D0070263

60
FARRAR
STRAUS
GIROUX

COMPETITION

COMPETITION

THE BIRTH OF A NEW SCIENCE

JAMES CASE

HILL AND WANG
A division of Farrar, Straus and Giroux
New York

Hill and Wang
A division of Farrar, Straus and Giroux
19 Union Square West, New York 10003

Distributed in Canada by Douglas & McIntyre Ltd.
Printed in the United States of America
First edition, 2007

Grateful acknowledgment is made for permission to reprint graphs and illustrations from the following publications:

Page 64: *Dynamic Models in Biology,* © 2006, Stephen P. Ellner and John Guckenheimer, p. 81. Reprinted by permission of Princeton University Press.

Page 77: *Economics and the Competitive Process,* © 1979, James H. Case, p. 4. Reprinted by permission of New York University Press.

Pages 117, 126, and 128: *Games of Life,* © 1993, Karl Sigmund, pp. 156, 42, and 43, respectively. Reprinted by permission of Oxford University Press.

Page 149: *Everyday Economic Statistics,* © 1995, Gary E. Clayton and Martin Giesbrecht, p. 115. Reprinted by permission of The McGraw-Hill Companies.

Page 164: *Irrational Exuberance,* © 2000, Robert J. Schiller, p. 186. Reprinted by permission of Princeton University Press.

Page 165: *The (Mis)Behavior of Markets,* © 2004, Benoit Mandelbrot and R. L. Hudson, p. 19. Reprinted by permission of Basic Books, a member of Perseus Books Group.

Pages 171 and 173: *Fortune's Formula,* © 2005, William Poundstone, pp. 269 and 238, respectively. Reprinted by permission of Hill and Wang, a division of Farrar, Straus and Giroux.

Page 179: *Miles to Go: A Personal History of Social Policy,* © 1996, Daniel P. Moynihan, p. 86. Reprinted by permission of Harvard University Press. Data provided by the Department of Commerce, Bureau of Economic Statistics; chart prepared by the Joint Economic Committee.

Page 256: *Butterfly Economics,* © 1998, Paul Ormerod, p. 6. Reprinted by permission of Pantheon Books. Data provided and chart prepared by Alan Kirman.

Page 272: *The Evolution of Cooperation,* © 1984, Robert Axelrod, p. 51. Reprinted by permission of Basic Books, a member of Perseus Books Group.

Library of Congress Cataloging-in-Publication Data
Case, James H., 1940–
Competition : the birth of a new science / James Case.— 1st ed.
p. cm.
Includes index.
ISBN-13: 978-0-8090-3577-9 (hardcover : alk. paper)
ISBN-10: 0-8090-3577-4 (hardcover : alk. paper)
1. Game theory. 2. Competition. I. Title.

HB144.C368 2007
338.6'048–dc22 2006036051

Designed by Patrice Sheridan

www.fsgbooks.com

1 3 5 7 9 10 8 6 4 2

Contents

Preface

Competition is a surprisingly difficult concept even to define. Modern lexicographers have failed repeatedly to improve on the definition given by Samuel Johnson in his 1755 work *Dictionary of the English Language*. In it he declares "competition" to be a noun meaning "the act of endeavoring to gain what another endeavors to gain at the same time; rivalry; contest." Yet even this definition is unacceptably narrow, since it limits to two the number of admissible competitors. Auctions, spelling bees, primary elections, horse races, and track and field events, as well as golf, tennis, and bowling tournaments, typically allow large numbers of hopefuls to compete. At the very least, Dr. Johnson's definition should be amended to read "what *others endeavor* to gain."

Like solitaire, crosswords, picture-, and other sorts of puzzles may be viewed as single-player games. All such games can, in principle, be solved mathematically. Although it might take the fastest conceivable computer millions of years to crack a particularly difficult puzzle, it is at least possible to define what is meant by a solution, and to specify the steps a computer would have to perform in order to find one. The same cannot be said of contests open to three or more competitors, since there is no universally accepted definition of a solution. As a result, leading authorities can disagree as to which, if any, of several proposed solutions for a particular many-player game is valid.

Two-player games can go either way. Those of the "zero-sum" variety, in which the winner wins only what the loser loses, are almost as solvable as puzzles and solitaire. Those that are not zero-sum are nearly as insoluble as many-player games. It is indeed unfortunate that the forms of competition that most directly affect human welfare—such as wars and commercial competition—are rarely zero-sum, and typically involve more than two competing factions. It is worth noting that, whenever many-player games such as Scrabble and Monopoly are contested at the tournament level, the rules are altered to transform them into two-player zero-sum games. Many-player and non-zero-sum games are simply too confusing for tournament play. With three or more players, there would be no end of complaints from alleged victims of collusion.

It was not until 1944 that John von Neumann and Oskar Morgenstern developed a truly workable definition of competition. It took them an entire chapter of their groundbreaking *Theory of Games and Economic Behavior* to explain what they had done. From the realization that a game *is*—for analytical purposes—nothing more than a book of rules, they distilled a concise (if forbiddingly technical) definition of a game. Only gradually did it become clear that all forms of competition can be made, with but little modification, to fit their definition. Indeed, more than sixty years after they wrote, no one has identified a form of competition that seems incompatible with the von Neumann–Morgenstern definition of a game. In 1950, Harold Kuhn developed the simpler and more graphic, yet logically equivalent, form of that definition to be found in Chapter 3.[1] Though dictionaries may never include the entry "com′pe▪ti′tion (kŏm′pê▪tĭsh′ŭn), *n.* See GAME," the emerging *science of competition* has yet to discover anything suggesting that such an entry would be misleading.

You won't find many departments of competition science listed alongside those of physics, chemistry, and food science in the catalogs of leading colleges and universities. There is as yet no *Journal of Competition Science*, nor any Society of Competition Scientists. Perhaps there never will be. Leaders in the field seem content to regard themselves as biologists, psychologists, computer scientists, statisticians, economists, mathematicians, and highway engineers. Yet all are keenly aware of the similarity between the mental gymnastics required of chess players, political consultants, military strategists, corporate planners, and others who compete for a living. Perhaps this book will alert them to the fact that the makings of a genuine and *very practical* science lie scattered throughout the literature of the decision sciences.

By far the most spectacular achievement of the new science has been the defeat of world chess champion Garry Kasparov by the IBM computer Deep Blue. There have, however, been plenty of other stirring victories. Computers now dominate the strongest human players of many familiar board games, as well as games encountered in the military, in commerce, and in government.

That's the good news. The bad news is that competition science is inescapably mathematical—all but impossible to explain in wholly nonmathematical terms. Though every effort has been made to minimize the number of pages infected with mathematical symbolism, I know not how to eliminate them all. What math remains is presented mainly in graphical form, since many readers find pictures less daunting than equations.

The most notable conquests of man by machine have made less use of mathematical game theory than of systematic experiment. Every computer chess, checkers, or backgammon tournament ever played, as well as every game of any kind between man and machine, has constituted a potentially telling experiment. It is impossible to count the number of such experiments conducted since the dawn of the computer age, or the number of once-plausible hypotheses rejected in the process.

Einstein once expressed the opinion that the "Development of Western Science is based on two great achievements: the invention of the formal logical system (in Euclidean geometry) by the Greek philosophers, and the discovery of the possibility to find out causal relationships by systematic experiment (Renaissance)."[2] Only after combining the two did Western Science initiate the four centuries of ever-accelerating material progress that history now records. And only after experimental methods began to complement a priori mathematical reasoning was significant progress made against the more challenging board games, like chess and checkers.

The first part of the book explains the nature and sources of man's existing knowledge of competition, while the second—lengthier and more controversial—explores economic competition. The latter, it will be argued, bears little resemblance to the carefully choreographed minuet described in books on the subject and too often relied upon to assert the will of the consumer. Indeed "consumer sovereignty" is but one of several colossal fictions enshrined in "mainstream economic theory." The central thesis of that vast oversimplification holds that something called "perfect competition" compels free markets to allocate "scarce resources" in a manner so "efficient" that all of mankind's conflicting wants and needs are resolved in the most satisfactory manner possible. On paper, all free markets work in more or less the same way.

In practice, most appear to work very differently, both from one another and from the way they are all said to work. Perhaps that's why so much "jawboning" is required to convince the public that things are not as they seem, but as free market logic commands them to be. Does it seem unwise to eliminate the inheritance tax on the superrich? Free market principles uphold the wisdom of such reform. Does it seem heartless to depress the minimum wage? Free market evangelists portray it as an act of compassion. Does it seem counterproductive to outsource American manufacturing jobs? Free market ideologues depict outsourcing as a subsidy to the lower middle class. And so on. It will be argued presently that the more preposterous claims of orthodox economic theory are direct consequences of expert willingness to conflate actual with perfect competition.

Unbeknownst to the public, numerous schools of "heterodox" economic thought dispute orthodox teachings. For a variety of reasons, they denounce mainstream (orthodox, neoclassical) economic thought as the worst kind of pseudoscience, in which the main conclusions were already in place by the dawn of the nineteenth century, leaving nothing more to modern scholarship than the discovery (read "fabrication") of ever more elaborate (read "abstract mathematical") justifications. The final chapters of the book will describe a few of the more active schools of heterodox economic thought, explain the faults they find with orthodox theory, and explore the policy implications of proposed amendments.

I hope the book will appeal to the entire scientific community, from career scientists to the most casual watcher of the Science and Discovery channels. Whereas the book's main purpose is to inform and entertain, I hope to encourage at least a few readers to join the growing chorus of agitators for open and public debate between orthodox and heterodox economists. As matters stand, those best able to judge such debates—namely experienced natural scientists—wish only to avoid involvement. Absent their participation, the issue will yet again be decided in the court of public opinion, presided over by a press corps notorious for its inability to distinguish real from pseudoscience.[3]

I would like to express my gratitude to those who, in various ways, have contributed to the completion of this project. My wife, Pat, as always, went above and beyond the call of duty by reading and criticizing every version of every chapter. Pat Scace read several chapters with equal—and much needed—attention to detail. Chih-Hung Cho (aka Samuel) transformed my awkward renditions into the clear and uncrowded line drawings to be found throughout. June Kim and the staff at Hill and Wang have been unfailing in their courtesy and attention to my occasionally idiosyncratic wants. Finally, my friend and editor Joe Wisnovsky exhibited the patience of Job with a multitude of unscheduled delays. His gentle reminders that this needed to remain a science book—as originally planned—kept me from wandering "off message" on numerous occasions. But for him, the following story would likely have remained untold. I alone am responsible for any errors or misrepresentations to be found within.

COMPETITION

CHAPTER 1

Man Versus Machine

On May 11, 1997, an IBM computer named Deep Blue defeated the reigning world chess champion in a six-game match. It was the first loss of Garry Kasparov's professional career and, with the possible exception of Bobby Fischer's victory over Boris Spassky in 1972, the most widely reported chess match in history. A pre-match press conference was attended by more than two hundred reporters from around the world, including TV crews from ABC, CBS, NBC, CNN, Fox, and several foreign networks. As the match progressed, and the likelihood of a Deep Blue victory increased, the number of reporters in attendance grew. Daily results were reported on the evening news, analyzed in the morning papers, and satirized by late-night TV comics. Not even the launch of the System/360 line of computers in 1964—by far the biggest public-relations event in IBM history—generated as much publicity.

While pundits differed as to the exact significance of Deep Blue's triumph, few doubted that an important milestone had been passed. This book will explore the implications of Deep Blue's achievement, and a growing list of others like it, as they contribute to man's understanding of competition, games, and the general welfare. For Deep Blue's triumph was by no means an isolated event. It followed hard on the heels of the 1994 defeat of Marion Tinsley—by all accounts the greatest checkers player of all time—at the hands of a computer program named Chinook, and it preceded by only three months the overwhelming (six games to none) victory of a program named Logistello over Takeshi Murakami, the then-reigning world champion player of Othello.

Chess, checkers, and Othello are but a few of the games at which computers now rival the strongest human players. Some of the others, like poker, bridge, backgammon, Monopoly, and Scrabble, will be familiar to most readers. Yet others, like awari, Kalah, dakon, draughts, Lines of Action, Nine Men's Morris, Qubic, shogi, Go-Moku, and Hex, may be less so.[1] Of particular interest is the ancient Asian game of Go, at which even the best computer programs are still no match for quite ordinary human players. Why do computers find Go so difficult?

The short answer to that question is *complexity*. A chess player has, on average, only twenty to thirty legal moves to choose from on any given turn, while checkers players seldom have more than a dozen. Go players, in contrast, have a hundred or more. This turns out to mean that brute-force searching—in which one simply tries every alternative—is virtually useless in Go, although it is an important component of every successful computer program for playing chess, checkers, and most of the other games at which computers currently excel. For any number of reasons, including the fact that at least two different measures of complexity have proven relevant for the analysis of computer games, a more complete explanation of Go's intractability will have to be deferred until later in the book. The reader should bear in mind that an entire science of complexity has emerged in recent years, to take its place beside the almost equally novel science of chaos. Both are at least peripherally related to the study of games and competition, and will recur in what follows.

Deep Blue's triumph was exceedingly hard won. After absorbing the lessons learned by previous students of computer chess, a three-man team still required twenty-eight man-years (beginning in or about 1985) to complete the project. Hardware designer Feng-hsiung Hsu alone devoted twelve years to the project. This lead threesome was assisted by an ever-changing cast of human chess experts, IBM project managers, and faculty members at Carnegie Mellon University, where the project began. Chinook was developed at the University of Alberta, under the direction of Professor Jonathan Schaeffer, beginning in 1989, and achieved its triumph within a "mere" five years. Logistello was built at the NEC Research Institute in Princeton, New Jersey, by a team led by Michael Buro. Hsu and Schaeffer have written books recounting the inside stories of their memorable campaigns.[2]

Computers exhibiting the ability to mimic aspects of human thought are said to possess artificial intelligence, or AI. A simple electronic calculator doesn't have AI, but a machine that can learn from its mistakes, or that exhibits the ability to reason effectively, undoubtedly does. As computers have grown more powerful, the dividing line between AI and "routine applications" has receded to the point where deeds once deemed indicative of AI are no longer so interpreted. Wags now suggest that a capability constitutes AI only as long as it retains the gee-whiz ability to delight and amaze. That said, computers able to play board games like chess and checkers at the highest human level, to compose arresting melodies, or to provide original proofs for known mathematical theorems have always been acknowledged to possess AI, and probably always will be. All this being the case, it is hardly surprising that most world-class game-playing computer programs have been designed by AI experts.

Arthur Samuel, the first man to program a computer to play checkers, once offered the following explanation for the scientific community's continuing interest in game-playing computers:

> Programming computers to play games is but one stage in the development of an understanding of the methods which must be employed for the machine simulation of intellectual behavior. As we progress in this understanding it seems reasonable to assume that these newer techniques will be applied to real-life situations with increasing frequency, and the effort devoted to games . . . will decrease. Perhaps we have not yet reached this turning point, and we may still have much left to learn from the study of games.[3]

Samuel's explanation rings almost as true today as it did when he first offered it. The "effort devoted to games" has actually increased in recent years, due in part to the advent of extraordinarily powerful desktop computers—a development not foreseen in 1960—and to the recent successes of Deep Blue, Chinook, Logistello, and other game-playing computer systems. His "turning point" is not yet in sight, and mankind still seems to have much to learn about competition and games. Nevertheless, his "newer techniques" have materialized and are being "applied to real-life situations with increasing frequency," not to mention admirable effect.

Nowhere are the effects more apparent than in the area of expert systems, also known as knowledge-based or rule-based systems. They consist of five component parts: an asset, a human manager, a book of rules, a compendium of facts, and an "inference engine." The latter three constitute a support system meant to help the human manager extract a high rate of return from the asset in question. Commercial inventory and revenue management systems are typical examples.

The facts stored in an airline's revenue management system concern the number of seats in each category that remain unsold for each scheduled flight, along with the economic data needed to estimate the expected demand for such seats. If an inordinate number of tickets for a given flight remain unsold as departure time approaches, the system will recommend that management begin discounting the more expensive ones. The inference engine will apply the rules contained in the system to the available facts to determine an effective schedule of discounts.

The facts stored in a motel chain's revenue management system concern the number of rooms of each size that remain unsold for each night in the coming year, along with the economic data required to predict unusual levels of demand. If an inordinate number of rooms remain unsold for a given date, the system will recommend that the manager begin discounting them, in order to stimulate sales. Alternatively, if unexpectedly many rooms have been sold, the system may recommend that the manager begin to encumber the remaining ones. If, for instance, Tuesday is the busiest night at a particularly business-oriented motel, the manager may refuse on Saturday, Sunday, and Monday to accept additional reservations for the coming Tuesday night, except from customers prepared to stay either Monday or Wednesday night as well. Again, the inference engine will apply the rules built into the system to the available facts to identify an effective program of discounts and/or encumbrances.

Complicated though commercial revenue management systems may sometimes be, their complexity pales by comparison with the knowledge-based systems contained in many game-playing programs. When Chinook reaches the stage at which eight or fewer pieces remain on the board—meaning that the pieces are deployed in one of exactly 443,748,401,247 possible configurations—it transfers command to its "endgame manager," an expert system of truly epic complexity. Game-playing computers stand at the forefront of expert system design.

The facts embedded in Chinook's endgame manager specify, for each possible configuration of eight or fewer pieces, whether it is a winning position for White, a winning position for Black, or a draw, as well as the maximum number of moves required to obtain the promised result. If, for instance, the system describes the current position as a W-43, meaning that White can win in forty-three moves, the inference engine will search the rule book for directions to a position designated W-42. Should Chinook's opponent play skillfully, the board will then be in a position designated W-41 when Chinook's turn comes to play again. But if the opponent makes a mistake, Chinook's next move may be made from a position designated W-40 or fewer. Only if Chinook's endgame manager contains a hitherto undetected programming error will it ever find itself in a nonwinning position after having occupied a winning one with eight or fewer pieces left on the board.

The lessons learned about the care and feeding of expert systems by the builders of Deep Blue, Chinook, Logistello, and other world-class game-playing programs all have been published in the open scientific literature, for the benefit of all. It would be surprising indeed if none of those lessons should prove applicable to future management systems. Samuel's "study of games" has proven more subtle, surprising, and fruitful than even he expected.

Fruitful though that study has been, it has failed to answer the most intriguing questions. It can be proven, for instance, as a mathematical theorem applicable to chess, checkers, and a host of other two-player games, that one of three mutually exclusive conditions prevails at any time during the course of play: either the player whose turn it is to move can win the game, or the opponent can win, or both can avoid losing, in which case the game will presumably end in a draw. Yet nobody knows which of those three alternatives applies to the opening move in chess, checkers, Othello, and other familiar board games currently ruled by machines.

The game plans executed by Deep Blue, Chinook, Logistello, and the like can all be beaten, and occasionally are. They are by no means the unbeatable game plans whose existence the above-cited theorem guarantees. Unbeatable game plans are known only for simpler—and less frequently played—games such as nim, tic-tac-toe, Qubic, Nine Men's Morris, mancala, Go-Moku, and Connect Four. Games for which unbeatable game plans are known are often described as "solved," "cracked," or "cooked." People soon lose interest in playing such games unless the unbeatable game plans are so complicated that only a computer can carry them out.

The trade-off between strategic complexity and effectiveness first became an issue during the 1960s, after Edward Thorp discovered that twenty-one (a.k.a. blackjack)—as then played in leading casinos—could be beaten by certain "card-counting" strategies.[4] Thorp, then a math instructor at MIT, used the school's state-of-the-art computer to demonstrate that the blackjack strategy that commits the house to "take a hit" on seventeen or less, but to "stand" on eighteen or more, is far from unbeatable.[5] He then published three increasingly sophisticated card-counting strategies and used the strongest to win thousands of dollars in casino play. As a result, he was allegedly asked to leave a number of such establishments and threatened with bodily harm should he return.

The simplest of Thorp's published strategies was not strong enough to beat the house strategy on a consistent basis. It merely allowed weekend warriors to get in on the action without undue risk. The second was a more or less break-even proposition requiring more skill on the part of the player. The third was a reliable cash cow but required users to practice several hours a day in order to realize its full potential. To defend their incomes from the army of card counters spawned by Thorp's book, and its eventual emulators, the casinos soon replaced the single deck with which blackjack had traditionally been played with a "shoe" containing no fewer than four identical decks, thereby increasing the number of cards to be counted seemingly beyond the limits of human mental arithmetic. After satisfying himself that he could clear perhaps $300,000 a year playing blackjack, Thorp abandoned casino gambling in favor of Wall Street—the biggest game of all—where he and his computer-generated strategies have prospered ever since.[6]

Chess, checkers, and Othello are games of "perfect information," meaning that anything either player might care to know about the current state of a game in progress is there on the game board for all to see. Because the players of Scrabble lack knowledge of the tiles held by their opponents, and the players of bridge and poker lack knowledge of the cards held by the other players, the players of those games are said to possess only "imperfect information." Games of imperfect information are conceptually more complex than those of perfect information.

Backgammon represents an intermediate level of complexity in that, although players possess perfect information, the randomness introduced by frequent rolls of the dice represents a definite complicating factor. It means, among other things, that the same two strategies, game plans, or computer programs for playing backgammon may have to oppose each other many times over before a clear winner can be determined.

The strongest backgammon-playing computer programs now perform at a superhuman level.[7] They differ from Deep Blue in that they employ machine-learning techniques to constantly improve their games. The idea that a computer program could teach itself to play board games by playing them over and over again against its own clone goes back to Arthur Samuel, as does so much of AI.[8] Samuel lacked only the computing power and machine-learning techniques to carry out his plan. The first program to do so successfully, called TD-Gammon, was constructed by Gerald Tesauro at IBM during the 1990s. The project has an interesting history.

In 1987 Tesauro, then at the Institute for Advanced Study in Princeton, New Jersey, collaborated with Terry Sejnowski, then at the Johns Hopkins University, to train a "neural network" to play backgammon. Neurons, the fundamental building blocks of the human nervous system, have been known to medical science for more than a century. During World War II, pioneers in AI invented an artificial (electronic) neuron, assembled small networks of them, and predicted that larger and speedier networks would one day emulate human intelligence. It took a long time to realize that prediction, but by the mid-1980s neural networks were becoming practical instruments of machine learning. Sejnowski was one of the key developers of neural network techniques.

Neural networks can often be trained to emulate human interpreters of complex information, whether that information pertains to potential oil-drilling sites, heart disease patients, credit applications, or baseball players. One trains such a network by supplying it with a library of stimulus-response data. For oil-drilling sites, the stimuli would include the seismographic and geological data available before drilling begins, while the responses would include the quantities and flow characteristics of whatever deposits were eventually found. For credit applications, the stimuli would include former applicants' financial status and histories, while the responses would include eventual default rates and recovery costs. Given an adequate supply of coherent data, neural networks (of which there are now many kinds) can often learn to interpret new data from the same source at least as accurately as human evaluators do, typically at a fraction of the cost.

To build their backgammon-playing program, Tesauro and Sejnowski supplied an off-the-shelf neural network with an expert's assessment of the best move from several thousand board positions.[9] The resulting program, which they called Neurogammon, could play at an intermediate level. They were pleasantly surprised by this, since backgammon permits more than 10^{20} possible board positions and Neurogammon was obliged to learn from a tiny fraction of that number. Yet somehow it was able to transfer what it had learned to unfamiliar board positions and to identify advantageous moves from most of them. But the level to which Neurogammon could hope to progress was limited: it could never hope to beat the expert from whom it had learned the game.

Some years later, after moving to IBM, Tesauro investigated a more ambitious approach to computer backgammon. He did so by combining Samuel's old self-teaching idea with a modern high-speed computer and a state-of-the-art machine-learning technique. The technique he chose is of interest in its own right.

Worker bees routinely travel several miles in a single foraging trip. They are instinctively attracted to blue flowers, but, being among the smartest of insects, they quickly learn to associate the color, shape, and fragrance of other local flowers with their nectar content. The more nectar they find in a given patch of flowers, the more likely they are to return to it. Only in 1995, after becoming the director of the Computational Neurobiology Lab at the Salk Institute in San Diego, did Sejnowski learn that a single neuron in the bee's brain underlies its odor-recognition capability.

This cell, called VUMmx 1, uses a substance called octopamine—which is chemically similar to the dopamine found in vertebrates—to influence the bee's decision-making system. Science has been aware of the dopamine system since the 1960s, when psychologists learned to implant electrodes in the dopamine-stimulating neurons of rats. The rats were then given access to a bar that would deliver a pulse of electrical current to those neurons whenever they pushed it, thereby releasing dopamine and activating the parts of the brain that respond to it. The rats promptly lost interest in food, sex, and whatever else ordinarily interests them. So anxious did they become to stimulate their dopamine systems that researchers claimed to have located the brain's "pleasure center."

The discovery of VUMmx 1 suggested that bees might be using their octopamine systems to improve their nectar-finding efficiency. If the location of an abundant source of nectar were to release an octopamine jolt into the bee's brain, causing it to identify flowers of that size, color, fragrance, and so on with raw pleasure, the entire hive would benefit. Although Pavlov and his successors had long ago demonstrated a link between learning and reward, modern psychologists tend to dismiss "reward learning" as an inferior variety. Yet it seemed to Read Montague and Peter Dayan—researchers in Sejnowski's lab—that simple creatures might use it to good effect. Accordingly, they built a computer model of a bee's octopamine reward system, based on the way the system appears to work in actual bees. Their simulated bee behaved remarkably like a real (worker) bee, exhibiting a number of known bee-havioral patterns, including an aversion to patches of flowers in which the nectar content fluctuates dramatically from one bloom to the next.

The success of their bee simulations led Montague and Dayan to speculate that the output of VUMmx 1 measures prediction error. When a bee visits a particular patch of flowers, in the expectation that nectar will be found there, the results either confirm or deny the expectation. If accurate expectations are rewarded with generous jolts of octopamine, while inaccurate ones are not, the reliability of a bee's expectations would seem likely to improve with practice. Such a mechanism could only enhance a bee's ability to teach itself what it needs to know about the world in which it lives.

Temporal-difference (TD) learning is an ingredient of many artificial learning systems. It compares the estimated chances of success in some multistage task—say, the building of a skyscraper—before and after the completion of each stage. If the estimated chances of completing the whole task on time and within budget don't change very much as stage after stage of the task is completed, the successive estimates were probably pretty good to begin with. But if successive best estimates vary all over the map, some of them have to be inaccurate. Thus it makes a certain amount of sense to award points to the estimator—as TD learning does—each time a new best estimate is in rough agreement with its predecessor.

Tesauro's familiarity with TD-learning techniques persuaded him to employ them in his implementation of Samuel's plan to build a self-teaching game-playing computer program. The result was TD-Gammon, the first backgammon-playing computer program to compete at the highest human level. After a few thousand games against itself, it began to play at a beginner level. After a hundred thousand games, it could beat Neurogammon. After a million games, it could compete with the best human players in the world, and was beginning to do so by 1995. The current version is said to play better than the best humans.

Perhaps surprisingly, the mathematical theory of games has played a relatively minor role in the development of world-class game-playing computer programs. That is in part because John von Neumann—the inventor of mathematical game theory—was primarily interested in games of imperfect information, such as Scrabble, poker, and bridge, in which it is possible to outsmart one's opponents by means of a successful bluff. He was an enthusiastic poker player himself and delighted in the diversionary aspects of the game. He and his immediate successors allowed the branch of game theory concerned with perfect information to languish to some extent. Only later did the developers of differential game theory, combinatorial game theory, and game-playing computer programs begin (each in their own characteristic way) to rectify this oversight.[10]

One reason for game theory's overall lack of success was the primitive state of electronic computers at mid-century. The computing power required to master games like chess, checkers, Othello, poker, backgammon, and Scrabble simply didn't exist until the late 1980s, when game-playing computers began (quite suddenly and unexpectedly) to challenge the strongest human players. The second reason for game theory's rather disappointing results has been that the most important forms of competition—the ones most directly affecting the general welfare—involve more than two players. This makes them significantly more complex than the two-player contests that computers seem destined to dominate.

Instead of appealing to game theory, or to any other theory for that matter, the "hackers" who have trained computers to play games at or above the highest human level have relied mainly on the experimental method.[11] Every contest between rival computer programs, or between a program and a human player, constitutes an experiment testing one or more hypotheses about what works and what doesn't. One expert has described the fifty-year effort to defeat a human chess champion as "the longest running experiment in the history of computer science."

Virtually every man-versus-machine contest has been exhaustively analyzed after the fact, prompting the reformulation of some hypotheses and the outright rejection of others. Hsu's book *Behind Deep Blue* is particularly informative in this respect, describing the many matches and tournaments played by Deep Blue's progenitors—including Deep Blue I, which lost to Kasparov in 1996—and the design changes made after each in response to lessons learned.

The five Computer Olympiads held in 1989, 1990, 1991, 1992, and 2000 were invaluable proving grounds for game-playing computer programs, and for the legion of working hypotheses incorporated into each one. The hackers who built those programs are not unaware of game theory. On the contrary, they know it well and apply it whenever it helps. Yet their victories owe far more to trial, error, and the experimental method than to any abstract theory.

It will be argued at length in what follows that the real significance of Deep Blue's triumph over Kasparov, and of the other defeats human champions have lately suffered at the hands of soulless machines, is that they mark the emergence of a genuine science of competition. Such a science differs from any mere theory of competition—such as the mathematical theory of games—in that it mines historical and experimental data to confirm, augment, and occasionally discredit conclusions reached long ago by unadorned logic.

While old sciences never really die, they are often eclipsed by newer and more dynamic ones. Many of today's most fruitful sciences, including space science, brain science, computer science, environmental science, and the sciences of chaos and complexity—not to mention the science of the human genome—were unknown a century ago. By the year 2100, the emerging science of competition may be as familiar and uncontroversial as highway science, poultry science, and dental science are today. If so, it will be a very influential science indeed, with implications for almost every aspect of public policy, including health, energy, environmental, education, immigration, military, and (above all) economic policy.

The early chapters of this book describe the current state of the fledgling science of competition—what it has discovered and what it seems destined to discover before long. The middle chapters attempt to identify the more important forms of competition in which mankind is currently engaged, and to explain what science can and cannot yet tell us about them. It is a book about science on the one hand and competition on the other. The common ground is surprisingly large and largely surprising.

CHAPTER 2

The Art and Science of Competition

The modern science of competition differs from the ancient art of competition much as modern medical science differs from the ancient art of healing. Generations of folk healers have used willow bark tea to relieve pain. It works because willow bark contains salicylic acid—aspirin. Tropical folk healers have long prepared tonics from the bark of the cinchona tree to comfort malaria sufferers. That works because cinchona bark contains quinine. English folk healers traditionally used extracts of foxglove to treat heart patients. That worked because foxglove contains digitalis. And so on. Although the vast majority of all folk remedies are either ineffective or counterproductive, many modern medicines are refined versions of effective folk remedies.

The science of competition has begun to function in much the same way, identifying, testing, and refining traditional competitive practices. A father advising his son to "keep your guard up" and "lead with your left" in school-yard fights is recommending the tried-and-true. A drama teacher advising a student to "speak to the back row" while auditioning for a role in the class play is doing likewise. A chess master comparing the merits of various opening gambits, an experienced salesman criticizing a trainee's performance, a 4-H leader explaining the rudiments of livestock judging, even a wrestling coach discussing takedowns, draw on a fund of well-tested practices.

Today there are instructional DVDs, schools, summer camps, and year-round training facilities to improve performance in golf, tennis, rodeo riding, ballet dancing, fiction writing, portfolio management, and almost every other known form of competitive activity. Each recommends some competitive practices and warns against others. Most maintain lists of satisfied customers eager to attest to the value of the training on offer. Most if not all of the same advice is available in book form. Such volumes as *The Art of Attack in Chess*, *The Art of Hitting* and *The Art of Pitching* (both about baseball), *The Art of Doubles* (tennis), and *The Art of Seduction* are relatively specific in their recommendations, while *The Art of the Deal*, *The Art of Leadership*, and *The Art of Winning* are meant to apply in almost equal measure to all forms of competition.

The Art of War by Sun Tzu may well have been the first book ever written about competition of any kind. War is a particularly ancient and well chronicled form of human competition. Indeed, if chimpanzees are any indication, warfare is even more ancient than mankind. Bands of chimps in the wild routinely seek to expand their territories by dispatching war parties against their neighbors.[1] Ordinarily the most vocal of creatures, chimps on the warpath steal silently through the forest to surprise the intended foe, picking up sticks and stones along the way for use as weapons. Since primitive peoples stage similar attacks, it seems likely that their common ancestors did so, as well. The skeletons of Neanderthals—many of which show the effects of wounds inflicted (with clubs, spears, axes, and the like) in battle—furnish additional evidence that warfare is not peculiar to *Homo sapiens*.[2]

Many peaceful forms of competition, including all but one of the events contested at the ancient Olympics, appear to have begun as training exercises for war. This is obviously true of boxing and wrestling—which exercise and reward the physical skills required for hand-to-hand combat—as well as running with and without armor, jumping, javelin throwing, and chariot racing. Indeed, of all the events on the ancient Olympic program, only the discus throw has no obvious connection with war. Even chess (the name of which comes from the Persian word *shāh*, meaning king) reputedly exercises the foresight and planning skills required of generals.

In addition to its influence on board games and athletic competition, war has inspired some of history's greatest books. Homer's *Iliad* and Tolstoy's *War and Peace* are cases in point. Military folk insist that the honor roll should include Clausewitz's *On War* and Sun Tzu's *Art of War*, both of which were written for the guidance of actual commanders in the field. Whereas the science of war concerns statistics, ballistics, logistics, and other matters safely entrusted to subordinates, the art of war concerns "strategy," a word derived from the ancient Greek στρατηγία, meaning "the art of the general." Even the most modern thinking on that age-old art borrows heavily from Sun Tzu.

Little is known about the life of Sun Tzu. If he ever lived at all, it was in China, between two and three thousand years ago. His book says far less about the usual elements of military strategy—time, space, and manpower ratios—than one might expect, and far more about morale, discipline, supply chains, propaganda, information, and especially the treatment of spies. "It is with their information," he reminds his readers, "that you move, act, and plan." He also described the "shock and awe" that fire—the most frightening weapon in the ancient arsenal—could inspire in an enemy, and distinguished five separate battlefield uses for it. He then warned that the time and conditions must be right for its use and that proper tools and training are required.[3] Last, he promised that leaders heedful of his advice "will not know defeat even in a hundred conflicts."[4]

Clausewitz made no such promises. In his experience, war was an inherently high-risk undertaking in which it is possible to make all necessary preparations, deploy one's troops impeccably, respond promptly and appropriately to every battlefield development, and still be defeated by an inferior force upon which fortune has chanced to smile. "In war," he warns, "more than anywhere else, things do not turn out as we expect."[5]

Far more is known about the life of Carl von Clausewitz (1780–1831) than about that of Sun Tzu. Clausewitz entered the Prussian military in 1792 and, after several years on active duty, was admitted to the Berlin academy for junior officers. Most of his writing was done after 1818, when—following extensive service in the Napoleonic Wars—he was promoted to major general and appointed director of the General War College (Allgemeine Kriegsschule) at Breslau. He returned to active service with the Prussian General Staff in 1830 and died a year later, of cholera.

Clausewitz's collected works were edited and published—in ten volumes—by his widow, with help from several military colleagues. The first three contain his magnum opus, *Vom Kriege* (*On War*), in which he develops his theory and philosophy of war. Subsequent volumes concern military history and leaders. Although Clausewitz apparently intended to write a book on the theory of war, and had begun to edit his papers accordingly, his untimely death terminated the project. As a result, the published version is little more than a draft—a lightly edited collection of essays written over at least twelve years, during which his thinking continued to evolve. Small wonder his book is proverbially difficult to read.

Instead of asking how wars may be won, Clausewitz asked what war really is: What are its enduring elements, how do they relate to one another, and how can the power of human reason be brought to bear on them? Despite its legendary lack of clarity, *Vom Kriege* continues to be read and quoted by political leaders, military thinkers, and captains of industry alike. Jack Welch—the legendary CEO of General Electric—summarizes Clausewitz's message to the business community as follows:

> Von Clausewitz summed up what it [his military career] had all been about in his classic *On War*. Men could not reduce strategy to a formula. Detailed planning necessarily failed, due to the inevitable frictions encountered: chance events, imperfections in execution . . . Instead, the human elements were paramount: leadership, morale, and the almost instinctive savvy of the best generals.
>
> The Prussian general staff, under the elder von Moltke, perfected these concepts in practice. They did not expect a plan of operations to survive beyond the first contact with the enemy. They set only the broadest of objectives and emphasized seizing unforeseen opportunities as they arose. Strategy was not a lengthy action plan. It was the evolution of a central idea through continually changing circumstances.[6]

Others have plumbed the analogy between war and commerce in greater depth. Al Ries and Jack Trout, CEO and president of Trout & Ries Inc., have written an entire book on the applications of military thought to marketing warfare, by which they mean the practice of "taking business away from somebody else."[7] They state without reservation that Clausewitz's *On War* is the best marketing book ever written, and that their own slender volume is intended only to demonstrate the applicability of his fundamental precepts to marketing issues. To that end, they distinguish four different kinds of marketing warfare and identify three winning principles for use in each.

The four kinds of marketing warfare are defensive, offensive, flanking, and guerrilla. In marketing as in conventional war, defenders enjoy a distinct advantage over their attackers. It is easier to defend a strong position—as at Thermopylae or Bunker Hill—than to attack one successfully. Their three winning principles for defensive warfare are (1) only the market leader should consider playing defense, (2) the best defensive weapon is the courage to attack yourself, and (3) strong competitive moves should always be blocked. While the first and third points seem largely self-explanatory, attacking yourself is not. By it they mean marketing a product that competes with your own flagship brand, in the way that Anheuser-Busch's superpremium beer Michelob was meant to compete with premium Budweiser. Though the device worked well for AB during the late 1960s and early 1970s, a similar move almost backfired on its leading competitor, Miller. It happened because Miller Lite took almost as much business away from Miller High Life (the flagship brand) as it did from rivals. Number two can get burned when trying to act like number one.

Ries and Trout offer similar winning principles for offensive marketing warfare, flanking marketing warfare, and guerrilla marketing warfare, along with a handful of more general principles loosely derived from their reading of Clausewitz. One parallels Napoleon's observation that "God is on the side of the big battalions" by asserting that "God smiles on the larger sales force." Not the better-trained, or more skilled, or better-motivated sales force, but the larger one. To expect anything else is to flout what they call the "better people" fallacy, which holds that you can beat more people with fewer, better ones.

The better people fallacy is not a logical fallacy but a practical one. Although the best team in the NFL might beat the worst even if the latter were allowed to field twelve men on every play—instead of the usual eleven—Ries and Trout deem it highly unlikely. Their feeling is that the twelfth man would be far too big an advantage for the stronger team to surmount, because even average NFL players are exceedingly talented. The fallacy becomes even more inescapable when teams, like sales forces, are composed of hundreds if not thousands of individuals, since it is unlikely that so many individuals could all be significantly "above average."

Similar to the better people fallacy is the "better product" fallacy, according to which the better product always wins in the end. This, they submit, does not accord with the verdict of history, in which the competition to produce a better product invariably ends in a tie, with several brands of beer, soap, soft drink, automobile, air conditioner, deodorant, candy bar, or razor blade being "parity products" asserting plausible claims to victory.

By that time, all of the inferior brands, along with some parity products, will have fallen by the wayside, leaving the market to be shared (unequally, of course) by those still standing. The leading share of that market will go to the firm that bamboozles the largest fraction of buyers into thinking its parity product more desirable than any other. In marketing as elsewhere, history is written by the victors. Only the losers have time to wonder if the best horse, man, team, army, product, or nation really won.

A good marketing strategy, say Ries and Trout, is "one that anticipates the competitor's counterattack." Too many marketing commanders, they maintain, draw up battle plans as if the enemy will take no notice of an aggressive act. In practice, enemies almost always react. "Cut your price in half," they warn, "and your competitor is likely to do the same." Or worse. Indeed, many of their principles for the conduct of marketing warfare already recognize (at least tacitly) the danger of a counterattack. Offensive principle number 2, for instance, states: "Find a weakness in the leader's strength and attack at that point." It is likelier that a leader will respond late, halfheartedly, or not at all to an attack on a minor profit center than on a major one.

Though Jack Welch would be the last to deny the importance of corporate strategy, he seems more interested in savvy, morale, and the other "human elements" of corporate leadership. He, and others like him, enable prominent football and basketball coaches to double and occasionally redouble their coaching salaries with product endorsements, speaking engagements, and books purporting to reveal the secrets of their success.

These men are acknowledged experts in the art of "competitive leadership," and their advice is eagerly sought. Such Hall of Fame coaches like Dean Smith, John Wooden, Red Auerbach, Paul "Bear" Bryant, and Bill Walsh have written such books, as have experts on golf, tennis, chess, checkers, backgammon, poker, bridge, marketing, litigation, and the dating game. Those books belong to the large and ever-growing literature on the art of competition. That literature contains, at any given time, an almost-up-to-date summary of the current state of that ever-evolving art. Yet one never quite knows how far to trust the advice offered in such books. By scientific standards, much of the content is vague, incomplete, and unsubstantiated. Advice based on an author's personal experience cannot, by its very nature, offer the detailed, step-by-step recipes for certain or highly probable success coveted by every serious competitor.

Not until the early years of the twentieth century did the idea begin to circulate that in certain situations, competition *can* be reduced to an exact science furnishing designated contestants with step-by-step instructions for achieving certain victory. That realization constituted a first important step toward the creation of a practical science of competition.

As it applies to tic-tac-toe, such a science can only confirm what every nine-year-old seems already to know—that each player can be furnished with a detailed game plan guaranteeing the possessor a draw, tie, or "cat's game."[8] Since such "unbeatable game plans" are available to both players, neither can reasonably hope to win. A draw is at once the most favorable outcome to which either contestant may realistically aspire and the least favorable either need accept.

Slowly at first, but then with increasing frequency, other unbeatable game plans began to surface. In time it was learned that such plans await discovery in a vast number of games, of which chess, checkers, and tic-tac-toe are merely the most familiar. For each such game, there must be either a winning game plan for White, a winning game plan for Black, or—as in tic-tac-toe—non-losing game plans for both players.

Among the first to publish a genuinely unbeatable game plan was one Charles L. Bouton of Harvard, who concocted a more or less complete theory of the game he called nim.[9] Played under different names in different places, Bouton's game was once a moderately popular pastime on college campuses and at street fairs in several parts of the United States. In the standard form of nim, a number of cards, matchsticks, coins, grains of corn, or other "tokens" are separated into several (often three) distinct piles on a tabletop. White (the first player to play) may remove some or all (but at least one) of the tokens from just one of the piles, leaving the other piles intact. Then Black may do likewise, and so on. The game continues, by turns, until all the tokens have been removed, the winner being that player who shall remove the last token or tokens from the table. Nim, rather obviously, can never end in a tie.

Bouton's relatively complete theory of nim involves binary numbers—those annoying little strings of zeros and ones that speak such volumes to computers and computer scientists while confusing everyone else. His game plan is explained and justified in any number of books on number theory and mathematical recreations.[10] A simpler theory applies to the single-pile version of nim, which is transparent unless a limit is imposed on the number of tokens a player may remove at each turn.

Suppose, for instance, that twenty-one pennies are placed in a single pile on a table and that each player may remove one, two, or three pennies from that pile at each turn. White can win in six moves by arranging—as plainly as she can—that the pile shall contain twenty pennies after her first turn, sixteen after her second, twelve after her third, eight after her fourth, and four after her fifth. The pile must then contain one, two, or three pennies after Black's fifth turn, allowing White to remove them all on her sixth.

Black can prevail at single-pile three's-the-limit nim only if White deviates from the indicated game plan, or if the original number of pennies in the pile is divisible by 4. In the latter event, Black can win by arranging that the number of pennies in the pile shall again be divisible by 4 after each of her own turns. Alternatively, if White should err at any time by allowing Black to reduce the number of pennies in the pile to a multiple of 4, Black can win by proceeding as if the game had begun anew following that blunder. Black can prevail at single-pile L's-the-limit nim only if the number of pennies in the pile is evenly divisible by $L + 1$, or if White allows Black to reduce that number to a multiple of $L + 1$ during the course of play.

In a book titled *The Mathematics of Games*, John D. Beasley explains the complete solution of multi-pile nim in particularly simple terms. He then describes several other games reducible to nim in the sense that anyone who knows how to solve nim can use that knowledge to solve the other games as well. Finally, he explains how all impartial games—in which the same moves are available to both players—which are guaranteed to terminate, reduce to nim in the foregoing sense.

Another solvable two-player game requires the two contestants to take turns placing nickels, one at a time, on a circular tabletop. The first player unable to find room for another nickel—without relocating any of those already in place—is declared the loser. White can win at what, for lack of a better name, may be called "nickeltop" by placing her first nickel at the exact center of the table, and then placing each subsequent nickel diametrically opposite the one most recently added by Black.[11] Because that location is necessarily unoccupied, White can't lose.

The age-old game of rock, scissors, paper is solvable in a more subtle sense. To play the game, each of the two contestants forms his right hand—on the count of three—into (1) a closed fist, signifying "rock," (2) an open down-turned palm, signifying "paper," or (3) a horizontal two-fingered peace sign, signifying "scissors." Barring duplication, the winner is determined as follows: scissors cut paper; rock breaks scissors; paper covers rock. Duplications may either be replayed or left as ties. A statistically unbeatable game plan is to toss a single die, out of sight of the opponent, and to react as follows: scissors on 1 or 2; rock on 3 or 4; paper on 5 or 6. The activation of a random device, such as the toss of a die, is a sure cure for predictability, a mistake to be avoided at all costs in many competitive situations.

Finger matching is solvable in much the same way. To play it, each player holds up—again on the count of three—either one finger or two. The challenger wins if both players show the same number of fingers, and loses otherwise. There are no ties. An unbeatable game plan is to toss a fair coin and show 1 on heads and 2 on tails. Here again, the activation of an appropriate random device (in this case, a coin toss) precludes predictability.

There are no guarantees in guessing games like finger matching or rock, scissors, paper. Even when following a statistically unbeatable game plan, one may lose time after time to an opponent upon whom fortune repeatedly smiles. Such plans are unbeatable only in the statistical sense that adherents will win about as often as they lose in any long series of contests.

In retrospect, it was soon realized that there is nothing particularly new about unbeatable game plans. Chess enthusiasts had been challenging one another for centuries to "solve chess problems" by devising just such plans. Classified as two-move problems, three-move problems, and so on, they are customarily posed pictorially, as in figure 2.1, over captions of the form "White mates in __ moves."[12] Readers are thereby challenged to solve the problem by explaining how anyone able to follow directions expressed in standard chess notation may (as White) unfailingly checkmate Black in the prescribed number of moves.

The Internet is currently awash with chess problems, checkers problems, bridge problems, Go problems, backgammon problems, and more. Each one represents a situation that could conceivably arise in the course of ordinary play and requires an unbeatable game plan for its solution. In times past, erroneous solutions could go undetected for years at a time. Today, however, published solutions are always computer verified and as nearly foolproof as things can be. The availability of computer verification has made possible the documentation of claims to the effect that White can win in *N* moves, from a depicted initial position, where *N* is a three-digit number.

Engineers often speak of the "state of the art" of building bridges, airplanes, TV sets, or computer chips. The states of most living arts are continually advancing. Computer chip manufacture is governed by

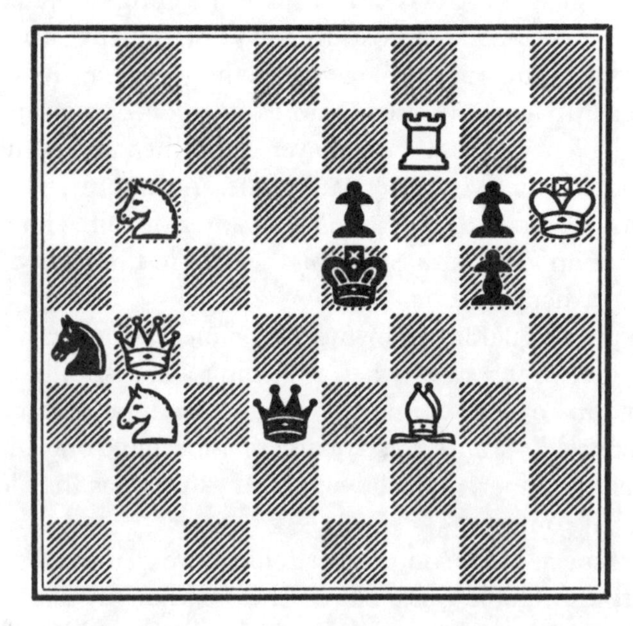

Figure 2.1. White mates in two moves. Dubuque Chess Journal, *December 1889.*

Moore's law, which asserts that the number of logic gates that can be built into a single silicon chip will double every eighteen months. That's been true since the 1960s, and promises to remain so for several years to come. The state of the art of playing chess is likewise advancing. Students of the game agree that today's best players would easily defeat the best of the World War I generation, who would in turn defeat Paul Morphy and the other nineteenth-century champions. State-of-the-art game plans for chess currently reside in electronic memory rather than the mind of man.

The same is true of checkers, Othello, backgammon, Scrabble, Monopoly, and quite a few other recreational games: state-of-the-art game plans for playing them now reside in computer memory. But better game plans will be devised, enabling the states of the art of playing them to continue their advance. The limit will not be reached until unbeatable game plans are discovered for each one, as in nim and finger matching.

Chess problems, as even the judgmental G. H. Hardy somewhat grudgingly conceded, are "genuine mathematics."[13] So are Go problems, checkers problems, and all the problems made solvable by Bouton's theory of nim. They may not be interesting mathematics, or important mathematics, but mathematics they surely are. Unbeatable game plans can therefore be guaranteed to produce, with "mathematical certainty," the advertised result: a tie in tic-tac-toe, a win for White in nickeltop, and an even chance of winning at rock, scissors, paper or finger matching.

At the time he concocted it, Bouton's theory of nim seemed little more than a curiosity. Yet by 1940, it had been recognized that the solutions of chess problems, checkers problems, nim problems, and the like belonged to a potentially vast branch of mathematics about which much remained to be learned. The more people found out about the subject, the more unanswered questions there seemed to be. Apparently unrelated discoveries began to merge into a respectable—perhaps even elegant—backwater of mathematical research. Then, quite without warning, everything changed.

The agent of change was a book, published in September 1944, by John von Neumann and Oskar Morgenstern.[14] Titled *Theory of Games and Economic Behavior*, it did for games what Clausewitz had done for war. Instead of asking how to win a particular game, or type of game, it began by asking what a game really is: What are its enduring elements, how do they relate to one another, and how can the power of human reason be brought to bear on them?

Most of the book's six-hundred-plus pages concern the manner in which human reason can be brought to bear on games. The unsurprising answer—given that von Neumann was one of the greatest mathematicians of the twentieth century, while Morgenstern was among the most mathematically inclined economists of his generation—was through mathematics. As a result, only a few of the book's many readers could understand more than a tiny fraction of it.

Fortunately, the authors were able to answer Clausewitz's most basic questions in a way that almost anyone could comprehend. This they did in the second of the book's twelve chapters, in which they explained "what a game really is." They did so by identifying the "enduring elements of a game" and revealing "how they relate to one another." They summarized their explanations in a pair of concise definitions—one for a game and one for a strategy, the latter being a game plan complete in every detail. Both definitions have stood the test of time.

Although other and sometimes better ways have since been found for bringing human reason to bear on games, no one has managed to improve on the definitions proposed by von Neumann and Morgenstern for strategies and games. Nor has anyone found—more than sixty years after publication, despite decades of intense research effort—a form of competition to which their definition of a game does not apply, either as written or after minor modification. For that reason alone, the book deserves an honored place in the annals of intellectual history. By answering Clausewitz's most basic questions for what they called games, the authors narrowly missed answering them for every conceivable form of competition.

In September 2004, Princeton University Press published a sixtieth-anniversary edition of *Theory of Games and Economic Behavior*. It included the entire third (1953) edition of the monumental work, along with perhaps a hundred additional pages of learned commentary concerning its overall legacy and lasting effect.

The commentators included Harold Kuhn, professor emeritus of mathematical economics at Princeton; Ariel Rubinstein, professor of economics at Tel Aviv and New York universities; and many of the book's original reviewers. Kuhn's new introduction lists a dozen reviews of the book, published between 1945 and 1950, which he describes as remarkable in both quality and quantity. Totaling 171 pages, all twelve appeared in leading scholarly journals, over the signatures of prominent or soon-to-be prominent mathematicians and social scientists, each of whom lavished praise on the book, the authors, and the boldness of their vision. Four of the chosen reviews are reprinted in the anniversary edition.

Leonid Hurwicz pointed out in *The American Economic Review* that the new methods were applicable to political science, sociology, and military strategy as well as to the economic problems and the "games proper" (primarily chess and poker) treated by the authors. In the *American Journal of Sociology*, H. A. Simon praised the new theory and urged his fellow social scientists to master it. In the *Bulletin of the American Mathematical Society*, A. H. Copeland began by suggesting that "posterity may regard this book as one of the major scientific achievements of the first half of the twentieth century," a breakthrough that might one day transform economics into "an exact science."

The years immediately following World War II were a time of progress, prosperity, and optimism, during which mathematics and the physical sciences reveled in postwar glory. Radar and the atomic bomb had brought the conflict to a successful conclusion months if not years ahead of schedule. No newsreel was complete without footage of helicopters, jet aircraft, and ballistic missiles. Cars were beginning to shift automatically and rooftops to bristle with TV antennas. Elected officials spoke of "harnessing the atom" and "conquering outer space." Who was to say what further wonders higher mathematics and "big science" might have in store? Why should there not be a single theory for all forms of competition? Why should it not turn the social sciences into exact ones?

At few other times in human history could such optimism have been taken so seriously. Von Neumann and Morgenstern had written the right book, with the right message and the right title, at the right time. By 1950, the U.S. Army, Navy, and Air Force had all put together high-powered teams to evaluate the potential of game theory. The mathematics departments at Princeton and the Rand Corporation (an independent, nonprofit research organization in Santa Monica, California, long funded by the Air Force) quickly emerged as the leading centers of game-theoretic research.

Perhaps the most telling comment in the anniversary edition came from the Nobel Prize–winning economist Paul Samuelson, who—sometime after 1960—described *Theory of Games and Economic Behavior* as "a work of genius" that "has accomplished everything except what it started out to do—namely, revolutionize economic theory." It would be hard to fault Samuelson's perception of the book's original purpose. Von Neumann and Morgenstern did indeed intend to revolutionize economic theory. Both were well read in the history of natural science, and both were keenly aware of the important part revolutions have played in its development. Even before they met, both believed a revolution in economic theory to be long overdue.

It will be argued presently that game theory should never have been expected to revolutionize anything of consequence, for the simple reason that game theory is only half a science. As its name implies, game theory is but a theory—or collection of theories—about events involving conflict. It can therefore constitute no more than the theoretical branch of an actual or potential science of those events.

Today there are experimental and theoretical branches of chemistry, biology, physics, geology, psychology, and almost every other science known to man. While cooperation between the two branches is closer in some instances than in others, there is never any doubt as to which is the controlling branch. In science, fact always trumps theory. It's been that way at least since Galileo—simply by dropping a light ball and a heavy ball of equal volume from the Leaning Tower of Pisa—discredited Aristotle's long-accepted theory that heavy objects fall faster than light ones.

Galileo's landmark experiment forced his contemporaries to decide for themselves (1) whether or not Aristotle's closely reasoned argument concerning the rates at which aerodynamic[15] bodies fall to earth yields anything more than a plausible hypothesis, and (2) whether or not Aristotle's hypothesis could still be taken seriously after failing to pass Galileo's test. Until a critical mass of learned Europeans shared Galileo's conviction that the answer to both questions was a resounding no, there could be no age of enlightenment.

Whether or not Galileo actually performed the famous experiment is almost beside the point. Until others replicated his (alleged) results, the results settled nothing. When others did obtain similar results, his findings (or alleged findings) became definitive. Aristotle was mistaken! Only gradually did observable fact supersede church-designated books and authors as the final arbiter of scientific truth.

Modern science was born in the early years of the seventeenth century, due in large part to the efforts of Galileo in Italy, Francis Bacon in England, and René Descartes in France. While sharing a common purpose, the latter pair differed as to method. Bacon was a confirmed empiricist who believed that the laws of nature would reveal themselves to any who gathered and digested sufficient quantities of data. Descartes, in contrast, was an armchair philosopher who proposed to "deduce" the laws of nature from the rules of logic (augmented only by the "fact" that God exists) while seated comfortably by the home fire. Where better to undertake the concerted thought that—in his considered opinion—was alone capable of unraveling the subtle laws of nature?

In isolation, neither fact-finding nor armchair philosophy has ever accomplished much. In combination, however, the two have generated more than four centuries of steadily accelerating scientific progress. Recently, as explained in Chapter 1, judicious combinations of elementary game theory and (massively computerized) experiment have enabled inanimate machines to defeat human champions at chess, checkers, Othello, backgammon, and other pastimes. So, even though game theory has failed in its original revolutionary purpose, the possibility remains that a practical science of competition—combining theory with observation and experiment in the manner of other sciences—may succeed where theory alone has so obviously failed. Indeed, the process is already under way.

Before examining the current state of the emerging science of competition, we should reflect in greater detail on the means by which Deep Blue's historic victory over Garry Kasparov was accomplished. They differ significantly from the means by which other soulless machines have been programmed to defeat other human champions. Specifically, we reflect on the fact that the Deep Blue team made surprisingly little use of formal experiments, relying instead on advice from experts in the art (as opposed to the science) of playing tournament chess. Perhaps the most remarkable aspect of the fifty-year campaign to defeat the reigning chess champion of the world is the manner in which team leaders exploited the art of competition to gain the long-sought victory. Chapter 3 will explain—among other things—how they went about it.

CHAPTER 3

Tree Games and Backward Induction

One fine day in 1926, at the tender age of twenty-two, John von Neumann unveiled his groundbreaking theory of games to an audience of professional mathematicians. Already a privatdozent at the University of Berlin—roughly the equivalent of an assistant professor in the United States—he was said to be the youngest man ever to hold that position. Simultaneously, he held a Rockefeller grant for postdoctoral study at the University of Göttingen, then a mecca for the world's mathematicians, logicians, and theoretical physicists. It was there, for instance, that he first met J. Robert Oppenheimer, with whom he would later collaborate on the Manhattan Project. The published version of his 1926 lecture is widely regarded as the founding document of modern game theory.[1]

For the purposes of this book, the words "game" and "competition" will be used interchangeably. On the one hand, every game is obviously a form of competition. On the other, nobody—more than eighty years after von Neumann's lecture, and sixty years after he last modified his definition of a game—has yet identified a form of competition that neither is nor closely resembles what he would have called a game. Although conflation of the two terms has lost whatever shock value it once had, it reminds one that the mental gymnastics performed by the players of chess, checkers, and penny-ante poker are closely akin to that required of lawyers, generals, financiers, and political consultants. There is an essential sameness in all activities for which the consequences of decisions taken by each of several independent decision makers are potentially important to all.

Von Neumann began his lecture by describing the class of games to which his theory would apply. It was later pointed out that all those games, along with quite a few others, may be played on "rooted decision trees" according to a simple protocol. Decision trees are but distant relatives of trees in the forest, having but one root, no central trunk, and numerous branches tipped with leaves. Because they have since proven useful in almost every branch of science or technology, an extensive vocabulary has grown up around them.

A particularly simple decision tree is shown in figure 3.1. The "root" or "origin" of the tree is designated *O*, while the (rectangular) "leaves" at the ends of the branches are labeled *L*, *L*′, *L*″, and *L*‴, respectively. In keeping with botanical usage, the places (here represented by circles) at which branches split are called nodes, while the lines connecting one node to another are called arcs or "edges." The root of the tree is sometimes called the initial node, while the leaves are referred to as terminal nodes. All other nodes are "internal" or intermediate. Figure 3.1 contains two internal nodes, labeled *A* and *B*.

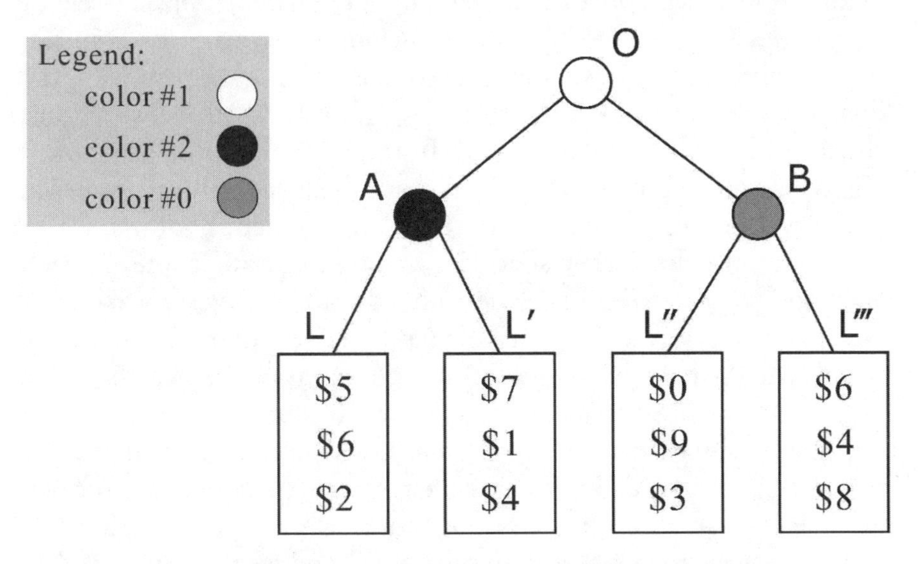

Figure 3.1. A small rooted tree, showing root, leaves, intermediate nodes, and connecting edges. The fact that each leaf contains three numerical rewards indicates that the game is played by three willful players, in addition to dispassionate chance.

Observe that in the present tree (as in all others) each leaf is connected to the root *O* by a single unbroken "path" of edges. Moreover, in any pair of nodes on such a path, one must lie closer to *O* than the other. The one nearer to *O* is said to lie upstream of the other, while the other lies downstream of it. If a pair of nodes is connected by a single edge, they are called neighbors in the tree. One neighbor necessarily lies upstream of the other and is called the upstream neighbor. The other, naturally enough, is called the downstream neighbor. All nodes, save the root and the leaves, have one upstream neighbor and several downstream neighbors.

The formal protocol for playing a "tree game" presumes the nodes to be color coded—each color corresponding to a distinct player—and each leaf to be inscribed with a definite reward for each player. That is the significance of the dollar rewards inscribed in the several cells of the leaves L, L', L'', and L''' of figure 3.1. The top reward in each leaf goes to player 1, in the event that the token ends up on that leaf, while the one beneath it goes to player 2, and the one beneath that to player 3. And so on.

The referee initiates play by placing a token (a coin, pebble, or kernel of corn will do) at *O*—which by convention is of color 1, typically white—and asks player 1 to select a downstream neighbor. The referee then slides the token to the chosen neighbor (either *A* or *B* in figure 3.1) and observes its color. If it is of color *K*, he asks player *K* ("the owner") to designate a downstream neighbor to which he would like the token moved next. And so on. In the fullness of time, the token will come to rest on a leaf, at which point the game will end and each player will receive the reward inscribed in his or her cell of that leaf.

Nodes of color 0 represent "chance moves" in the game. To each such node there corresponds an urn filled with routing slips to downstream neighbors. Instead of asking dame chance where she wants the token sent, the referee draws a slip at random (as in a lottery) from the corresponding urn and slides the token to the designated neighbor.

The fact that there is no internal node of color 3 in figure 3.1 means that player 3 can never be called upon to make a decision. Such an "inactive player" need do nothing more, in von Neumann's conception of a game, than collect a reward at the end of play. The fact that they make no decisions of their own does not prevent such players from influencing the decisions made by others, as will presently be seen. Each decision-making player presumably tries, throughout the duration of play, to steer the token toward a leaf containing an attractive reward for him- or herself.

Because chance is regarded as "player 0," it may be said that every nonterminal node in a game tree represents a situation in which one of the players may be called upon to make a decision. The edges connecting a node with its downstream neighbors then represent options among which the rules of the game permit the owner to choose. By committing the tree of a game like nim or tic-tac-toe to paper, one is actually drawing the rules of that game. It is, upon reflection, astonishing that such a thing is possible. Although the trees corresponding to games like chess, checkers, backgammon, and Go are far too large to be drawn on a single sheet of paper—or etched on a single parking lot, for that matter—it remains instructive to think of the corresponding game trees as *depictions* of the rules of those games.

Whereas human patience and reliability are limited, computers can be programmed to execute the foregoing protocol to near perfection, even on trees containing hundreds of thousands of paths from root to leaf. Indeed they do exactly that—and more—during computer chess tournaments, computer checkers tournaments, and any number of other tournaments organized to challenge the designers of game-playing computer programs.

The root of the chess tree has twenty downstream neighbors, each of which has twenty more, and so on. Whereas the nodes farthest from the root have few downstream neighbors, many of the centrally located ones have forty or more. Combined with the fact that many of the paths from root to leaf consist of more than a hundred edges, the consequent multiplicity of downstream neighbors suggests that the chess tree is vast and complex. Even the reduced tree for tic-tac-toe contains 764 internal nodes, 26,830 leaves, and (as a result) 26,830 distinct paths from root to leaf.[2]

There is a simple step-by-step process for identifying at least tentative solutions of tree games. It works by asking what it is worth to the several players to have the token at rest on a particular node. The answer is called the value of that node. If each player were certain to collect $2 at the end of the game, given that the token currently rests on node P, it could be said that the value of P to each player is $2. It is typically the case, however, throughout the course of play, that different players expect to collect different rewards at the conclusion of play. If players 1, 2, and 3 were confident of collecting \$7, \$8, and \$9 at the end of the game, those sums would constitute the values (plural) of node *P* to the several players.

Should the token ever come to rest at *A* in figure 3.1, player 2 is all but certain to send it on to *L* since the \$6 reward she can obtain by so doing exceeds the \$1 she could earn by dispatching it to *L*′, her only alternative. Thus, when the token is at rest on *L*, the players may confidently expect soon to collect \$5, \$6, and \$2, respectively, a fact that may be recorded by writing VAL(*A*) = COL(\$5, \$6, \$2), where COL(*x*, *y*, *z*) is nothing more than a typographically convenient abbreviation for the column of alphanumeric symbols in which *x* stands above *y*, which stands above *z*.

The conclusion that player 2 will choose *L* over *L*′, if and when the token arrives at *A*, is particularly firm in the present case because the extra \$2 that players 1 and 3 would each earn if the token were to finish at *L*′ instead of *L* does not add up to the \$5 player 2 would forgo by allowing that to happen. The situation would be quite different if the middle reward in *L* were \$2 instead of \$6, for players 1 and 3 could then—by contributing a dollar each—offer player 2 a \$2 bribe to send the token on to *L*′. Each of the three would gain a dollar by collusion. On the other hand, players 1 and 3 need offer player 2 only \$0.51 apiece for the agreement to benefit player 2, or player 2 might hold out for \$1.99 from each in order to go along. In real life, deals often go unmade because the potential beneficiaries are unable to reach agreement on a division of the spoils.

By sending the token to *B*, player 1 turns the remainder of the game into a lottery in which player 2 may win either \$4.00 or \$9.00, player 3 may win either \$8.00 or \$3.00, and she herself may win either \$6.00 or nothing. If she knew the urn corresponding to node *B* to contain equally many slips directing the token to *L*″ and *L*‴, she could calculate that a ticket to participate in the proposed lottery would be worth \$3.00 to herself, \$6.50 to player 2, and \$5.50 to player 3.

The more slips the urn contains directing the token to *L*″, the more such a ticket is worth to player 2, and the less it is worth to players 1 and 3. When the probabilities of the potential outcomes in a lottery are known, it is a simple matter to compute the "actuarial value" of a ticket to participate as player 1, player 2, or player 3. Should the urn contain eleven slips directing the token to *L*‴ for every one directing it to L'', tickets to participate as player 1, player 2, or player 3 in the lottery at B would be worth \$5.50, \$4.42, and \$7.58, respectively. In the former case, where chance was equally likely to dispatch the token to either L'' or L''' from B, VAL(B) = COL(\$3.00, \$6.50, \$5.50). In the latter case, in which L''' is by far the likelier destination, VAL(B) = COL(\$5.50, \$4.42, \$7.58).

When choosing her opening move, player 1 faces a situation akin to that depicted in figure 3.2. The two edges emanating from the root node O lead to "pseudo leaves" A^* and B^*, upon which are inscribed the values of nodes A and B in figure 3.1, as deduced above. A^* and B^* are not true leaves, because the rewards inscribed on them are not guaranteed by the rules of the game. They are conditional on the assumption that player 2 will indeed choose L over L' if and when given the opportunity to do so, and that the actuarial values of tickets enti-

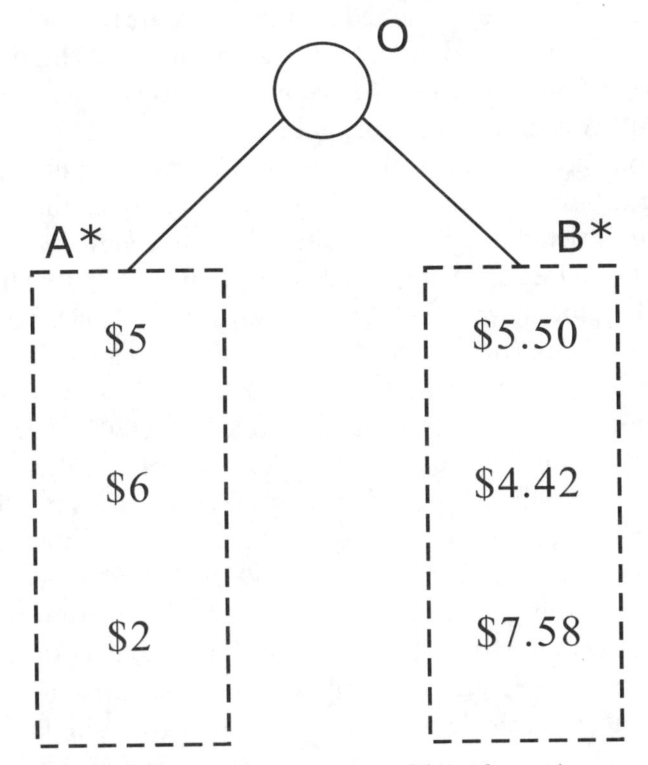

Figure 3.2. The reduced tree obtained from the one in figure 3.1 by turning A *and* B *into pseudo leaves* A* *and* B*, *containing pseudo rewards inferred from the actual rewards in 3.1.*

tling the bearers to assume the roles of player 1, player 2, and player 3 in the lottery to be held at *B* fairly represent their respective prospects in that contest. The pseudo leaves *A** and *B** are bounded by dotted lines instead of solid ones to emphasize that they contain only pseudo rewards rather than actual rewards guaranteed by the rules. Since $5.50 exceeds $5.00, it seems reasonable to suppose that player 1 will use the opening move to send the token to *B*. Hence VAL(*O*) = VAL(*B*) = COL($5.50, $4.42, $7.58).

VAL(*O*) is often called the value of the entire tree game depicted in figure 3.1. The conclusion that VAL(*O*) = VAL(*B*) is conditional on the assumption that player 1 will prefer eleven-to-one odds on winning $6 to the (all but) certain $5 to be had by sending the token to *A*. The process by which that conclusion was reached is known as backward induction, since it reasons from the leaves of the tree back toward the root, in the upstream direction, opposite to the direction in which the token is constrained to travel. Despite the many assumptions involved, backward induction is a very useful technique. Without it, Deep Blue could never have beaten Kasparov, and games of pursuit and evasion would be insoluble.

One of backward induction's most attractive features is that it works on arbitrarily large trees. To see why, consider the somewhat larger tree depicted in figure 3.3. The values of nodes *R*, *Z*, *U*, *V*, and *W* can all be deduced exactly as were VAL(*A*) and VAL(*B*) in figure 3.1, because they, too, have only leaves for downstream neighbors. When that is done, the portions of the tree lying downstream of those nodes may be replaced by the pseudo leaves *R**, *Z**, *U**, *V**, and *W**, containing the values of the newly evaluated nodes, exactly as *A* and *B* were replaced by *A** and *B** in figure 3.2. Next, the values of *X*, *Y*, and *Q* may be deduced, then *P*, and finally VAL(*O*). It makes no difference whether *X* is evaluated before or after *Y*, but it is essential that *Z* be evaluated before *X*, and *W* before *Y*. The value of each node is deduced in turn by analyzing a series of mini trees that differ from the one in figure 3.2 only in the (frequently large) number of edges that emanate from the root node, and the (typically rather small) number of rewards inscribed on the leaves and/or pseudo leaves with which those edges connect.

The root node of any one-move game is easily evaluated. If it is a chance node, it assumes its actuarial value, as did node *B* in figure 3.1.

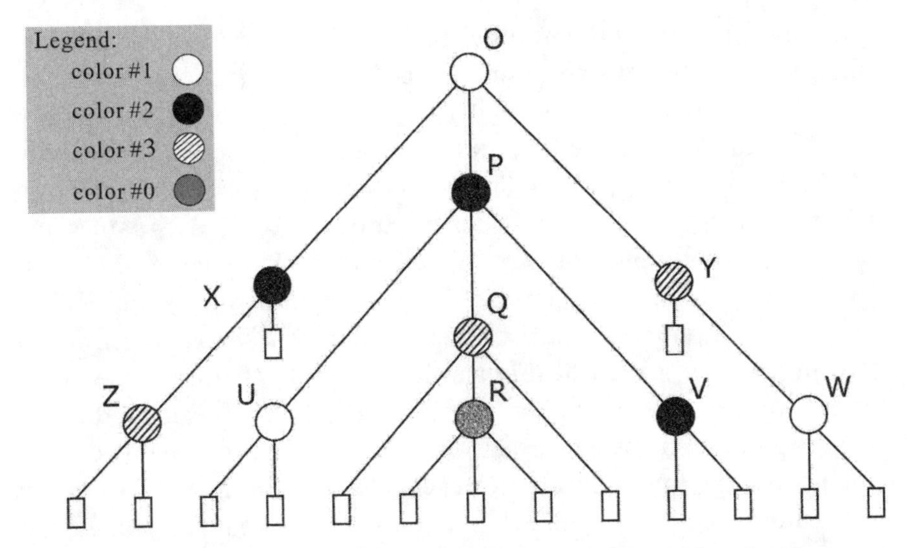

Figure 3.3. A slightly larger rooted tree, in which several successively reduced trees must be considered before the root node can be evaluated.

If it is a player node, it assumes (as did both *A* and *O* in figure 3.1) the column of values contained in the leaf or pseudo leaf which rewards the decision maker most generously.

When player 1 has no opponents, as in Sudoku, the method associates a single number (column of height one) with each node of the tree in question. VAL(*O*) then represents the largest reward White can obtain. To earn it, White need only steer from node to node, destined always for the downstream neighbor of greatest value. When (as in solitaire) chance is White's only opponent, the method identifies the largest reward White can reasonably expect to obtain. Single-player tree games are often called multistage decision problems, or problems of "dynamic programming." Since about 1950, computer scientists have studied such problems in fine detail.

It makes little difference, for the method of backward induction, how many players are involved. The calculations and bookkeeping don't change all that much. But the results are far more convincing in some games than in others. That is because opportunities for collusion, as discussed in connection with node *A* in figure 3.1, tend to invalidate the solutions obtained. Because backward induction ignores the possibility of collusion, and because opportunities for collusion tend to proliferate as the number of decision makers grows, solutions obtained thereby for games involving three or more players must be regarded as tentative at best.

Whereas single-player games such as solitaire are conceptually simple, games involving three or more players—complicated as they are by the possibility of collusion—can be complex. Two-player games may be either. The simplest of these are called zero-sum games. Mutually beneficial collusion between the players in such a game is impossible.

A two-player tree game is of the zero-sum variety if and only if the two numbers in each leaf differ only in sign. In that case, the upper and lower rewards in a given leaf add up to zero, meaning that the winning player must win exactly what the losing player loses. In all such cases, the lower number in each leaf is redundant. Accordingly, the value VAL(*P*) of a node *P* in a zero-sum tree game is adequately described as a single number, with the understanding that it represents the amount Black (player 2) should expect to owe White (player 1) when the game is over. If it is a negative number, White must expect to owe it to Black. By that convention, the value attributed to *O* by the method of backward induction represents a reward that is at once the largest White can single-handedly obtain and the least he or she need accept.

The innate complexity of many-player games is regrettable, since the most important forms of competition, from the standpoint of human welfare and the common good, are of the latter sort. Because the possibility of bribes, side payments, and spontaneous cooperation limits the utility of backward induction for solving multiplayer and non-zero-sum tree games, two different versions of many-player game theory have evolved. While John (the beautiful mind) Nash, Reinhard Selten, John Harsanyi, and no few others were busy developing "the noncooperative theory of many-player games," in recognition of which those three shared the 1994 Nobel Prize in Economics, the theory developed by von Neumann and Morgenstern was coming to be known as the "cooperative theory." In principle, the two complement each other and represent distinct parts of a whole that is greater than either part. In practice, disputes often arise as to which theory applies in a given situation.

Perhaps the most notable feature of the tree game concept is its versatility.[3] A remarkable variety of practical real-world situations are fruitfully compared to simple tree games. This has been demonstrated on countless occasions, by a legion of books and authors. Of these, few are more dramatic than Marek Kaminski's recent book *Games Prisoners Play: The Tragicomic Worlds of Polish Prison.*

Kaminski's book chronicles the bizarre, terrifying world of rapes, knife fights, suicides, blunt talk, and self-inflicted injuries that existed behind Polish prison walls during the waning years of communist rule. Apprehended in his native Warsaw, during March 1985, in a car filled with anticommunist (Solidarność) literature, Kaminski was sentenced within days to a year in jail. As a student of social science, he quickly resolved to devote his time behind bars to the most scientific study possible under the circumstances of Polish prison culture.

The central thesis of Kaminski's book is that prison life, by its very nature, obliges inmates to function as indefatigable cost-benefit analysts immersed in a seemingly endless succession of subtle non-zero-sum games. Because the contests he observed—and occasionally took part in—involve small numbers of choices from short lists of alternatives, Kaminski is able to construct their entire trees, and so to show how clever, well-informed inmates can sometimes reason their way around misfortunes that their less cerebral companions learn to avoid only through sad experience.

On another front, Henry Ford, acting for the Ford Motor Company (FMC), and Alfred P. Sloan, acting for General Motors (GM), were the key participants in a tree game of great significance during the formative years of the automobile industry. John McDonald's 1975 book, *The Game of Business*, describes the contest between them.[4] As Sloan's ghost biographer, McDonald was the ideal choice to chronicle their conflict.

McDonald spent several decades as a writer and editor at *Fortune* magazine, in which capacities he both pioneered the effort to explain von Neumann and Morgenstern's theories to the public and began his involvement with Sloan.[5] He recounts that in the early spring of 1921, a team of GM executives led by Sloan spent several weeks deciding when and how to invade the market for low-priced automobiles then dominated by Ford's Model T. At the time, GM controlled about 10 percent of the American automobile market, while Ford controlled more than half. Yet because GM cars were more expensive—ranging from the mid-priced Chevrolet to the high-priced Cadillac—GM had finished a relatively close second to Ford in dollar sales the previous year.

In May, Sloan's team submitted a daring plan to the GM board of directors, who adopted it the following month. The plan was predicated on a perceived gap in the product line being marketed to the American public. There were no cars in the price range between the Ford Model T at $415 and the Chevrolet touring car at $815. While Ford was without competitors in the $400–$500 price range during the spring of 1921, Chevrolet had several in the $800–$900 range.

Instead of challenging the legendary Model T head-on, the Sloan team proposed to market a slightly less austere form of "basic transportation" at a slightly higher price. While the Model T had to be hand-cranked, for instance, the Chevrolet came equipped with an automatic starter. Moreover, in an age when flat tires were commonplace, the Chevy boasted a spare tire mounted in advance on an "interchangeable rim," greatly facilitating tire changes.

In September 1921, the U.S. economy was barely beginning to emerge from the harsh recession of late 1920. GM chose that moment to cut the price of the Chevrolet touring car to $525. It was done in the expectation that by such action, GM could draw customers away from both the lower-priced Ford and the somewhat more expensive cars that the Chevy had previously competed with on roughly equal terms. McDonald estimates that GM needed to sell between 150,000 and 200,000 Chevrolets by the end of 1922 in order to sustain its attack.

Had they wished to familiarize themselves with the strategic possibilities of the struggle they were about to initiate, GM executives might have constructed a "management-training game." Though common today, such games were all but unheard of in 1921, save in the military. McDonald suggests that an adequate training game might have been laid out on a game board containing the tree depicted in figure 3.4. The referee would have begun play by placing a token on the board at "Start" and sliding it downstream to *O*, simulating GM's initial price reduction. The four edges of the tree emanating from *O* represent possible reactions by Ford. He could have (1) simply held his price (*hp*) where it was while awaiting further developments; (2) cut his price (*cp*) immediately; (3) devised a souped-up version of his Model T—call it a Mercury Model T (*mt*)—to compete directly with the Chevrolet in both price and amenity; or (4) done both (*cp* + *mt*), cutting the price of the standard Model T *and* introducing a souped-up version to deprive Chevrolet of needed sales volume. Had Ford felt unready to produce such a vehicle himself, he might have purchased an existing firm such as Hup, Franklin, Reo, or Essex to produce it for him. Scores of

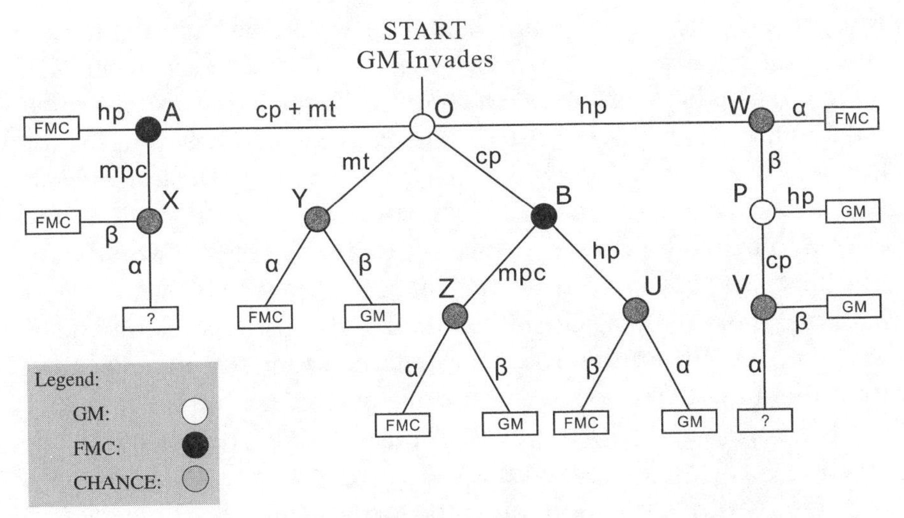

Figure 3.4. Tree-game conceptualization of the contest between Ford and General Motors, in which an early blunder by Ford enabled GM to gain a toe-hold in the (then huge) market for low-price cars during the early 1920s.

automobile companies were still in operation in 1921, and Ford had the resources to acquire any one of them.

The black interior nodes (*A* and *B*) in figure 3.4 correspond to situations in which it is GM's turn to move, while the white ones (*O* and *P*) correspond to Ford's moves. The unmarked ones (*U*, *V*, *W*, *X*, *Y*, *Z*) represent chance moves corresponding to random changes in the health of the national economy. The eventual outcome was highly dependent on economic conditions. Indeed, GM's gamble would surely have failed had the recession of late 1920 continued into—or resumed during—1922. The Greek letters alpha (α) and beta (β) denote "complementary probabilities" (meaning that together they add up to 100 percent) of which β is the larger.

There is no real reason to suppose that all the small probabilities (alphas) in the figure, nor all the large ones (betas), were equal. Yet the fact that McDonald neglects to specify which of the large probabilities exceed which others would seem to suggest that their absolute—as opposed to their relative—sizes are of little consequence. Readers uncomfortable with algebraic symbolism should read ⅓ (33 percent) for α and ⅔ (67 percent) for β. An honest die will show either 5 or 6 about one time in three.

Had the Sloan team member assigned to represent Ford in the

training game chosen $cp + mt$, the referee would have pushed the token downstream to A, a move belonging to GM, and asked the GM representative how he wished to continue. If the latter responded by holding the Chevy price constant, the referee would have slid the token farther downstream along the edge marked hp to the leaf marked FMC, signifying an immediate victory for Ford. But if the GM rep elected to match Ford's every price cut (*mpc*), the referee would have slid the token downstream to the chance move X. There, he would have cast the die and again slid the marker downstream along one of the two edges emanating from that node, choosing the one marked α if the die showed 5 or more, and the one marked β if not.

The edge marked β ends at a leaf marked FMC, representing victory for Ford, while the one marked α leads to a leaf containing a question mark, signifying an outcome too close to call. There the referee must toss a (fair) coin to select a winner. The choice $cp + mt$ thus enables Ford to win for sure if GM is foolish enough to select hp at A, and with probability $\beta + \frac{1}{2}\alpha$ if not. The latter exceeds 75 percent for any admissible values of α and β, and exceeds 83 percent if $\alpha = \frac{1}{3}$ and $\beta = \frac{2}{3}$. The decisions made by every practice player during every practice session, along with the resulting outcomes, would naturally have been recorded in meticulous detail for later analysis.

With repetition, the executives involved would likely have become proficient. Although they wouldn't have done so as quickly or as thoroughly as a computer-learning program of the sort that conquered checkers and backgammon during the 1990s, they would surely have discovered that Ford had no better opening move than $cp + mt$, because the gap in the industry's product line was just too wide to allow Chevy to occupy it alone.

In the event, Ford was slow to appreciate the breadth and significance of the gap GM was moving into. He made no effort to produce a souped-up version of his Model T and was slow to cut his price for the standard version. That enabled Chevrolet to sell 200,000 cars—roughly 13 percent of the market for low-priced cars—by the end of 1922. Chevy sales increased to 18 percent in 1923, allowing GM to expand and strengthen its chain of dealerships at Ford's expense. By decade's end, after a second round of strategic interaction, GM's share in every segment of the American automobile market was regularly exceeding Ford's.

McDonald has written the definitive account of GM's successful campaign to wrest the mantle of leadership in the U.S. automobile market from Ford's once firm grasp. It is far fuller than the foregoing summary and defends—among other things—the thesis that GM's survival in the game of 1921–23, and its success in the subsequent battle for dominance, were due as much to luck in the matter of economic growth rates as to Sloan's clear understanding of the strategic possibilities of the U.S. automobile market as then constituted.

Never again would the market for low-priced new cars be as important as it had been in the early years of the industry, in part because used cars were beginning to satisfy much of the demand for "basic transportation." Buyers of limited means were increasingly inclined to purchase previously owned Cadillacs, Lincolns, Packards, and Pierce-Arrows instead of new Model Ts. That understanding—which seems to have originated with Sloan—enabled GM to best Ford in a game combining the effects of both strategy and chance.

It should not go unmentioned that McDonald's management-training game is easily solved by backward induction. Since the method works with numerical labels only, the would-be solver must begin by replacing the abbreviations inscribed in the leaves of figure 3.4 with numerical values. The obvious expedient is to replace each FMC label with a 1, symbolizing a victory for Ford, each GM label with a -1, symbolizing a defeat for Ford, and each question mark with a 0. Such replacement seems appropriate whenever the victor's reward is independent—or largely so—of the margin of victory.

The reader will observe that nodes X, Y, Z, U, and V are immediately eligible for evaluation, and that evaluating them makes A, B, and P eligible as well. Next comes W, and finally O. The fact that $\mathrm{VAL}(O) = \beta$ turns out to mean that Ford should expect to win with probability $\beta + \frac{1}{2}\alpha$, and to lose with probability $\frac{1}{2}\alpha$. But Ford should not expect to win unless it exercises option $cp + mt$ at O, because the unserved segment of the market between the Model T and its lowest-priced competitors was too extensive for even the dominant Ford to concede.

Claude Shannon of Bell Laboratories was the first to propose a practical method of enabling computers to play such games as chess and checkers, which generate almost inconceivably large trees. In 1950, at the dawn of the computer age, he noted that all the nodes, edges, and leaves downstream of a given node in any game tree constitute a "sub-tree" of the original, and proposed that one should be able to play tolerably well by performing backward induction on the "pruned sub-trees" obtained by cutting away all but the first few arcs of each path in the sub-tree rooted at the current board position. Deep Blue II, for instance, seems to have terminated such paths after at most eleven moves.[6] The result was a class of pruned sub-trees small enough (if only barely) to be treated by backward induction once the ends of the foreshortened paths were equipped with "pseudo leaves" containing appropriate numerical rewards or, as Shannon called them, "scores."

Shannon proposed a relatively simple way of assigning an appropriate score to any conceivable board position. He began by subtracting the number of black pawns on the board from the number of white ones. Then he subtracted the number of black bishops and knights from the number of white bishops and knights, multiplied the difference by 3, and added the result to the previous total. Next he subtracted the number of black rooks from the number of white ones, multiplied the difference by 5, and added that result to the previous total. Finally he multiplied the difference (if any) in queens by 9, the difference (if any) in kings by 200, and added those as well to the previous total. A negative score indicates that the position favors Black. The "weights" 1, 3, 3, 5, 9, and 200 were intended to reflect the relative values of pawns, bishops, knights, rooks, queens, and kings in the game.

The result turned out to be a surprisingly accurate indicator of the strength of White's position on the board relative to Black's. Indeed, such "material evaluations" frequently suffice to determine which player holds the momentary advantage. But Shannon wasn't finished yet. He next proposed to subtract the number of "poorly positioned" black pawns from the number of poorly positioned white ones, divide the difference by 10, and subtract that result from the previous total. He also proposed to subtract the number of moves currently available to Black from the number currently available to White, divide that difference by 10, and add the result to the previous total. These corrections for "pawn placement" and "current mobility" represent significant improvements over the "raw score" provided by material evaluation alone. In later years, Shannon proposed yet other modifications.

Modern chess-playing computer programs employ far more elaborate scoring systems than any proposed by Shannon. Advantages of only a few hundredths of a pawn are now considered significant. If game-playing programs were books, state-of-the-art scoring systems would occupy the most important chapters. Detailed accounts of important man-versus-machine matches—including the ones between Garry Kasparov and versions I and II of Deep Blue—invariably describe last-minute changes made by the program's handlers to their ever-evolving scoring systems, in an effort to correct weaknesses exposed during the latest encounter. In so doing, they typically formulate, debate, and test a considerable variety of hypotheses concerning the cause of the observed weakness and the relative effectiveness of proposed remedies. They behave, in other words, like working scientists.

As a result, the contrast between their undeniable success and the disappointing results achieved by pure theory is turning the study of competition—slowly but surely—into a mainly experimental science. The study of board games like chess, checkers, Scrabble, Monopoly, and backgammon is merely the most easily explained aspect of this propitious trend.

Except for the first five to fifteen moves made in a particular game by a typical chess program, and possibly a few at the very end, every move such a program makes in a match is determined by the scoring system. After each move made by its opponent, the program generates the pruned sub-tree that begins at the current board position and ends a short distance downstream. It then appends a pseudo leaf to the downstream end of each foreshortened path, employs the scoring system to insert appropriate scores into the pseudo leaves, and performs a quick backward induction before making a single move and awaiting the reply.

The first five to fifteen moves in a game of computer chess or checkers are made quite differently. The programs involved are invariably equipped with extensive built-in libraries of board positions likely to occur near the beginning of a game, together with preselected moves to be made from each such position. These libraries are known as "opening books." Human opponents often strive to confuse their inanimate rivals by taking them "out of their opening books" as quickly as possible, even if such behavior courts disaster.

There also exist libraries of known endgame positions, such as the one in figure 2.1, for which unbeatable strategies are known and cataloged. Should such a program recognize the current board position among the ones tabulated in its "closing book," it has no further need to strategize. Its closing book reveals the means by which the current game can be brought to the most favorable conclusion still possible.

In addition to permitting the solution of simple little games, such as McDonald's formulation of the game between Sloan and Ford, and in addition to suggesting partial-solution procedures for games (such as chess and checkers) too large to be solved completely, tree games provide a context within which it seems possible to analyze any imaginable form of competition.

The definition of a tree game actually evolved through three distinct stages. Von Neumann's 1928 paper, like his 1926 lecture, made no mention of contests in which one or more of the players may lack potentially relevant information when called upon to move. Known as "games of imperfect information," these include all versions of hide-and-seek, most of cops and robbers, and any card game in which cards are dealt facedown. Another such game is the one von Neumann and Morgenstern described as "double-blind chess," or Kriegspeil. Devised as a training exercise for military schools, the earliest version of Kriegspiel was played on a map of the French-Belgian border divided into thirty-six hundred squares.[7] Game pieces advanced or retreated from sector to sector like real armies. The game soon came to the attention of the Prussian high command, which promptly supplied each regiment with an official book of rules and issued standing orders that every officer should learn to play. The kaiser himself attended Kriegspiel tournaments in full military dress. Growing up in Budapest, the young von Neumann played an improvised version of the game with his brothers, especially during World War I. Today, a streamlined version of Kriegspiel is played with three chessboards arranged in the following manner: ●□|□|□○. The black and white circles represent the two players, while the three squares represent chessboards. The vertical slashes between the boards represent screens that prevent either player (though not the referee) from seeing any board other than the one in front of her.

The middle board belongs to the referee, who knows the current position of every piece of either color. The players, in contrast, know only the positions of their own pieces. The black chessmen on White's board represent only White's best estimate of the current position of Black's men, and vice versa. Such estimates are reasonably accurate near the beginning of the game but rapidly deteriorate as the game unfolds. Not knowing where his opponent's pieces actually are, each player is obliged to obtain the referee's permission before making any move. As one might expect, many requests are intended merely to furnish information. To listen to a game in progress is to hear the referee—often corrected by bystanders—issue one lengthy sequence of denials after another, interspersed with occasional grants of permission.

Not until 1944, after Morgenstern had revived his interest in the subject, did von Neumann expand his definition of what he called "games in extensive form" to include games of imperfect information. He and Morgenstern devoted an entire chapter of their 1944 book to explaining what a game is, before offering their formal definition. Finally, in 1950, Harold Kuhn gave a vastly superior—yet logically equivalent—form of the von Neumann–Morgenstern definition. Since then, due to its visual character and ease of application, the von Neumann–Morgenstern–Kuhn definition has remained the gold standard.[7] The main elements have already been described. They include (1) the rooted tree itself, (2) the complete list of final scores (rewards) inscribed in each leaf, (3) the color scheme dictating which players are entitled to make which routing decisions as the token progresses from root to leaf, and (4) the urns associated with each uncolored (chance) node, from which the referee is obliged to draw random routing slips. They also include (5) a division of nodes of a common color into what Kuhn called "information sets," among the elements of which the owner (a.k.a. the player corresponding to the color in question) is unable to distinguish.

Figure 3.5 contains four information sets, one of which belongs to White, one to Black, and two to Gray. The fact that Black's information set—like both of Gray's—contains multiple nodes means that neither Black nor Gray will know the exact whereabouts of the token when his or her turn comes to play. Informed only that the token is temporarily at rest in information set IV, for instance, Gray is left to wonder whether it is at *R* or *S*. She is therefore unable to choose—as plainly she would if she could—the right fork (as seen from upstream) when the token is at *R* and the left one when it is at *S*. Gray cannot do as she chooses, because she is allowed to know only that the token is at *R* or *S*. Such is the significance of information sets—each represents a decision to be made without exact knowledge of the token's current position.

The rules of the game do not allow Gray to predicate her decision on the token's position within IV, because they do not allow her to know which of the two positions (nodes) in IV it occupies. Assuming she guesses right as often as she guesses wrong, Gray can expect to win about $1.50 every time the token lands in IV. She may therefore be willing to pay as much as $1.50 to learn the token's whereabouts within IV. Information has measurable cash value when playing for money.

The players in a tree game are said to possess perfect information whenever each information set contains but one node. The fact that backward induction can be used to evaluate every node in the tree of any game of perfect information, including the root node, is known as the perfect information theorem. The original version was proven in

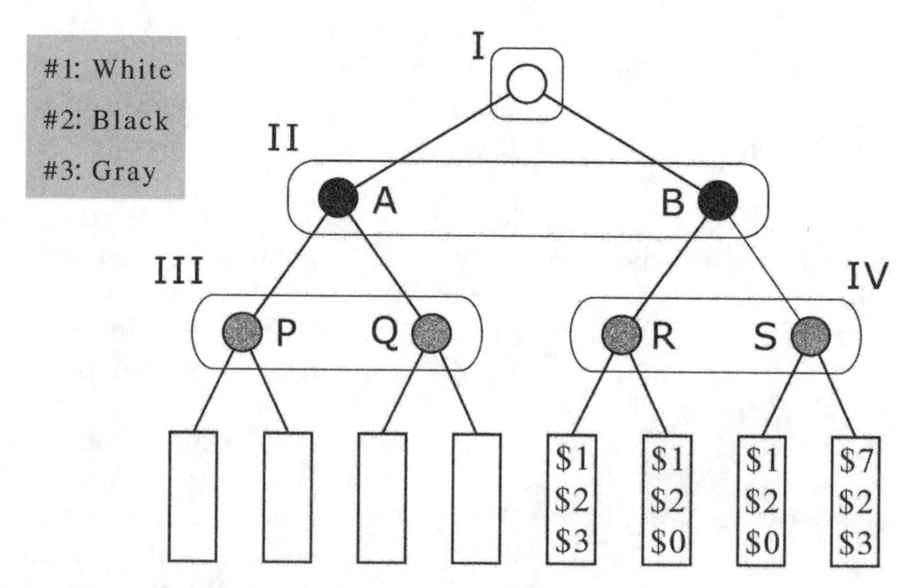

Figure 3.5. The tree of a game of imperfect information, showing information sets I to IV.

1913 by a mathematician named Ernst Zermelo, whose primary interest was in chess.

The fact that the trees corresponding to chess, checkers, and Go contain no chance nodes, together with the fact that every leaf in those trees contains exactly one of the numbers 0, 1, and -1, implies that the value of any node in any of those trees is equal to one of those three numbers. The nodes P for which VAL(P) $= 1$ are called winning positions for White, while those for which VAL(P) $= -1$ are called winning positions for Black. The ones for which VAL(P) $= 0$ are known as drawn positions. Although computers now play chess and checkers more skillfully than any human, the values of the root nodes of those games (like that for Go) remain unknown.

Imperfect information is the rule rather than the exception in the more common forms of high-stakes competition. Football coaches are often obliged to prepare game plans without knowing whether the opposing team's star running back will be able to play. A military commander may have to issue orders while his unit is under attack, without knowing the attackers' real strength. Bidders at an auction for yearling racehorses cannot know how fast a particular colt or filly will be able to run as a three-year-old. A Kriegspiel player denied permission to advance his queen's rook more than four squares knows that the way ahead is blocked by an opposing piece but cannot tell whether the blockage is due to a pawn, bishop, knight, or queen. Nor can he tell what if any additional opposing pieces lurk beyond the roadblock. And so on. Participants in almost all forms of competition are regularly obliged to make decisions based on an incomplete or unreliable knowledge of the facts.

Kuhn's information sets completed the five-part definition of a tree game. The parts are:

1. a rooted tree upon which the token is required to move;
2. a complete schedule of rewards inscribed on each leaf;
3. a color scheme indicating which players are entitled to make which routing decisions;
4. a collection of urns from which the referee shall draw routing slips at random;
5. a partition of nodes of a common color into information sets as indicated in figure 3.5.

Although more inclusive definitions have since been proposed, the von Neumann–Morgenstern–Kuhn version remains the gold standard. Its narrowness is mitigated by its pristine clarity, and by the previously mentioned fact that *no one has yet described a form of competition that neither is nor closely resembles a tree game*. Even the simplest forms of competition, such as sprint racing and contests of strength, involve some strategic decisions. Each sprinter must weigh the need for a fast start against the risk of a false one and decide whether to retaliate when bumped and whether to avoid injury by easing up near the finish line. Weight lifters must likewise consider the fact that extra effort increases the risk of a foul and/or an injury. Although it seems unlikely that anyone will become a more successful track or field athlete by learning game theory, the fact remains that even the simplest forms of competition do involve strategic decisions.

Among von Neumann's more inspired choices was his decision to identify each game with "the totality of the rules which describe it." Thus, if asked the difference between baseball and softball, between chess and checkers, or between soccer, rugby, and American football, he could in each case have given the same two-word answer: "the rules." Indeed the rules of any organized sport describe virtually everything about it, including the space in which it is to be played, the equipment to be used, and the means by which the final score shall be determined. The rules of golf, tennis, baseball, chess, backgammon, and the like contain only minor ambiguities, which future editions of the official rule books will strive to eliminate. A game tree is nothing more than a visual presentation of the rules of a particular game. Posterity may well regard the von Neumann–Morgenstern–Kuhn definition of a game as the foundation upon which all future studies of competition should rest.

Among the central theses of this book is the assertion that, whereas purely theoretical investigations have relatively little to teach about competition, impressive progress has been achieved by blending a small amount of theory—such as the method of backward induction—with an abundance of factual evidence accumulated through trial and error. On the other hand, pure game theory has enjoyed genuine moments of triumph, a few of which will be described in chapters 5 and 6, following a brief digression on the nature of scientific inquiry.

CHAPTER 4

Models and Paradigms

Every year, over a weekend in early February, several hundred three-member teams of college, university, or (in some cases) high-school students from around the world take part in the Mathematical Contest in Modeling (MCM). First held in 1985, the contest is designed to offer students of applied mathematics a competitive challenge comparable to that which the older William Lowell Putnam Competition has long offered to students of pure mathematics.

Applied mathematics deals with mathematical models of the world in which we live. Climate models, population models, contagion models, traffic-flow models, and econometric models—to name but a few—are increasingly in the news. They influence the most important decisions federal, state, and local officials are called upon to make. The inability of existing climate models to demonstrate conclusively that global warming is under way has long impeded efforts to forestall it, as the inability of existing econometric models to predict the onset of recessions impedes efforts to prevent them.[1]

MCM teams are allowed to see the contest problems at about six o'clock on Thursday evening and have until the same time Monday to submit their solutions. There are always two problems—designated A and B—to choose from, and each team must begin by deciding which to work on. Teams may consult as many books, papers, and Internet sites as they wish and may download any software they deem relevant. The object is to submit as complete a solution as possible to the chosen problem in the time allotted.

A typical MCM problem can be posed in a paragraph or two of plain English, not unlike a "word problem" in high-school algebra. Contestants must construct and at least partially solve a "mathematical model" of the chosen problem. To that end, they must identify the numerical variables implicit in the problem statement, specify ranges over which those variables may reasonably be expected to vary, and supply a list of conditions—usually in the form of equations ($x = y$) and/or inequalities (x may not exceed y)—that the values assigned to the problem variables must satisfy in order to constitute a solution of the model.

To "solve" such a model, the students must produce a set of numerical values that, when assigned to the corresponding variables, meet the stated conditions. Then they must extract from their (numerical) solutions a plan of action addressing the given problem and explain why their plan is trustworthy. This final "verification" step is frequently the most important, because much detail is necessarily lost in translation from a question posed in plain English to a concise yet solvable mathematical model. That is why the history of applied mathematics is replete with instances in which the solution of this or that elegant mathematical model failed miserably to solve the practical problem of interest. Sometimes these failures can be remedied by tweaking the model imperceptibly, and sometimes they require a wholesale return to the drawing board.

Perhaps the most remarkable thing about the modeling process—amply reflected in the annals of the MCM—is the variety of models that can shed light on a single practical problem. Every veteran judge of the contest (the author included) can recall models of which he or she would never have thought, yet which permitted less than obviously relevant mathematical methods to be brought to bear on the problem at hand. Sometimes different models yield starkly different solutions, in which case at least one of them must be wrong. More often than not, however, quite different models yield similar conclusions concerning the practical problem at hand, so that each confirms the validity of the other.

The concept of a mathematical model is of relatively recent vintage. Seventeenth- and eighteenth-century scientists saw little need to distinguish between natural phenomena and the models they themselves were constructing of those phenomena. Their nineteenth-century successors were less fortunate. The various crises that shook mathematics and physics during their careers forced them to ponder such distinctions at length. Yet it was not until the early years of the twentieth century, as physics absorbed first relativity and then quantum mechanics—both of which offer alternatives to the traditional Newtonian-Galilean model of the universe—that leading geometers and physicists arrived at a clear understanding of the matter. Practitioners of the other sciences learned the subject from them. Today, colleges, universities, and even a few elite high schools offer courses in mathematical modeling.

G. H. Hardy cut to the heart of the matter in his 1940 memoir *A Mathematician's Apology*. In it he discussed "a truism which a good many people who ought to know better are apt to forget."[2] Even astronomers, he began—who make extensive use of mathematics in their work and are well trained in the subject—have been known to announce "a mathematical proof" that some part of the physical universe must behave in a particular way. But, he went on, *"all such claims, if interpreted literally, are strictly nonsense."* That in a nutshell is Hardy's truism. *"All such claims . . . are strictly nonsense."* They cannot be otherwise, because "it cannot be possible to prove mathematically that there will be an eclipse tomorrow." No such proof can ever exist "because eclipses and other physical phenomena do not form a part of the abstract world of mathematics; and this, I suppose, all astronomers would admit when pressed, however many eclipses they may have predicted correctly."[3]

The "abstract world of mathematics" to which Hardy refers is made up of lines, circles, numbers, and the like, none of which exist in nature. This is perhaps clearest in the case of geometric figures because—as teachers of the subject never tire of pointing out—the points, lines, and circles one draws with pencil on paper, or with chalk on a blackboard, are but imperfect realizations of the abstractions they are intended to represent. The same is true of ordinary numbers like 3, 7, and 11, along with exotic ones like π, $\sqrt{2}$, and $i = \sqrt{-1}$—all of which fail to exist anywhere in the physical universe. Although everyone knows what is meant by such phrases as "three strikes and you're out" and "ten minutes to train time," few can explain what 3 and 10 actually are. Numbers (like sets and geometric figures) seem to exist only in the mind of man.

Even though sets, numbers, and geometric figures exist only in the imagination, they and other inhabitants of the "abstract world of mathematics" are the only things about which mathematics can actually prove anything. Hardy could assert with "mathematical certainty" that 8,712 and 9,801 are the only four-digit numbers that are exact multiples of their "reversals," in the sense that

$$8{,}712 = 4 \times 2{,}178 \quad \text{and} \quad 9{,}801 = 9 \times 1{,}089.$$

Likewise, there are just four numbers greater than 1 that equal the sum of the cubes of their digits, namely

$$153 = 1^3 + 5^3 + 3^3; \qquad 370 = 3^3 + 7^3 + 0^3;$$
$$371 = 3^3 + 7^3 + 1^3; \qquad 407 = 4^3 + 0^3 + 7^3.$$

Never can it be said, with the same sort of certainty, that there will be an eclipse tomorrow. On the contrary, an astronomer who announces a proof that an eclipse is imminent necessarily overstates his case. Strictly speaking, he can mean only that such an event will occur in the model or models with which he is familiar, models consisting—as all mathematical models do—of numbers, sets, and other figments of the imagination.[4] The fact that predictions based on this or that model have never been wrong in the past is beside the point. Hardy's truism asserts that neither tides, eclipses, and electrical phenomena, which are very well predicted by mathematical models, nor earthquakes, weather conditions, and stock market fluctuations, which still defy every effort to predict them, can ever be predicted with truly mathematical certainty.

Not once in discussing his truism did Hardy use the word "model." Few scientists did until after World War II. Whereas physicists, geometers, and philosophers of science clearly grasped the concept, they had yet to develop the terminology needed to communicate their understanding to a broader audience. Social and biological scientists, who made scant use of mathematics before World War II, had little occasion to concern themselves with the limitations of mathematical methods and models. Hence 1940 represents a time of transition, when the idea of a mathematical model had at last been fully digested by those with a "need to know" but had yet to reach a broader audience. Today, the subject is routinely taught to college, university, and even precocious high-school students, all of whom soon learn to respect Hardy's truism.

An early use of the term "mathematical model," along with a clear explanation of its meaning, is to be found in a 1943 article by the Norwegian economist Trygve Haavelmo titled "The Statistical Implications of a System of Simultaneous Equations."[5] With few exceptions, colleges and universities did not begin to offer courses in model building until the 1960s, and books for use in such courses remained scarce until about 1970. Henry Pollak, a leader in the drive to bring model building into the classroom, has contributed a chapter to a history of mathematics in the schools describing the campaign in which he participated to add modeling to the curriculum.[6] In it he stresses the need to acquaint students not only with the abstract world of higher mathematics—which traditional courses of study do rather well—but with the process and pitfalls of constructing analogies between the mathematical world and the "real" one.

The most successful model ever devised consists of Isaac Newton's three laws of motion, together with his "inverse square" law of gravitation, and certain primitive concepts previously identified by Galileo. Thus equipped, Newton was able to explain the motions of the planets, the ebb and flow of the tides, and all manner of previously unexplained experiments. In the process, he did as much as anyone to bring about the age of enlightenment and to convince the world that natural phenomena are governed by laws that—with sufficient thought and experimentation—may be discovered and applied to the benefit of all.

So successful did Newtonian methods become in the hands of his successors that by the middle of the nineteenth century, they were widely regarded as absolute truth, valid in any and all circumstances. Not until 1882 did A. A. Michelson perform an experiment that could not be explained within the Newtonian-Galilean model.[7] And so disturbed was Michelson by his seemingly inexplicable findings that his wife attempted to have him committed to an insane asylum. Yet no flaw could be found in his results, which were eventually published and replicated by others. Not until 1905 would Einstein explain that the Galilean principles underlying Newtonian physics are only approximately correct as they apply to objects moving at ordinary speeds, and grossly incorrect as they apply to objects moving at almost the speed of light. In short, the Newtonian-Galilean model of particle motion is of limited validity. However reliable it may be within established limits, it is not to be trusted beyond them.

Later it was learned that very small particles, such as the electrons surrounding the nucleus of an atom, also violate Newton's laws. Newtonian methods suffice to design earthquake-resistant skyscrapers, to build passenger aircraft capable of transporting hundreds of people thousands of miles without refueling, and to put men on the moon. Yet even the most successful models have limited ranges of validity, and as every student of mathematical modeling soon learns, any model can be "tortured" by being pushed beyond those limits. Einstein's theory of relativity, together with the advent of quantum mechanics, forced scientists to acknowledge the limits of the Newtonian-Galilean model of particle motion.

Geometry itself is a model—a model of the space in which we live. From the time of the ancient Greeks until the early years of the nineteenth century, Euclid was thought to have constructed the one and only logically coherent model of space. Then, almost simultaneously, János Bolyai in Hungary, Nikolai Lobachevsky in Russia, and Carl Friedrich Gauss in Germany arrived at the opposite conclusion: other geometries are logically possible, and may even furnish more accurate models than Euclid's of the space around us. Such a possibility is implicit in the realization that Euclid's notorious fifth postulate—the one concerning parallel lines—could not be deduced from genuinely self-evident truths and could conceivably offer a dangerously inaccurate description of the space around us. The possibility that the space in which we live may be non-Euclidean was one of the most sobering mathematical discoveries of all time, and one whose implications required several generations to digest.

Legend has it that Gauss reacted to his disturbing discovery in a particularly dramatic fashion. He is said to have instructed three assistants to place lanterns, in the dark of night, atop three of the highest towers in the district, that each might measure the angle whose arms extended from his own lantern to the other two. Because the sum of the angles in this giant triangle (the towers were located between seven and twelve miles apart) did not differ appreciably from 180°—by more, that is, than could be accounted for by measurement error—it afforded Gauss no grounds for rejecting the hypothesis that the total is exactly the 180° (two right angles) predicted by Euclidean geometry.

Lobachevsky later confirmed—at least to his own satisfaction—that the same was true when the legs of the triangle were of lengths comparable to the distance between the earth and the sun.

Had the results been different, Gauss and Lobachevsky would have been the first to discover that the space in which we live is non-Euclidean. As it was, they were merely the first to realize that it doesn't have to be Euclidean. It is an experimental fact rather than a logical necessity that the space in which we dwell is at least approximately Euclidean. To geometers, it meant that the centuries-old initiative to deduce Euclid's dubious fifth postulate from genuinely self-evident truths was futile. To mankind, it meant that logic alone is an unexpectedly feeble instrument of discovery. Unaugmented by experimental evidence, it fails even to distinguish between the geometry of the space in which we live and all the other geometries human ingenuity has since devised. Hardy's truism applies to space itself. Mathematics can prove that the models men make of it possess certain properties, but not that physical space must possess them, too.

There are several different kinds of non-Euclidean geometry.[8] The simplest in which the angles of a triangle add up to less than two right angles is called hyperbolic, while the simplest in which they add up to more than two right angles is called elliptic. For consistency, Euclidean geometry is occasionally described as parabolic. There are plenty of other ways—in addition to the sums of the angles in a triangle—in which the several geometries differ. Yet that is beside the point here. Gauss's three-towers experiment establishes that surveyors do not torture the Euclidean model when they employ the method of triangulation over distances typical of the early nineteenth century, and Lobachevsky's astronomical measurements confirmed that astronomers too may rely on the method of triangulation.[9] Euclidean geometry is adequate for most if not all practical purposes.

Elliptic geometry is not as unfamiliar as its name might suggest. In fact, it is nothing more than a formal version of the geometry of global navigation, in which "great circle routes" furnish the shortest paths between pairs of points on the earth's surface. It does not differ all that much from Euclidean geometry, since the points and great circles of elliptic geometry interact in almost the same way as the points and straight lines of Euclidean geometry. The most notable exception, of course, is the Euclidean axiom asserting the existence of parallel lines.

It is violated in elliptic geometry by the fact that no two great circles can be parallel, since every pair meets in exactly two (diametrically opposite) points.

The sphere, together with its points and great circles, is said to provide a Euclidean model of elliptic geometry. It proves, by its very existence, that there is nothing inherently illogical about a geometry in which no two lines are parallel. The same thing can be done for hyperbolic geometry, although the requisite models are somewhat more complex. They teach us that there is nothing inherently illogical about a geometry in which many lines parallel to a given line pass through a single point. Remarkably, the axiom of parallels turns out to be the only axiom of Euclidean geometry that the points and lines of hyperbolic geometry fail to satisfy. Elliptic geometry, in contrast, fails also to satisfy the Euclidean axiom whereby exactly one of any three distinct points on a line must stand between (a.k.a. "separate") the other two.

All this concern with the logical structure of the subject goes back to Euclid himself. He made his reputation by arranging the geometric truths known during his lifetime in an order permitting the more complex to be deduced from the less so. No such organizational task can continue indefinitely because—sooner or later—the search for ever-simpler truths must grind to a halt. Euclid terminated his quest after identifying five "postulates" so simple they seemed to require no proof, and from which he believed it possible to deduce all other geometric truths. He himself deduced more than four hundred, and others have since deduced more. What he called postulates would today be called axioms, while their logical consequences would be called theorems instead of propositions.

It has long been apparent that Euclid overestimated the significance of his five postulates, since the list he gave is both redundant and incomplete. Archimedes, less than a generation younger than Euclid, was the first to identify an axiom neither acknowledged by Euclid nor deducible from his five. While the fourth of Euclid's axioms is a consequence of the first three, many of his proofs are invalidated by the fact that they have been found to rely on "hidden assumptions" neither included among his five postulates nor deducible from them. Although none of Euclid's theorems are known to be false, it was not until 1882 that Moritz Pasch succeeded in producing a complete list of the assumptions needed to justify them. As refined by David Hilbert, the complete list contains no fewer than twenty distinct, independent, and necessary axioms.[10] An axiom is deemed necessary if, by replacing it with something else, one creates a different geometry.

To the ancient Greeks, Euclidean geometry became a model in another sense. So impressed were they by its successes that they made it the model for all other sciences.[11] Thus they predicated their astronomy on the "self-evident" axioms that (1) the earth stands motionless at the center of the universe and (2) whereas the earth is corrupt and ever changing, the heavens are immutable and perfect. Then, because they had long considered the circle to be the most perfect of curves, it seemed to follow that the heavenly bodies must move in circles centered on the earth.

As navigators, calendar makers, and other scrutinizers of the night sky accumulated evidence to the contrary, ancient astronomers were forced to concede that certain heavenly bodies might—like moons orbiting invisible planets—move in circles about points moving in circles about the earth. Around A.D. 150, Claudius Ptolemaeus (Ptolemy) formalized these "epicycles" into what came to be known as Ptolemaic cosmology, a system later endorsed by the Catholic Church. Awkward though it was, the Ptolemaic system could be used to navigate ships, predict eclipses, and perform other useful tasks. It went all but unchallenged until the time of Copernicus.

In similar fashion, Aristotle espoused a theory of particle motion predicated on "self-evident" axioms that included an assertion that the speed of an object in free fall is proportional to its weight. This is precisely the assertion discredited by Galileo, when he observed that a middleweight ball and a heavyweight ball of equal volume dropped from atop the Leaning Tower of Pisa both struck the ground simultaneously. Did Aristotle even consider such an experiment before proclaiming his axiom self-evident?

On another occasion, Aristotle claimed to have proven—by the impeccable logic for which he was famous—that men have more teeth in their mouths than women do. For nearly two thousand years after the golden age of Greek civilization, learned men were content to answer virtually any question concerning the material universe by appeal to the writings of Aristotle, Ptolemy, or Euclid. It just didn't occur to them to perform even the simplest corroborative experiments, such as dropping balls of different weights from on high or counting the teeth in the mouths of a few men and women. Galileo, Kepler, and Copernicus were among the first to test the reliability of the ancient authorities.

In or about 1645, a group of English gentlemen began to hold informal meetings, replicate Galileo's experiments, and perform new ones. In 1660, the group was granted a royal charter by King Charles II. Members of this royal society performed physical experiments, exchanged letters in English describing their results, and openly disputed each other's conclusions. By so doing, they formed one of the world's first scientific communities. Yet they did not immediately command the respect of their learned contemporaries in England or beyond. Not until Newton joined the society, fresh from his landmark investigations, did informed opinion change. The power and beauty of the Newtonian-Galilean system, which enabled all who mastered it to deduce empirically verifiable truths from a handful of plausible assumptions, proved quite irresistible. Since that time, the principle that contrary fact trumps even firmly established theory has not been seriously challenged in the West.

For the purposes of this book, it will be necessary to keep a few simple models in mind. A computer is a machine that *accepts* certain inputs and *returns* specific outputs. The typical desktop computer—like the original Apple computer designed by Steve Wozniak—accepts inputs typed in at the keyboard and returns outputs on a video screen. Printers were a later refinement. Some computers require as inputs a date, a point of departure, a destination, and a range of departure times to return a list of compatible airline flights. Others accept locations, distances, and lines of business (such as hotels, restaurants, or dealers in plumbing supplies) as input to return a list of firms within a given distance offering the desired good or service. Any desktop can connect to such computers via the Internet.

Even the simplest pocket calculators accept pairs of numbers and, at the mere push of a button, return their sum, difference, product, or quotient. More expensive models can also provide square roots, logarithms, reciprocals, and a host of trigonometric treasures formerly accessible only by table lookup. Most computers are quasi universal in the sense that they can be *programmed* to do almost anything—given time—that any other computer can.

Computer programs come in all sizes and shapes and may be written in any of several languages. Lisp, a contraction of "list processing," is among the simplest languages and also one of the oldest. Designed by John McCarthy at MIT during the late 1950s and early 1960s, the original version was meant to facilitate research in artificial intelligence. Every Lisp expression is a list to be evaluated by interpreting the first name on the list as an operation to be performed, and all subsequent names as raw material upon which to perform it. Thus (÷ (+ 3 5) 2) is a program for dividing the quantity (+ 3 5) by 2, where (+ 3 5) is a subprogram for adding 3 and 5. The value of the entire expression is 4. Lisp aficionados—of whom there are many—approve of the fact that the language represents data and programs in the same way, since that seems to encourage the development of programs that employ other programs as subprograms. In theory, there is no limit to the number of subprograms that can nest, one inside the other, within a single master program. The downside of the nesting process is an aggravating profusion of parentheses.

For computers or computer programs that accept as input—and return as output—only a single number, it is possible to represent the entire input-output "map" as a graph on a sheet of paper. Figure 4.1, for instance, displays the input-output map of a program that accepts

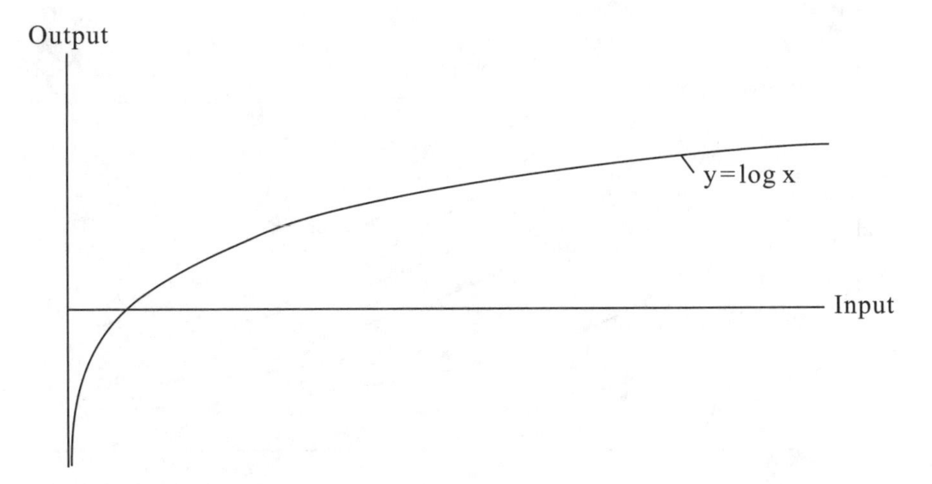

Figure 4.1. The input-output map for a computer program (or subprogram) accepting positive numbers as inputs and returning their logarithms as outputs.

positive numbers as input and returns their logarithms as output. When furnished with 0 or a negative number as input, such programs return error messages reminding the user that only positive numbers have logarithms.

Although it is mere convention that inputs are always measured along the horizontal axis, while outputs are measured in the vertical direction, serious confusion can result from an interchange of the two axes. Programs that accept two numbers as input, while returning only a single number as output, can be represented graphically by contour maps, as in figure 4.2. The "contour" labeled 4 at the lower right contains all the input pairs (x, y) that, when multiplied together, return the output 4. And so on. The input-output map for a program that accepts a single input, t, and returns two outputs, x and y, would consist of a single curve in the xy-plane, each point (x, y) of which corresponds to a particular value of the input t. And so on. Reference is often made to input-output maps corresponding to programs that accept entire tables of numerical input and return comparable tables of outputs, even though there is no way to draw such maps with pencil and paper.

Random phenomena are everywhere. The toss of a coin that starts a football game, the roll of the dice in Las Vegas, and a turn of the

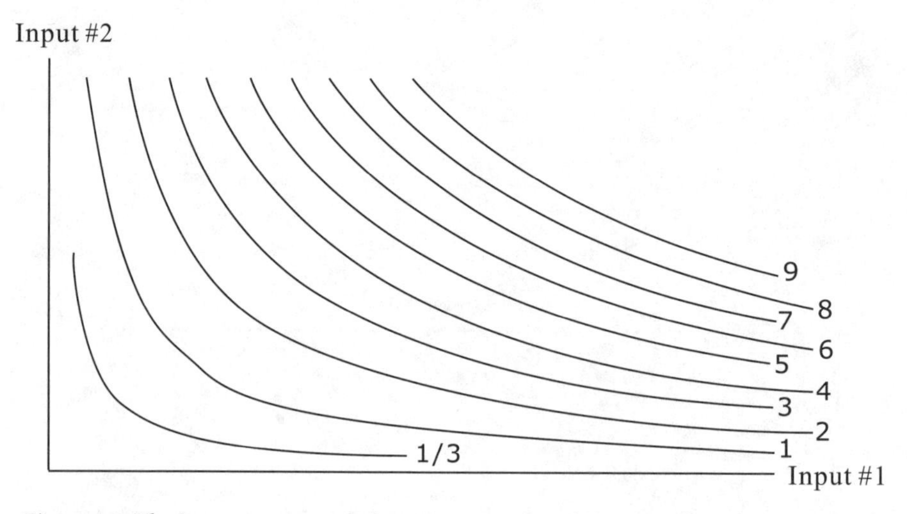

Figure 4.2. The input-output map for a program accepting pairs of positive numbers as inputs and returning their products as outputs.

wheel of fortune at a Rotary Club meeting are all performed for their unpredictability. Each one, in isolation, is indeed unpredictable. Yet if any is repeated many times, certain altogether predictable patterns will emerge. A fair coin will come up heads about half the time. An honest pair of dice will come up six, seven, or eight more than 44 percent of the time, seven being slightly likelier than the other two. Each number on a wheel of fortune—or on a roulette wheel—will come up about equally often. And so on. Likely events are said to have a high probability, while unlikely ones have low probabilities. Girolamo Cardano (1501–76) wrote a book titled *Liber de ludo aleae* (*The Book on Games of Chance*) that pursued the study of probability far enough to enable the reader to win—or at least bet advisedly—at various card and dice games.

If one tosses a coin twice in succession, one may obtain two heads, two tails, or one of each. To Jean Le Rond d'Alembert, a leading figure in eighteenth-century science, that made it obvious that the three were equally likely and that the probability of each must therefore be one in three. So confident was he of his conclusion that he included it in his famed encyclopedia. But people who actually conducted such trials began to notice that mixed outcomes tended to occur about as often as pure ones, whereas, by d'Alembert's logic, they should have been outnumbered two to one. Upon further reflection, it became clear that nature perceives four equally likely outcomes of a "d'Alembert trial": two heads, two tails, a head followed by a tail, and a tail followed by a head. This accounts for the rough equality of pure and mixed results in small-scale trials and demonstrates that even the most pristine a priori logic requires experimental verification when applied to natural phenomena.

Experimenters at Cornell University recently conducted an unusually large-scale test of elementary probability theory.[12] Stephen Ellner and John Guckenheimer programmed a computer to perform five thousand runs of one hundred machine-simulated d'Alembert trials each. An individual trial was deemed successful if it resulted in at least one head, and a failure otherwise. The "score" for each hundred-trial run was the number of successes achieved. Because the chance of success in each individual trial was about three in four, the scores should average about 75. In fact they did, while ranging from a low of 60 to a high of 91. Nearly eighteen hundred of the runs produced scores of 74, 75, 76, or 77, while all other scores were less frequent. Figure 4.3 contains a complete tally of the results. The asterisks mark—for each k between 60 and 91 inclusive—the number of scores equal to k, while the circles directly above or below the asterisks record the number of such scores predicted by exact theory. Though far from perfect, the agreement between theory and experiment is unmistakable. There is no surer indication of falsified experimental data than suspiciously close agreement between prediction and observation.

The smooth curve in the figure is known as a Gaussian bell curve, in honor of Carl Friedrich Gauss (1777–1855), arguably the greatest mathematician of all time, who pioneered its use in statistical analysis. Such curves arise in various contexts throughout mathematics, physics, biology, and the social sciences. It is firmly established, for instance, that the heights of recruits in various armies cluster about their average values, much as the foregoing scores cluster about theirs, and agree even more closely with Gaussian bell curves when displayed

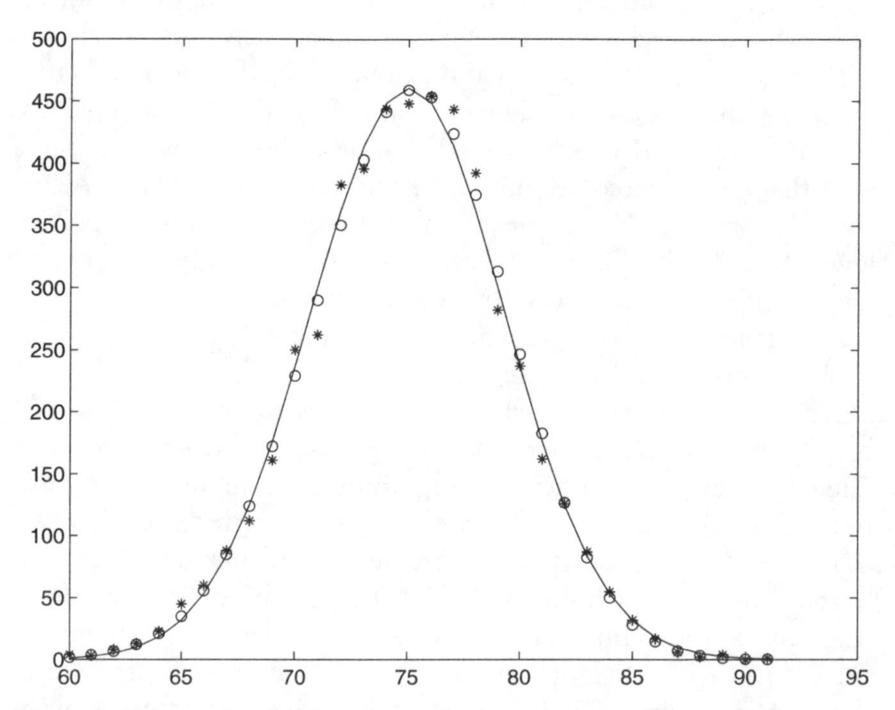

Figure 4.3. The results of a computer experiment consisting of 5,000 runs of a hundred d'Alembert trials each, showing the extent to which the observed (∗) and predicted (∘) numbers of successes fell along a Gaussian bell curve.

in the foregoing manner. Scores on such standardized tests as the SATs and ACTs taken by aspiring college students are said to do likewise, as do the values obtained by repeating a single physical measurement several times over, whether it be obtained from a surveyor's transit, a truck scale, a volt meter, a yardstick, a Geiger counter, or any of a host of other familiar measuring devices.

Gaussian bell curves are by no means the only bell-shaped curves known to science. Yet because they are so important in so many branches thereof, and because they recur repeatedly in this book, it seems advisable to digress briefly on the geometry of such curves. The area beneath any Gaussian bell curve is one square unit. Given a horizontal line on a piece of paper, and a point *P* somewhere above that line, *P* is the peak of exactly one Gaussian bell curve asymptotic to the given (horizontal) line.[13] Figure 4.4 displays a few of the curves peaking above a common point marked on a horizontal. Given the horizontal and vertical coordinates of the peak of a Gaussian bell curve, one can discover anything else one might wish to know about the curve with a few keystrokes on a pocket calculator.[14]

As a practical matter, the altitudes of bell-shaped curves are seldom mentioned in the technical literature, because the "widths" of such curves—which are inversely proportional to their heights—are of greater practical significance. An unambiguous measure of the width of a bell curve can be obtained from the observation that the highest part of each such curve is "cap shaped," like an inverted letter

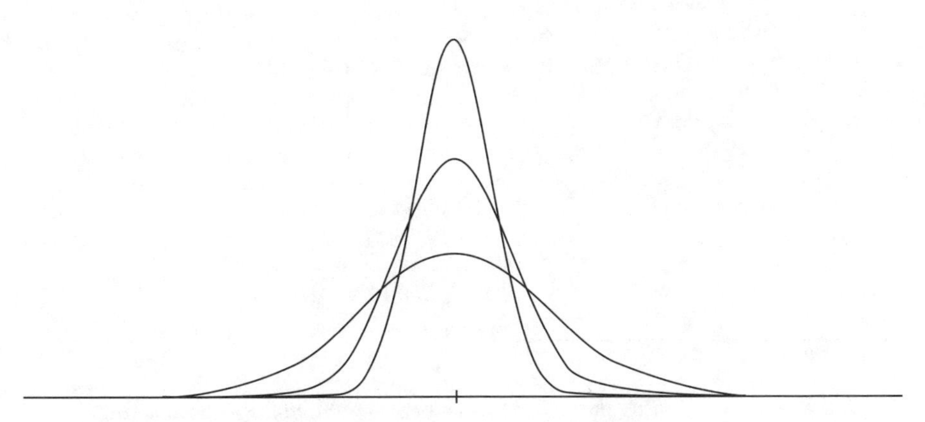

Figure 4.4. Exactly one Gaussian bell curve peaks at a given distance above a distinguished point P *on the horizontal axis of coordinates.*

U, while the lower portions constitute "tails," extending indefinitely to the left and right. The cap-shaped portions of the curves all bend inward, toward their (vertical) lines of symmetry, while the tails bend outward, away from those lines. The precise points at which a bell curve ceases to bend inward, toward its line of symmetry, and begins to bend outward, away from that line, are called inflection points. Each bell curve has two of them, located at the bottom of its cap-shaped part. The distance between the inflection points of a particular Gaussian bell curve is the conventional measure of the width of that curve. For reasons soon to be made clear, the half width of a Gaussian bell curve is denoted by σ, the lowercase Greek letter sigma.

Each of the five thousand scores from the five thousand runs of one hundred d'Alembert trials constitutes a "population" of numbers. Moreover, because the asterisks in figure 4.3 lie as close as they do to the solid curve, it may be said that the population at least approximates a "Gaussian normal population." Tall and slender bell curves correspond to tightly clustered Gaussian normal populations, while wide-bodied ones correspond to looser clusterings.

A population of numbers is said to be Gaussian normal if the fraction of its members lying between a lower limit *a* and an upper limit *b* is at least approximately equal to the area between a Gaussian bell curve and the interval extending from *a* to *b* on the horizontal asymptote (axis) as shown in figure 4.5. Another way to say the same thing is that if each of the numbers in the population were written on a raffle ticket—duplicates included—and all the tickets placed in an urn, the

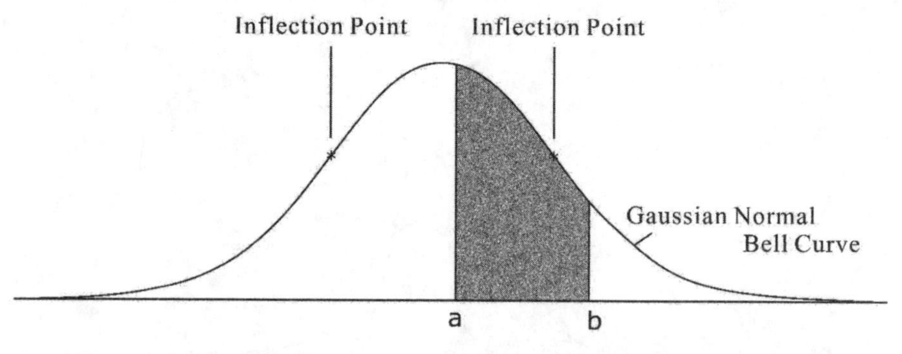

Figure 4.5. The probability that a number chosen at random from a (Gaussian) normal population of numbers shall exceed a *but not* b *is equal to the shaded area.*

probability that the number on a ticket drawn at random from the urn would lie between *a* and *b* is approximated by the shaded area.

Gaussian normal populations cannot be small populations. Figure 4.6 displays one consisting of only 1,024 whole numbers, ranging from 0 to 15 inclusive, that is about as close to being Gaussian normal as a small population can get. For the record, it contains 1 zero, 5 ones, 15 twos, 35 threes, 65 fours, 101 fives, 135 sixes, 155 sevens, 155 eights, 135 nines, 101 tens, 65 elevens, 35 twelves, 15 thirteens, 5 fourteens, and 1 fifteen. The reason an approximately normal population of numbers has to be so large is that the most common members (in this case sevens and eights) have to be so many times more numerous (here 155 times) than the least common ones (zeros and fifteens).

Given a population of numbers large enough to be potentially Gaussian, it is a simple matter to identify the particular Gaussian bell curve that best describes the manner in which the members cluster about the average of the numbers in the population. By convention, that "population average" is ordinarily called μ, the lowercase Greek letter mu. It is the horizontal coordinate of the peak of the bell curve. Once we know μ, the vertical coordinate may be obtained as follows: First, calculate all the deviations between the individual members of the population and μ, the population average. Second, square each deviation and add the results. Finally, divide the total by the size of the population, and extract the square root of what you get. The result is

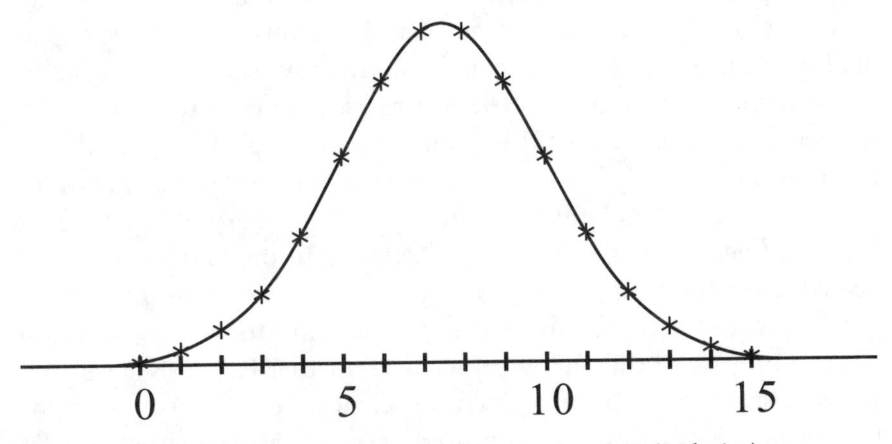

Figure 4.6. A population of 1,024 numbers between 0 and 15 (inclusive) can approximate a normal population of numbers surprisingly well.

called the standard deviation from the population mean μ and is customarily denoted by σ, the lowercase Greek letter sigma. The vertical coordinate α of the peak will then approximate $0.4 \div \sigma$.

The term "paradigm" is a relatively recent addition to the vocabulary of science. The 1961 edition of *Webster's New World Dictionary* describes it as a little-used synonym of "model," which was "common now only in its grammatical sense of an example of a declension or conjugation, giving all the inflectional forms of a word." Thomas Kuhn popularized the term among scientists and philosophers of science with his 1962 book, *The Structure of Scientific Revolutions*, in which he argued that science operates by accepting models (a.k.a. paradigms) of how the world works. "Normal science" tends only to elaborate a currently accepted model, or paradigm, by answering the sorts of questions to which it calls attention. Only rarely does "revolutionary science" cause an established model, or paradigm, to be abandoned in favor of another. Kuhn's most telling example was the "Copernican Revolution," which caused the centuries-old Ptolemaic model—in which the earth stood motionless at the center of the universe—to be abandoned in favor of a particular "heliocentric" alternative.

Primatologist Shirley Strum, herself the instigator of one such revolution, is well qualified to explain the process. She did so in her 1987 book, *Almost Human*, by pointing out that "once a paradigm is in place, it is very hard to dislodge, and when discrepancies are found they are either ignored or written off as the results of bad methods or 'bad science.' But these discrepancies eventually become overwhelming, and another version, model, or paradigm replaces the old one." It happened in the ancient world, when accumulating evidence to the effect that the heavenly bodies did not move in circles about a stationary earth obliged Ptolemy to put forward his theory of epicycles. It happened again during the Copernican revolution, and has been repeated many times since.

Few would deny, for instance, that Darwin's theory of evolution has revolutionized the biological sciences, or that Pasteur's germ theory of disease revolutionized medical science. The discovery of non-Euclidean geometry revolutionized not only mathematics but epistemology as well, since the demonstration that Euclidean geometry does not constitute absolute knowledge went a long way toward convincing the scientific community that absolute knowledge of the physical world is a pipe dream. The (thankfully remote) possibility always exists that tomorrow's news will bring word of an experiment disproving even the most firmly established scientific principle. Gödel's incompleteness theorem later established limits on what can be achieved even within the imaginary world of mathematics.

The atomic theory of matter—doubted by many leading physicists as late as 1900—has revolutionized physics and chemistry. The discovery by Watson and Crick of the molecular structure of DNA has revolutionized biology. Alfred Wegener's theory of continental drift, later superseded by the theory of tectonic plates, has revolutionized geology. Maxwell's incorporation of electricity, magnetism, and optical phenomena within a single dynamic model revolutionized nineteenth-century physics, even as Einstein's theory of relativity, together with quantum mechanics, did twentieth-century physics. Some expect chaos theory to do the same for twenty-first-century physics. Countless lesser discoveries—such as Strum's observation that male dominance is not the universal law of the jungle—have revolutionized other branches of science. At least in the "hard" physical sciences (in which he himself was trained) Kuhn noted that revolutions are often initiated by outsiders—agnostics uncommitted to the current model, which can narrow a scientist's field of vision like blinders on a horse.

The tree games of Chapter 3 furnish a remarkably adaptable model, or paradigm, for strategic competition. Though designed to comprehend only recreational games like chess, checkers, and poker, along with a few simple instances of economic competition, that definition (a.k.a. model) has since been modified to include almost every conceivable form of strategic competition. Indeed—if one is prepared to accept at face value the testimony of sprint racers, weight lifters, and field athletes to the effect that even those elemental forms of competition are partly strategic—the tree game paradigm could be called universal. No one has yet identified a form of competition to which that model fails, perhaps after superficial modification, to apply.

The need to modify the von Neumann–Morgenstern–Kuhn definition of a game was not foreseen in 1950, when it assumed its final form. Only gradually did the need to consider situations in which competitors confront an infinitude (let alone a continuum) of options make itself felt.[15] Yet numerous forms of competition do offer infinitely many options. In an automobile race, each driver must at each instant choose a setting for his steering wheel between −1 (hard left) and +1 (hard right). Likewise, he or she must choose a setting for the accelerator between 0 (foot off the pedal) and 1 (pedal to the metal). Ditto the brake. Moreover, drivers must make these and other decisions at every instant between time zero ("Gentlemen, start your engines") and the finish line. Although there isn't time to think separately about each individual decision, drivers are responsible for all of them, and discharge that responsibility by skillful movements of the hands and feet.

Adaptation is commonplace. The ax may have been invented for chopping wood, but it was soon adapted to warfare. The knife may have been invented for cutting meat, but it soon grew into the sword. Wheels may first have carried crops from the field, but they soon appeared on the chariots that—for a time—put fear in the hearts of foot soldiers around the world. And so on. What serves one purpose well is soon put to another.

The bottom line would seem to be that models play an increasingly conspicuous role in science, and that von Neumann, Morgenstern, and Kuhn made possible a science of competition by furnishing a surpassingly versatile model for the purpose. Even von Neumann might be surprised to learn that eighty years after he first proposed it, and more than sixty years after he last helped update it, no one has yet identified a form of competition to which that model cannot—with suitable modification—be made to apply. The opportunity their model represents has been exploited to the full in the realm of two-person zero-sum competition, as will be seen in Chapter 5, but has been less availing in the more complex and consequential sphere of many-player competition.

CHAPTER 5

Two-Sided Competition

Although history may remember his concise definition of a game as von Neumann's most important contribution to human understanding, it was by no means the high point of his 1926 lecture. Neither was his new and improved version of Zermelo's perfect information theorem. The pièce de résistance was an entirely new theorem concerning two-player zero-sum games in which each contestant has but one information set. Figure 5.1 displays the tree of one such game. Each nonterminal node in the figure has three downstream neighbors, and each individual leaf is inscribed with the dollar amount Black must pay White if the token should come to rest on that leaf. Black (player 2) is obliged to choose among the right fork (R), the left fork (L), or the middle fork (M) without knowing whether the token is currently at rest on A, B, or C. Both the perfect information theorem and the method of backward induction are inapplicable to such games.

Observe that if White sends the token to B, Black will have to pay her \$3, \$2, or \$4, the least of which is \$2. Likewise, if Black chooses option M, she will have to pay White \$1, \$2, or \$0, the greatest of which is \$2. Hence \$2 is at once the least payment White need ever accept from Black and the largest Black need ever make to White. Consequently, VAL(O) = \$2, and the choices B for White and M for Black constitute simple yet unbeatable game plans for obtaining that value. So far, so good. But how might one discover all this? A surprisingly

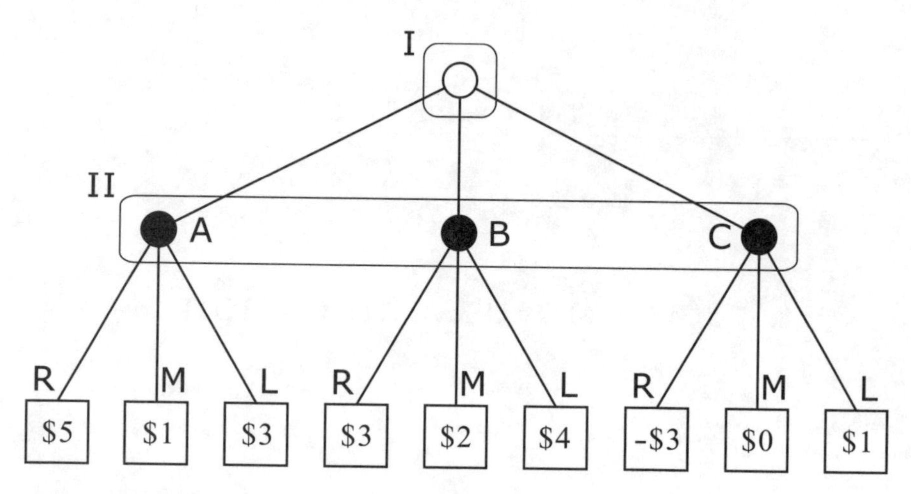

Figure 5.1. The tree of a two-player zero-sum game in which each player has but one information set.

simple procedure frequently suffices, even for tree games with many, many leaves.

Let the contents of the leaves in figure 5.1 be arranged in the tabular (a.k.a. matrix) form shown in figure 5.2. Notice that the rewards in the (horizontal) rows of the matrix correspond to the ones in the leaves upon which the token may come to rest if White (the payee) chooses *A*, *B*, or *C*, respectively. Similarly, the rewards in the (vertical) columns of the matrix correspond to the ones in the leaves upon which the token may come to rest if Black (the payor) selects *R*, *M*, or *L*. This confirms that figures 5.1 and 5.2 represent the same game, albeit in different formats.

In each (horizontal) row of the completed matrix, mark the reward that is least desirable to the row chooser (a.k.a. payee) by enclosing it in a circle. Then, in each (vertical) column, mark the reward that is least desirable to the column chooser (a.k.a. payor) by enclosing it in a square. In case of ties, mark all. Any number that ends up enclosed in both a circle and a square must be the value of the game, since the payee can obtain at least that amount by choosing the row in which the doubly marked number lies, while the payor can avoid paying more by choosing the column. Such a cell (the center cell of the matrix in figure 5.2 being one example) is called a saddle point of the ma-

		PAYOR		
		R	M	L
PAYEE	A	5	1	3
	B	3	2	4
	C	-3	0	1

Figure 5.2. The reward matrix corresponding to the game of figure 5.1.

trix because the numbers in the matrix rise and fall like the surface of a saddle, being high in front of a rider seated at the saddle point facing either left or right, and also behind, while being low on either side so that her legs may hang down in comfort.

A matrix may contain many saddle points or none. If there be many, the numbers within must all be equal, since no one of them (being the smallest in its row and the largest in its column) is capable of exceeding any other.[1] The payee may obtain the value of the game—and possibly more should the payor behave foolishly—by choosing any row containing a saddle point. Likewise, the payor can avoid paying more by choosing any column containing one. Matrices containing no saddle points at all, such as the matrix for finger matching displayed in figure 5.3, require further analysis.

Only two years before von Neumann's lecture, a prominent French mathematician named Émile Borel had managed to solve a few such games by means of "mixed," or "randomized," strategies. Noting that each player is free to toss a fair coin in secret before playing finger matching, and to show one finger (I) on heads and two (II) on tails, he observed that such behavior—which might be considered a third alternative and denoted $M = \frac{1}{2}I + \frac{1}{2}II$ to indicate that both options should be exercised with equal probability—could be expected to produce roughly equal numbers of wins (1) and losses (-1) against I, II, or (in consequence) M after repeated play. Hence the

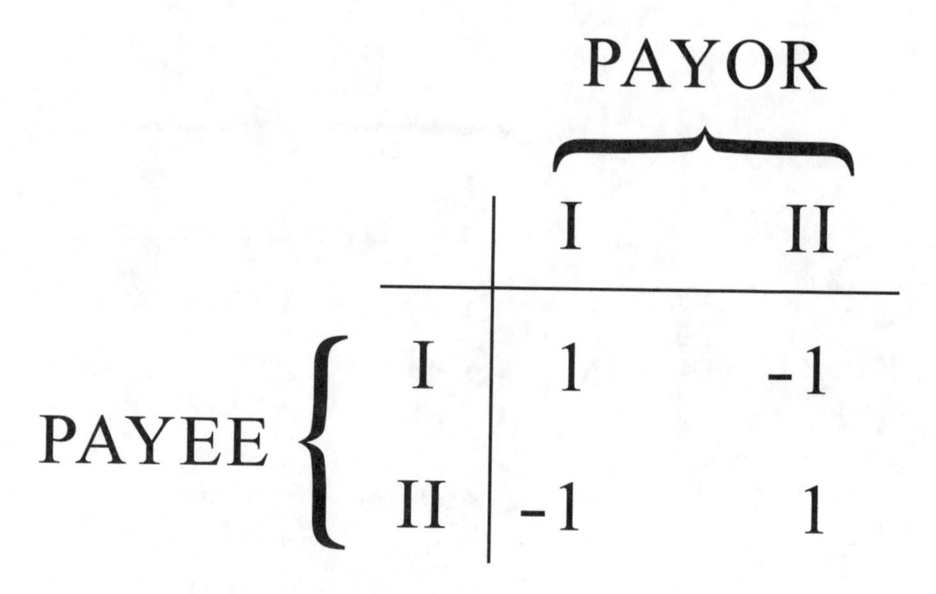

Figure 5.3. The reward matrix for finger matching.

"augmented matrix" displayed in figure 5.4 exhibits the strategic possibilities of finger matching even more explicitly than does figure 5.3. In particular, it contains a saddle point in the lower-right-hand corner, since 0 is both the smallest number in its row and the largest in its column. Hence 0 is the value of the game, and either player can do at least that well by playing the randomized strategy *M*.

The same trick works for the game of rock (*R*), scissors (*S*), paper (*P*) mentioned in Chapter 2. The obvious randomized strategy is $M = \frac{1}{3}R + \frac{1}{3}S + \frac{1}{3}P$, since it promises to produce roughly equal numbers of wins (1), losses (–1), and draws (0) in the long run against each of the original three options, *R*, *S*, and *P*, and therefore against itself as well. The corresponding augmented reward matrix is shown in figure 5.5. It, too, exhibits a saddle point in the lower-right-hand corner, indicating that the value of the game is 0 and that either player can do as well by adhering to the randomized strategy *M*.

The high point of von Neumann's 1926 lecture—as of his 1928 paper—was a concise mathematical proof of the fact that any matrix game, no matter how large, can be solved in terms of these "mixed" or randomized strategies. In other words, every matrix game has a numerical value, and neither player need accept anything less. About twenty years later it was discovered that the strategies in question can,

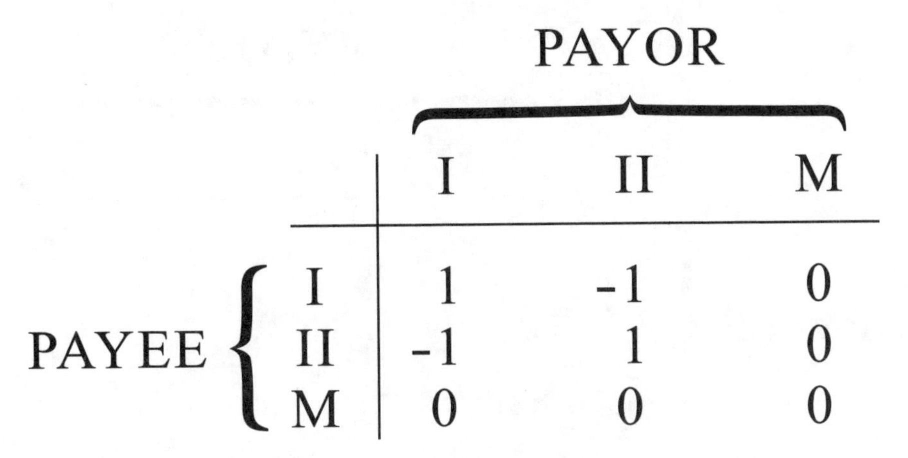

		PAYOR		
		I	II	M
PAYEE	I	1	-1	0
	II	-1	1	0
	M	0	0	0

Figure 5.4. The augmented reward matrix for finger matching.

along with the value of the game, be calculated by a process known as linear programming. That process is among the most successful inventions of the computer age and works well even on matrices containing tens of thousands of rows and columns.

Von Neumann went on to point out that any two-player zero-sum game can—at least in theory—be converted into a matrix game. The result will contain a row corresponding to each "pure strategy" available to White (the payee) and a column corresponding to each pure strategy available to Black (the payor). A mixed strategy is a random combination of pure strategies, such as the mixtures *M* employed above to solve finger matching and rock, scissors, paper. A pure strategy for player X in a tree game is a complete and unambiguous list of instructions of the form:

When in information set ___, exercise option ___.

Such a list is complete only if it contains an instruction for every information set of color X, and is unambiguous as long as it includes no conflicting instructions for any such set. Strategies differ from game plans only in that strategies must be complete and unambiguous: a strategy for X must specify one and only one option to be exercised in every conceivable situation (a.k.a. information set) in which X might conceivably be called upon to take action. A mere game plan, in contrast, may allow the executor some latitude in certain situations

		PAYOR			
		R	M	L	L
PAYEE	A	0	1	-1	0
	B	-1	0	1	0
	C	1	-1	0	0
	M	0	0	0	0

Figure 5.5. The augmented reward matrix for rock, scissors, paper.

and/or include a "contact number" to be called should the token come to rest on an information set for which the plan's designer left no instruction.

Because every two-player zero-sum tree game (with or without perfect information) can be converted into a matrix game, von Neumann's theorem applies to all such games. They all have values, which the players can obtain with the help of (potentially computable) mixed or randomized strategies. That's the good news. The bad news is that the matrices corresponding to even rather simple tree games with perfect information can contain trillions—instead of tens of thousands—of rows and columns, placing them beyond the reach of all known means of computation.

The applications of two-player zero-sum game theory are, for the most part, to military, athletic, or recreational competition. The contest between Ford and GM described in Chapter 3 is a rare instance of commercial relevance. Baseball furnishes a wealth of athletic applications, of which we need consider only one. In it, the primal conflict is the one between pitcher and batter. Assume for simplicity that a three-ball, two-strike count has already been reached and that the pitcher has sufficient control to hit either the low-outside (LO) or the high-inside (HI) corner of the strike zone with his fastball at will. Or he can throw the ball down the middle (M), should he choose to do so. He might in practice consider other options (curves, sliders, and the low-inside corner, for example), but it suits our present illustrative purposes to limit him to the foregoing three.

The basic reward structure is indicated in figure 5.6. The batter is skilled at hitting balls thrown down the middle, as indicated by the .400 batting average shown in the middle (M) portion of the strike zone. It is to be interpreted as an assertion that the batter will hit safely with 40 percent probability if the pitcher chooses M while he is in his ordinary batting stance. He also does well against HI pitches, since his average against them is .300, but fares less well against LO pitches, where his average is only .200. These probabilities are in rough accord with those suggested for himself by Ted Williams.[2]

The batter's alternatives are guesses at where the pitch will be thrown. If he guesses LO, he can bend over and become a .400 hitter against LO pitches. But he does so at the expense of becoming only a .300 hitter against M pitches, and a .000 hitter against HI ones. Similarly, if he guesses HI, he can rise up and become a .400 hitter against

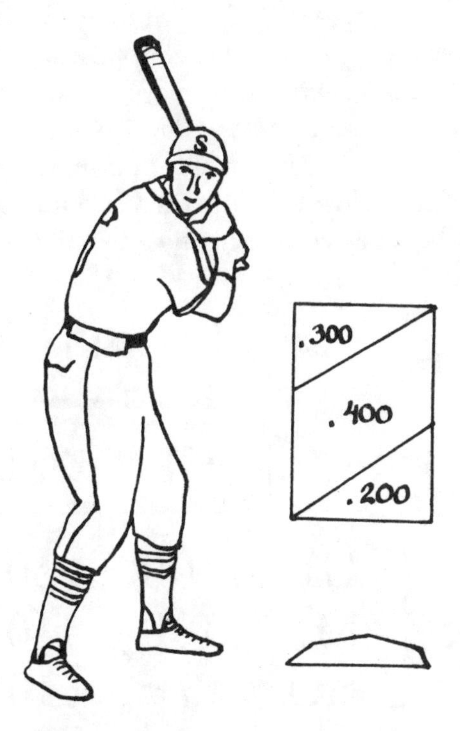

Figure 5.6. No batter can cover every part of the strike zone equally effectively.

HI pitches, though at the expense of becoming a .200 hitter against M pitches and a .000 hitter against LO ones. Hence the two are engaged in the matrix game shown in figure 5.7. The matrix entries are probabilities that the batter (or payee) will hit safely, which he would like to maximize by his decisions, while the pitcher is trying to minimize them by his. It can be shown, by appending an additional row and column to the foregoing matrix, that the value of the game is 24 percent and that the pitcher and batter can each avoid any less desirable outcome by choosing the strategies $\frac{2}{5}$HI + $\frac{3}{5}$LO and $\frac{4}{5}$M + $\frac{1}{5}$LO, respectively. For the record, the row and column to be added are respectively 24 percent, 38 percent, 24 percent, 24 percent, and COL 16 percent, 24 percent, 24 percent, 24 percent.

The foregoing strategies may be discovered by linear programming—a standard mathematical technique well suited to electronic computation—or in any of several other ways. Among the alternatives, the method of "fictitious play" deserves special mention. Suppose the game is to be played many times in succession, with each player keeping a running tally of the number of times each opposing option has been exercised to date. Suppose, too, that each invariably chooses, in the next game to be played, that option which *would offer* the most attractive reward *if* the opponent could surely be counted upon to exercise each available option with probability equal to the fraction of times he or she has exercised it in the past. Were the above process to be repeated sufficiently often, the relative frequencies

		PITCHER		
		HI	M	LO
BATTER	HI	.400	.200	0
	M	.300	.400	.200
	LO	0	.300	.400

Figure 5.7. The reward matrix awaiting the batter depicted in figure 5.6.

with which the individual options have been exercised would necessarily come to approximate the probabilities with which the players must exercise them in order to play the game as well as it can be played.

The most remarkable feature of the saddle-point solution is the extraordinary predictive burden of which it relieves the players. No longer must the batter reason that "most of his pitches are LO, so I should guess LO, but knowing that I know this, the pitcher will doubtless throw HI, so I shall . . ."—a cycle from which there is no escape. Since von Neumann, the argument goes, the batter need only observe that an "optimal" mix of pitches exists and his sensible opponent will surely adhere to it. Hence nothing can be lost by using his own "optimal" mixture $\frac{4}{5}$M + $\frac{1}{5}$LO, and perhaps something will be gained, should the pitcher (inexplicably) interject a few Ms into his mixture. The pitcher, too, may engage in such reasoning.

On the field, however, the foregoing argument loses much of its force. Since the entries in the matrix of figure 5.7 are only estimates inferred from the record of past encounters, and since the pitcher and batter may draw quite different inferences from those encounters, the batter's assumption that the pitcher will throw what the batter takes to be an optimal mixture is just a guess, like any other, and may even prove a bad one.

If, for instance, the pitcher persistently includes Ms in his mixture, it is probably because he doesn't believe the .400 shown in the middle of figure 5.6. In that case, the batter should doubtless eliminate LO from his mix until either the Ms disappear or he, too, is obliged to revise his opinion about the .400. It is foolish to behave as if the pitcher were using the mixture $\frac{2}{5}$HI + $\frac{3}{5}$LO in the face of continued evidence to the contrary.

In short, the remarkable feature of the saddle-point solution discussed above is largely illusory here. The players are *not* relieved of the need to observe their opponent's behavior and to predict his future actions from the record of his past. This is due to the shortage of available information concerning the game's reward structure, and also to the fact that it is played not once but many times.

Football, too, can be analyzed as a matrix game. There the decisive conflict occurs on third-down-and-long-yardage situations. The quarterback (QB), or one of his coaches, must choose one of several available offensive plays, while the opposing team's defensive captain (DC) must select an appropriate defensive alignment. Assume for simplicity that the QB has only two plays to choose between, a run and a pass, while the defensive captain must decide whether to stack the defenders near the line of scrimmage in anticipation of a run or to spread them out in anticipation of a pass. In reality, of course, the range of options available to each decision maker is far broader.

If the rewards of the game are taken to be the probabilities of gaining the yardage required for a first down, the game matrix could be the one shown in figure 5.8. It may again be shown, by appending an extra row and column to the given matrix, that the value of the game is 45 percent. Moreover, $\frac{5}{12}$ run + $\frac{7}{12}$ pass and $\frac{1}{2}$ stack + $\frac{1}{2}$ spread are saddle-point strategies for the QB and DC, respectively. Finally, the row and column to be added are (45 percent, 45 percent, 45 percent) and COL(45 percent, 45 percent, 45 percent).

Here, too, the matrix entries have to be inferred independently by the respective coaching staffs from the record of past encounters. Quarterbacks and defensive captains—with and without their coaches—spend endless hours watching game films in order to accomplish roughly that. Hence the considerations limiting the relevance of saddle-point solutions for baseball limit their relevance for football as well. In particular, the assumption that one's opponent will employ what one believes to be a saddle-point strategy is only that—an assumption, which may prove either right or wrong. Yet some as-

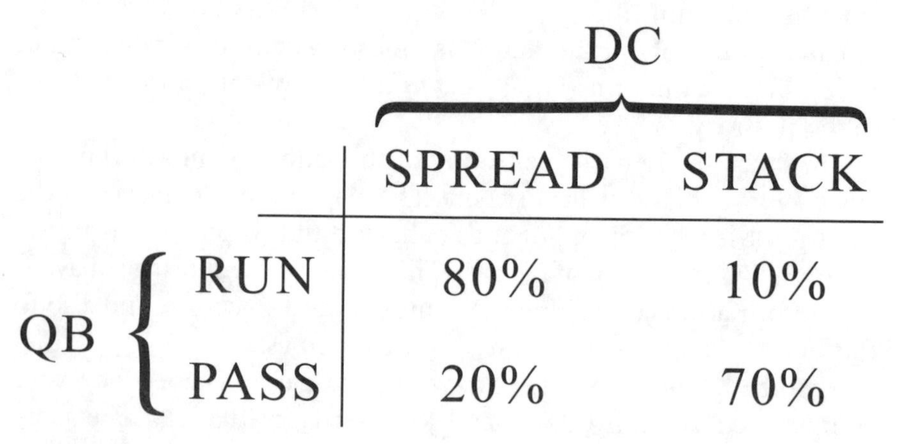

		DC	
		SPREAD	STACK
QB	RUN	80%	10%
	PASS	20%	70%

Figure 5.8. The reward matrix for a simple game within a game of football.

sumption must be made, and the theory of matrix games furnishes a plausible candidate.

The people who operate football and baseball teams have long been aware of game theory—there was even a rock band by that name during the 1980s—yet make little if any use of it. Why should they? They play frequently, and they have ample opportunity to observe what they call the tendencies of their opponents. They know what opposing players and teams have done in the past, in critical situations, and expect them to do more or less the same again. Occasionally they do get fooled by a pitcher who throws a pitch he's never thrown before, or a football team that runs a play it's never run before. But they seldom lose games for that reason.

Military organizations, in contrast, have little opportunity to "scout" their opponents. Wars are infrequent, and no two are quite alike. A professional soldier can spend an entire career preparing for a war that never comes. That is why war gaming (a.k.a. Kriegspiel) has so rich and storied a history. Denied the opportunity to observe the opposing force in action, military leaders are forced to rely on estimates of its manpower, weaponry, and political objectives, and to assume that the opposing leaders will deploy their assets more or less as they themselves would do in similar circumstances. As a result, the U.S. and Soviet military establishments funded extensive research, during the Cold War years, on the applications of game theory to combat.

Perhaps the simplest application is to games of the so-called Colonel Blotto type. The colonel, with a force of several regiments, is instructed to attack a number of militarily valuable objectives—such as border crossings or mountain passes—defended by a somewhat weaker force. Both commanders are free to divide their forces as they see fit, to secure a favorable outcome. The forces are to occupy their assigned positions during the night, for attacks to commence at dawn.

Assume for simplicity that the colonel has four regiments under his command, while his opponent has only three, and that there are only two equally valuable objectives called A and B. Since Blotto can assign zero, one, two, three, or four regiments to objective A, and is under orders to hold none in reserve, he has just five pure strategies, while his opponent has only four. If the colonel assigns more regiments to a given objective than does his opponent, he is assumed to capture the defending regiments as well as the objective, which is deemed equal in

		DEFENDER			
		0	1	2	3
BLOTTO	0	4	2	1	0
	1	1	3	0	-1
	2	-2	2	2	-2
	3	-1	0	3	1
	4	0	1	2	4

Figure 5.9. The reward matrix for a simple version of Colonel Blotto.

value to one regiment. If he assigns fewer regiments to a given objective than does his opponent, he is assumed to lose both his regiments and the objective. If each side assigns equal numbers of troops to an objective, the attack ends in a draw and nothing is won or lost.

All this suggests that A and B are extraordinarily soft targets, since the defender is without his usual advantage. Despite the fact that the Princeton mathematicians John Tukey and Charles Winsor devoted considerable effort during World War II to formulating more realistic versions of Colonel Blotto, the present example will serve to illustrate the solution process. Since the defender always loses exactly what Blotto gains, the contest between them reduces to a matrix game of the sort shown in figure 5.9. The solution is moderately surprising: Blotto should usually (with 89 percent probability) attack either A or B with all four regiments, otherwise attacking both A and B with two regiments each. Never should he send three regiments against one objective and just one against the other. The defender, in contrast, should usually (with 89 percent probability) assign two regiments to one objective and the remaining one to the other. Otherwise (with 11 percent probability) he should assign all three of his regiments, at random, to either A or B. The value of the game is $14 \div 9 = 1.556$, as can be verified by augmenting the matrix with an additional row and column in the now familiar manner.

Another application of matrix game theory is to the general problem of destroying a single hidden object. A defender has something of great military value, such as a command and control center, which he may conceal in a variety of locations, some being more secure than others. The attacker must continue to attack the possible locations until the target is destroyed. In which location should the defender hide the target, and which locations should the attacker attack first? The solutions of such problems invariably involve mixed strategies in which the more secure locations are the more likely to contain the target and to come under attack. Typically there will be low-security locations to which the optimal strategies assign zero probability of containing the object and/or of coming under attack.

The theory of games can be adapted to deal with the timing of competitive decisions. Such problems are not immediately convertible to tree games, because the passage of time is continuous. If each player is entitled to act at any time between, say, dawn and dusk, each would seem to have an infinite number of alternatives. Duels, being among the simplest games of timing, have been studied in exhaustive detail.[3] The simplest ones involve single-shot dueling pistols of dubious accuracy. The duelists are presumed to start out of range and to walk toward each other until one elects to stop and fire. If the first bullet finds its mark, the shooter's opponent is dead or disabled. If it misses, the opponent advances to point-blank range before firing. If both duelists fire simultaneously, either, both, or neither may be hit. These were not the rules in force during the early years of the nineteenth century, when Alexander Hamilton and Aaron Burr fought the most famous duel in American history, but they are well suited to precise analysis.

To decide when to shoot, both duelists have to know both their own and their opponent's accuracy profile, a typical pair of which is displayed in figure 5.10. Both men are 100 percent accurate at point-blank range, but their accuracies decline with the distance between themselves and their targets, one of the two being visibly more accurate than the other at middling distances. Both should hold their fire until the other is within their effective firing range, but neither should hold for too long.

The game between the two duelists has a saddle-point solution in pure strategies. In other words, each has a single optimal firing distance, rather than a range of potential firing distances among which

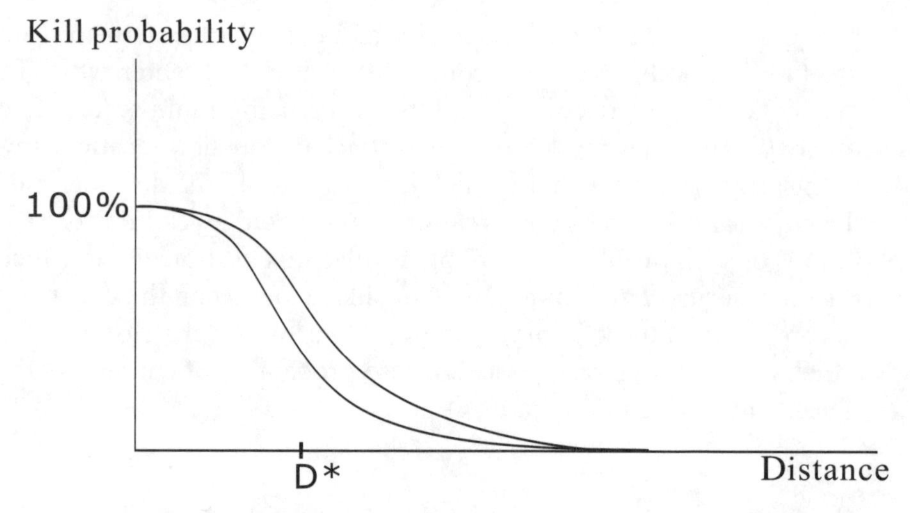

Figure 5.10. The kill probability for any duelist diminishes with the firing distance. The higher curve belongs to the more accurate shooter.

each should pick at random. Moreover, despite the difference in their marksmanship, the optimal firing distance is the same for both contestants. That is because each is able to calculate the other's optimal firing distance and has no call to fire at any greater distance than that. The common optimal firing distance is denoted by D^* in figure 5.10 and is characterized by the fact that at that distance, the two individual kill probabilities add up to 100 percent. Clearly there can be but one such distance.

The foregoing duel is presumed noisy, in the sense that each duelist can hear his opponent's gun go off and is free to advance to point-blank range if and when the gentleman should miss. Game theorists have also investigated silent duels, in which the contestants never know whether the opponent has yet fired. The saddle-point strategies for such games are typically randomized in the sense that there is a largest and a smallest distance at which it is advisable to fire, and a diminishing probability that each will hold his fire until the distance separating them is less than a given intermediate value. There are also multi-bullet versions of both silent and noisy duels, with presumed military applications, and a host of other variations on the underlying theme.

Duels might seem to disprove the assertion that every form of competition either is or closely resembles a tree game. The noisy duel

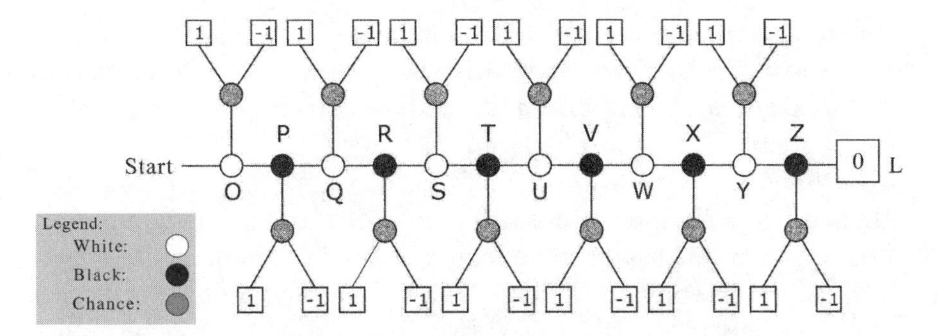

Figure 5.11. The tree of a duel in which the contestants take turns advancing three paces before deciding whether to fire at the current distance or to allow the opponent to advance an equal distance before facing a similar decision.

described above clearly is not such a game, and its resemblance to one is less than obvious. So consider the following rooted tree. It has the twelve player nodes *O*, *P*, *Q*, *R*, *S*, *T*, *U*, *V*, *W*, *X*, *Y*, and *Z* located on the horizontal path from the root node *O* to the leaf *L*, along with twelve chance nodes located immediately above and below the player nodes. It corresponds to a duel in which the contestants take turns advancing three paces toward each other before deciding whether or not to fire. When one contestant advances and elects not to fire at the new distance, it becomes the other's turn to advance, decide, and so on. Should White elect to fire at node *S*, he need only push the token to the chance node located directly above *S* and instruct the referee to spin the wheel of fortune calibrated to decide whether he shall hit or miss his mark. Should one duelist fire and miss, the other may advance to point-blank range before firing. It is clearly a big mistake to fire too soon.

Because the guns are noisy, figure 5.11 depicts a game of perfect information. Each participant knows at all times the distance between himself and his opponent, and whether or not the opponent still has a bullet in his gun. Hence the game can be solved by backward induction, once the kill probabilities are known at each permissible firing distance. Let the ones in the upper row of chance nodes be 3 percent, 6 percent, 13 percent, 32 percent, 70 percent, and 97 percent—reading from left to right—while those in the lower row are 4 percent, 9 percent, 20 percent, 49 percent, 88 percent, and 99.9 percent. These were arrived at by supposing the firing distances to be evenly spaced along the (horizontal) distance axis in figure 5.10. The 1s inscribed in some of the leaves are to indicate wins by White, while the -1s indicate wins by Black, and the 0 in leaf L betokens a tie in which the combatants fire simultaneously at point-blank range.

Observe to begin with that all the chance nodes are immediately eligible for evaluation, since their only downstream neighbors are leaves. When this has been done, it is possible to evaluate the player nodes in reverse order. Indeed VAL(Z) $= -99.9$ percent, since Black is almost certain to win by firing at three paces, should White be foolish or unfortunate enough to allow him that opportunity. Likewise, VAL(Y) $= 95$ percent, since White is better advised to fire at six paces with 97 percent accuracy than to allow Black to fire at three paces with 99.9 percent accuracy. And so on. The ultimate conclusion is that neither should fire at any distance greater than fifteen paces, while each should seize any opportunity to fire at that or any shorter distance. The fact that VAL(O) $= 2$ percent indicates that White is slightly more likely than Black to prevail in this particular approximate version of the noisy duel depicted in figure 5.10.

Fifteen paces is obviously a crude approximation to the critical distance D^* indicated in the figure, which is roughly equal to 16.77 of the paces employed above. More accurate estimates could be obtained by constraining the duelists to advance in increments of a single pace, or even a single foot or inch, between firing opportunities. But the main point has already been made: Games with a continuum of decision points can be approximated by tree games. Moreover, the accuracy of the approximation can be increased by increasing the number of decision nodes. The process by which this is done is known as "discretization," since it replaces an infinite number of objects of interest—such as the firing distances depicted in figure 5.10—with a discrete sample of those objects that can be evaluated one by one.

A familiar example of the discretization process is the method by which one estimates the area of an irregularly shaped pond on a map. After superimposing a grid of regularly spaced horizontal and vertical lines upon it, one may obtain an underestimate of its area by counting the cells that lie entirely inside the pale blue region, and an overestimate by counting the cells that lie *on or* inside the boundary of that region. If the grid is sufficiently fine, the resulting estimates will be almost equal to each other and will bracket the area of interest.

Wholly supported for many years by the U.S. Air Force, the Rand Corporation naturally devoted much time and effort to problems of air warfare. Many such problems are fruitfully regarded as two-player zero-sum games between the opposing commanders. Indeed the commanders themselves often think and speak in such terms. A typical situation endows each with a force of multipurpose aircraft, to be depleted by enemy fire and—if the campaign is to be of long duration—more or less simultaneously augmented by replacement craft.

Imagine for simplicity that each aircraft must be assigned either to (1) counter air operations, designed either to destroy enemy aircraft on the ground or to incapacitate the airports from which they fly, or (2) air-defense operations, designed to impede the enemy's counter air operations, or (3) ground-support missions against enemy troop concentrations and fortified positions for the benefit of allied ground forces. Although the ultimate goal is success on the ground, tasks 1 and 2 are necessary to sustain the ability to render services of type 3. Each day, during a given campaign, each commander must divide his forces among the three tasks. Each such decision may be represented as a three-sector pie chart depicting the fraction of available aircraft assigned to each task. The pies in question will expand and contract as the number of aircraft available to each side waxes and wanes.

Barring successful espionage, each commander is obliged to make his daily allocations in ignorance of the corresponding ones being made more or less simultaneously by his opponent. So even if each knows how many aircraft he and his opponent have available on a given day, the two are perforce engaged in a game of imperfect information. The precise nature of the game depends on a variety of factors, including the prevailing weather conditions, the expected duration of the campaign, the bombing and air-to-air accuracy of the aircraft involved, the experience level of the pilots, and so on. Nevertheless, the saddle-point strategies for different games of the foregoing type have much in common.

For one thing, both leaders devote the final phase of the campaign exclusively to ground-support missions, regardless of the relative strengths of their forces. At all earlier times, the commander of the stronger force should behave quite differently from his opponent, since he alone has an optimal pure strategy for dividing his forces among tasks 1, 2, and 3. The commander of the weaker force needs to confuse his opponent by concentrating his forces on a single task, to be chosen at random by drawing instruction slips from an urn whose contents change as the campaign progresses.

There is a critical force ratio of about 2.7 such that, if the stronger force outnumbers the weaker by anything less than the critical ratio, it should refrain entirely from dispatching ground-support strikes during the initial stages of the campaign, to concentrate on counter air and air-defense strikes. But if the stronger force should outnumber the weaker one by more than the critical ratio, it is better advised to dispatch missions of all three types until the final stage of the campaign begins. In either case, the optimal division of forces among tasks 1, 2, and 3 changes gradually from day to day, until the time comes to concentrate exclusively on ground-support missions.

The weaker force should engage in the same two or three tasks as the stronger one during the early stages of the campaign, but should choose a single (randomly chosen) one of them each day, so as to keep the opposition guessing. In particular, the commander of the weaker force should—with high probability—dispatch his entire force on a counter air strike early in the campaign. And knowing this to be so, the commander of the stronger force should assign a fraction of his aircraft roughly equal to the whole of the weaker force to air defense during the early stages of the conflict. Only later should his focus shift from defensive to offensive measures.[4]

Games of pursuit and evasion have been investigated in even greater detail. A typical example pits a fast but somewhat ungainly pursuer (P) against a slower but more agile evader (E). Suppose, for simplicity, that the autos driven by P and E both have full tanks of gas and that the chase is to take place in a large, flat, empty parking lot devoid of obstacles. If the "fitness ratio" for each vehicle is obtained by dividing its top speed by the square of its turning radius, it may be proven as a mathematical theorem that the driver of the fitter vehicle is capable of winning the contest.[5]

Should the vehicle in question belong to the pursuer, P can "capture" E by crashing his own car into the one driven by E. Or, should E be blessed with the fitter vehicle, he can avoid a chase-ending collision indefinitely. In the latter case, E need only wait until the faster P is bearing down on him from behind, then—perhaps after flipping a coin—veer either hard right or hard left as a certain critical distance between the two is approached. Cursed with an overlarge turning radius, P will go barreling on past E without making contact. The maneuver can be repeated as often as necessary. When P has the fitter vehicle, a relatively complex series of preparatory maneuvers may be necessary before capture becomes inevitable.

Games of pursuit and evasion are discussed at length by Rufus Isaacs, in his book on *differential games.*[6] They can be extremely difficult to solve, especially if they take place in three-dimensional space rather than on a flat parking lot, if momentum is an issue, if there are obstacles to be avoided, and/or if they must contend with limited supplies of fuel or ammunition. Pursuit and evasion games with imperfect information are even more problematic. Research in the field rose to a peak during the 1960s and early 1970s, before declining due to a perceived lack of progress.

It is no accident that the foregoing applications of two-person zero-sum game theory have focused on military, athletic, and recreational forms of competition. The theory has but few applications to social, political, and commercial conflicts because, even when such conflicts involve only two parties, they are rarely zero-sum. Two-person non-zero-sum games are significantly more complex than their zero-sum relatives.

Among the simplest two-player non-zero-sum games is the so-called prisoner's dilemma, involving a pair of criminals apprehended while in possession of stolen goods, far from the scene of any crime. Each will be interrogated separately by the authorities, in an attempt to obtain a confession. Should neither talk, both will be found guilty of possessing stolen goods, and each will serve a year behind bars. Should one talk while the other remains silent, the confessor will receive a minimal sentence of three years for breaking and entering, which will then be suspended by the judge so that no time need actually be served, while the book is thrown at the silent offender, who will then serve five of the maximum eight years prescribed by law for breaking and entering. Should both talk, both will spend one year behind bars. The foregoing information can be summarized in the tabular (a.k.a. matrix) form shown in figure 5.12. Only the penalties (negative rewards) accruing to ME are shown. The penalties THEE

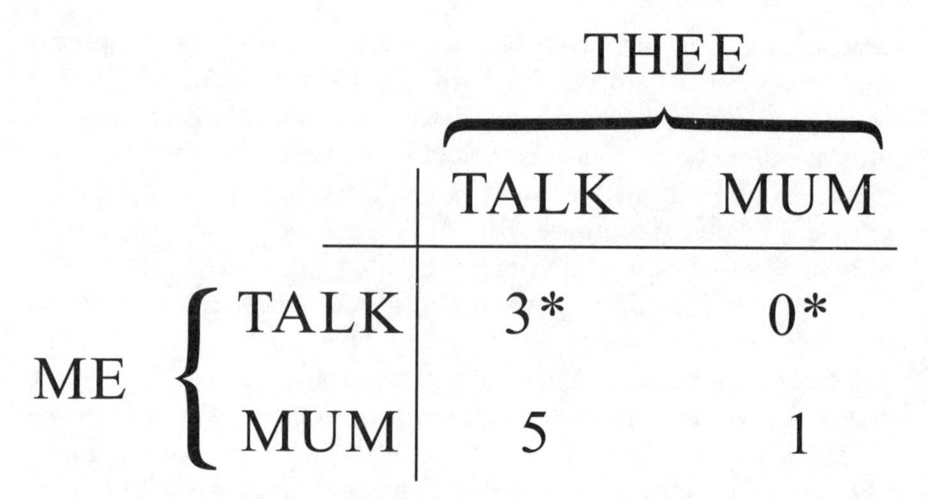

Figure 5.12. The reward matrix for ME in prisoner's dilemma. The one for THEE is obtained by interchanging the numbers 0 and 5.

faces can be recovered—in the present symmetric case—by interchanging the numbers in the lower-left- and upper-right-hand corners of the matrix. They are not to be obtained—as in the case of zero-sum (matrix) games—by inserting a negative sign in front of each individual penalty. Because two matrices would be required to display the complete reward structure of a non-zero-sum game in the absence of symmetry, such games are often called "bimatrix" games.

The asterisk on the 3 is to indicate that if THEE elects to talk, the best thing for ME to do is to talk as well, since three years are fewer than five. Similarly, the asterisk on the 0 indicates that if THEE elects to stay mum, the best thing for ME to do is still to talk, since zero years are shorter than one. No matter what THEE elects to do, the best thing for ME is to talk. And no matter what ME elects to do, the best thing for THEE is to talk. The talk strategy may thus be said to dominate the mum strategy in the sense that it yields a more attractive reward against every opposing strategy.

Since both players are free to reason in the foregoing manner, it seems only natural that both should elect to talk, and to serve three years each for their crime. On the other hand, it would not be all that surprising if the pair contrives to serve a total of only two years by staying mum rather than the six collective years they get by talking.

Both would surely stay mum if questioned together in the same room, and it is by no means inconceivable that "brothers in crime" might achieve the same result even in isolation. To some, it may seem paradoxical that the best collective result should be obtained when both participants employ a dominated strategy. To others, prisoner's dilemma merely illustrates the fact that selfish behavior can well prove harmful to all concerned.

Studies of actual criminal behavior have been inconclusive. Some talk, while others don't. When both detainees are members of the same street gang, it is often expected that the higher-ranking member will talk while his subordinate remains silent. In some cities, elaborate "don't snitch" campaigns discourage cooperation with authorities by circulating DVDs showing the corpses of recent talkers. Laboratory experiments are similarly inconclusive, in that some pairs of subjects manage to cooperate, even when denied the opportunity to communicate, while others fail even when permitted to do so. William Poundstone reports that the first prisoner's dilemma experiment was conducted at the Rand Corporation in 1950 and that the U.S. Air Force sponsored an extended series of such experiments at Ohio State University during the late 1950s and early 1960s.[7] He also quotes Anatol Rapoport to the effect that between 1965 and 1971, the results of perhaps two hundred such experiments were published.[8]

The foregoing plethora of (largely inconclusive) experiments dramatizes the world of difference between zero- and non-zero-sum competition. Once solved, zero-sum games tend to stay solved. Experiments pitting human subjects against a computer programmed to implement an unbeatable game plan in nim, tic-tac-toe, or finger matching would be pointless, because unbeatable really means unbeatable in zero-sum situations. Non-zero-sum games, in contrast, are seldom if ever definitively solved. The discovery of an effective game plan for one of them does not halt the search for better ones, just as the discovery of an effective treatment for joint pain or high blood pressure does not halt the search for even more effective treatments. Chapter 6 will summarize existing knowledge of many-player and non-zero-sum competition.

CHAPTER 6

Many-Sided Competition

The many-sided version of the prisoner's dilemma is known as the tragedy of the commons. It takes its name from a disaster that befell the inhabitants of small English villages during the eighteenth century. Even as they continued to graze their cattle in common unfenced pastures, the villagers sought gradually to increase their herds. Each additional cow entitled its owner to more butter, more milk, more cheese, more offspring, and (eventually) more meat and hides, at the cost of a barely perceptible degradation of pasture quality. While the benefits accrued exclusively to the extra animal's owner, the tradition of unrestricted access obliged his or her neighbors to share the cost. So, as one farmer after another grew more prosperous, one village common after another became overgrazed. Garrett Hardin has famously argued that similar forces are at work today on population growth, resource depletion, and environmental degradation.[1]

Any competitive venue can generate a tragedy of the commons. The (highly competitive) annual free-agent drafts conducted by the National Football League, the National Basketball Association, the National Hockey League, and other professional athletic leagues are particularly susceptible. Although the draft system has proven highly successful in maintaining competitive balance between rich and poor teams in a variety of professional sports, it is not an unmitigated blessing.

Every annual player draft operates in more or less the same fashion. First, the team with the worst record in the most recently completed season is allowed to pick from a pool of players new to the league. Next, the team with the second-worst record is allowed to pick from the pool, and so on, until every team has chosen one new player. Then the process is repeated until each team has chosen two new players, then three, and so on, through the appointed number of rounds. Pool remnants become "undrafted free agents," eligible to negotiate with any team.

For illustrative purposes, Philip Straffin asked his readers to consider a hypothetical three-team, two-round draft in which the pool of new players is only six deep.[2] Call them A, B, C, D, E, and F, and suppose the competing teams rate them in descending order, as shown in figure 6.1. If each team always chooses the highest-rated player left on the board, according to its own ranking, the first five players chosen will be A, E, C, B, and F, with each team getting its highest-rated player in the first round and its second-rated player in the second. All that harmony will come to an end, however, with the sixth and final choice, by which time the Raptors' second and third choices—namely, F and E—will already have been chosen, leaving the Raptors no better option than D, their fourth-ranked player.

Might the Raptors not have been better advised—suspecting that the Titans are more anxious to acquire F than C—to pick F in the opening round? That would oblige the Titans to choose D in the second round—since A, B, E, and F will all be gone by then—leaving C, secretly the Raptors' most wanted recruit, for them to claim with the final pick. Such "insincere drafting" might enable the Raptors to obtain their first- and second-rated players, instead of their first and

Giants	Titans	Raptors
A	E	C
B	F	F
C	B	E
D	A	D
E	D	A
F	C	B

Figure 6.1. The orders in which three teams rank draft-eligible players A to F.

fourth choices. But the Titans, too, may draft insincerely. By passing over E and F in the first round in favor of B, they can hope to obtain their first and third choices instead of their first and fifth. Yet they can do so only if the Giants draft sincerely in the first round.

Any player draft constitutes a tree game, to which backward induction applies. By carrying it out, the Giants can decide which of the original six players to choose with the first pick and which to select from any remaining threesome with the fourth. Likewise, the Titans can decide which to choose from any remaining fivesome with the second pick and from any remaining pair with the fifth. Finally, the Raptors can decide which to select from any remaining foursome with the third pick overall. There is no decision to be made concerning the sixth pick, since only one player will then remain.

When a team has decided what to do in every possible situation, it has formulated a strategy. When all three teams have done so, the outcome of the game is determined. If each team's strategy is the best possible response to the choices made by the other two teams, its chosen strategy is in Nash equilibrium. Each of the strategies identified by backwards induction turns out to be optimal against the rest.

If the teams draft insincerely, the players in the pool seem likely to be chosen in the order C, E, F, A, B, D, leaving the Giants with their first and third choices, the Titans with their first and third, and the Raptors with their second and fourth.[3] That constitutes an undeniable tragedy of the commons, since it obliges every team to accept a lower-rated pair of players than it would have obtained by picking sincerely throughout. Insincere drafting can be a way for the teams to outsmart themselves.

The Titans are likely to draft insincerely because they fear that the Raptors will. The Giants are likely to do so because they fear that the Titans will. And so on. Actual player drafts are far more complicated, since they typically involve thirty or more teams, seven or more rounds, hundreds of aspiring players, and scores of unscrupulous agents. Moreover, teams make every effort to disguise their sincere rankings of the available talent. Yet two key features of Straffin's miniature example are common to real and hypothetical drafts—the strategic thinking that permeates all such contests and the (minor) tragedies that occur.

Raised as they invariably are on the teachings of Adam Smith, modern policy makers seem baffled by the tragedy of the commons.[4]

So ingrained is their faith in the essential goodness of what Smith described as "the uniform, constant, and uninterrupted effort of every man to better his condition" that they regularly overlook the conflict between that effort and the public purpose. While conceding that "it is his own advantage, indeed, and not that of the society, which he [the decision maker in question] has in view," Smith insisted that "the study of his own advantage naturally, or rather necessarily leads him to prefer that employment [of whatever resources he can command] which is *most advantageous* to the society."[5] As a result, the story goes, every effort to influence such a man to employ his resources in a way *more advantageous* to the society is doomed to fail. It cannot succeed, because, left to his own devices, such a man is automatically guided by Smith's "invisible hand" to employ "whatever capital he can command" in the way most advantageous to the society. Does the tragedy of the commons precipitated by the teams' decision to draft insincerely not contradict Smith's assertion?

Although it is impossible, at this late date, to know how steadfastly Smith would have defended the foregoing pillar of the faith that has grown up around his teachings, it is abundantly clear that many of his current followers take him at his literal word. Did a slip of the pen cause him to write "most advantageous" when he meant only "advantageous"? Or did he write what he meant to write? The consequences are not insignificant.

It is one thing to concede that an individual's self-interested actions are advantageous to the society in which he or she lives. It is another to insist that no other course of action could possibly be more so. Under the first (less optimistic) interpretation of Smith's declaration—which he repeated on various occasions in slightly altered form—there is plenty of scope for government programs designed to divert private capital from socially advantageous activities into socially more advantageous ones. Under the second, there are no more socially advantageous ones. The English villagers' experience during the eighteenth century, as well as Straffin's analysis of a free-agent draft, would seem to discredit the more optimistic interpretation.

Insincerity is by no means uncommon. Elections involving more than two candidates, such as the American presidential election of the year 2000, positively invite it. Bush opponents rightly foresaw that that year's contest would be close and made every effort to minimize the impact of the Nader candidacy on Gore's chances. Some organized vote exchanges in which Gore voters from states Gore was sure to win were paired with Nader voters in "swing states." If each promised to vote for the other's candidate of choice, Nader's chances of winning 5 percent of the popular vote—so that a Green Party candidate would be eligible for federal campaign funds in 2004—would be undiminished, while the odds against Gore would shorten.

On November 3, 2000—Election Day—Gary Trudeau used his popular cartoon strip, *Doonesbury*, to exhort readers to help defeat Bush. Each of the first three panels showed a different character listening to the radio as the announcer cataloged fears aroused by Bush:

- "If you'd like to see abortion re-criminalized . . ."
- "If you're for unrestrained logging and drilling, and for voluntary pollution control . . ."
- "And if you favor more soft money in politics, then the choice today is clear . . ."
- "Vote Nader."

Trudeau foresaw—correctly, as it turned out—that unless the voters who preferred Nader to Gore to Bush could be prevailed upon to vote for Gore, Bush would surely win.

Surprisingly little "strategic insincerity" would have changed the result. In New Hampshire, where Nader received twenty-two thousand votes, Bush beat Gore by fewer than eight thousand votes. In Florida, where Nader received ninety-seven thousand votes, Bush beat Gore by even fewer. If a few more Nader voters in those two states had been prepared to vote strategically, Bush would have lost in the Electoral College, as well as in the popular vote. This, together with the actions of the Supreme Court, brought forth louder than usual calls for the abolition of the Electoral College. Others argued for the adoption of a more discriminating method of counting votes.[6] It has long been known that "plurality voting"—whereby each voter merely identifies a first-choice candidate—is by no means a reliable method of discovering what voters actually want. Is there a better way?

The first serious study of voting procedures was undertaken by members of the French Academy of Sciences shortly before the French Revolution. Foreseeing that the nation's most critical decisions might soon be made in this new and unfamiliar way, they wondered how trustworthy majority rule might actually be. Their conclusions were far from reassuring. Although plurality voting seems to produce the desired result in two-candidate elections, it can easily malfunction in more complex situations. Perhaps the most striking—and decidedly the best known—of its shortcomings was discovered by Marie-Jean-Antoine-Nicolas de Caritat Condorcet, a mathematician, philosopher, politician, and savant of the revolutionary era.

Because he did trust two-candidate elections, Condorcet believed that every issue should be decided by pairwise comparisons between comparable alternatives. With the 5 alternatives A, B, C, D, and E, there are 10 pairs to compare—namely, AB, AC, AD, AE, BC, BD, BE, CD, CE, and DE. With 10 alternatives, there are 45 pairs; with 15 alternatives, 105 pairs. There is no need to conduct a separate election between every pair of candidates, since the same information can be acquired from a single election in which every voter is required to arrange all candidates, from first to last, in order of preference.

Condorcet was well aware that such elections are likely to produce ambiguous results. They will whenever candidate A beats candidate B, who beats someone else, who beats someone else, who . . . beats candidate A. Yet it doesn't have to be that way. A single candidate can outrank every opponent on a majority of the ballots cast. And surely, said Condorcet, such a candidate represents "the people's choice." Donald Saari refers to such candidates as "Condorcet winners" and to candidates who rank below each opponent on a majority of ballots as "Condorcet losers."[7] Condorcet proposed that any acceptable voting procedure should, at the very least, elect the Condorcet winner when there is one. Indeed, he seemed to regard the failure to elect a Condorcet winner as a (potentially major) tragedy, comparable in many ways to the tragedy of the commons. So imagine Condorcet's surprise upon learning that plurality voting can fail to elect such a winner.

To demonstrate his landmark discovery, Condorcet considered a three-candidate election in which thirty voters prefer *A to B to C*, one prefers A to C to B, ten prefer B to C to A, twenty-nine prefer *B to A to C*, ten prefer C to A to B, and only one prefers C to B to A. Here the three orderings in which A precedes B have been underlined, while the three in which A precedes C are italicized. It is hoped that these visual aids will encourage the reader to verify that (1) A beats B by a score of 41 to 40, (2) A beats C by a score of 60 to 21, and (3) B wins the plurality election with 39 votes to A's 31 and C's 11, making A a Condorcet winner unelected by plurality voting. This is the voting paradox forever associated with Condorcet's name. Something like it can occur whenever voters are asked to choose among three or more candidates. The failing is by no means peculiar to plurality voting.

Condorcet's investigation of voting procedures seems to have been motivated by a desire to discredit a scoring system devised by a certain Count Borda—also a member of the French Academy—that was used for some time to elect new members. Borda proposed that in an election in which each voter is required to rank all candidates in order of preference, each candidate should receive from each ballot the number of points obtained by subtracting his or her rank on that ballot from the total number of candidates. Thus, in a five-candidate election, a given ballot is worth four points to the candidate ranked highest on it, three points to the candidate ranked second, two points to the candidate ranked third, one point to the candidate ranked fourth, and none to the candidate ranked last. The one with the most points wins. Condorcet was delighted to discover that Borda's scoring system also fails to elect A, the people's choice.

Borda's system—known today as the Borda count—is but one of many ways to distribute a fixed number of points among some or all of the candidates in a multicandidate election. In a five-candidate election, the Borda count distributes ten points among the top four candidates on a given ballot. In an eight-candidate election, it distributes twenty-eight points among the top seven candidates. And so on. Ordinary plurality voting awards a single point to the top-ranked candidate. Saari coined the term "positional methods" to describe scoring systems that distribute a fixed number of points, in a fixed manner, among a fixed number of the highest-ranking candidates on each ballot.[8] He has discovered a truly impressive number of pitfalls that the Borda count—alone among positional methods—manages to avoid. Yet neither the Borda count nor any other positional method manages to avoid the Condorcet paradox. Any one of them can fail to elect a Condorcet winner. So, too, can "approval voting," one of the few scoring systems for multicandidate elections that is not a positional method.

Approval voting was apparently first proposed by Robert Weber in his Yale thesis of 1971. It allows each voter to check off as many names as desired on his or her ballot, thereby indicating both approval of those candidates and disapproval of all others. Each checkmark is then worth one point to the candidate beside whose name it appears, and the candidate with the most points wins. Approval voting is not what Saari would call a positional method, since the number of points to be distributed by a single ballot among the several candidates can vary. Yet it, too, can fail to elect a Condorcet winner. Approval voting has much to recommend it in primary elections, held to decide which candidates deserve to be included in a final runoff. An entire book has been written about approval voting.[9]

When called upon to choose between three or more courses of action, legislatures—possibly influenced by Condorcet—often vote on them two at a time. Thus, after choosing between option 1 and option 2, they choose between the winner and option 3, then between that winner and option 4, and so on. The order in which the options are introduced is called an agenda, and whoever controls it is in an excellent position to influence the outcome. In March 1988, the U.S. House of Representatives dealt in such fashion with two quite different plans to provide humanitarian aid to the so-called contra rebels in Nicaragua. Three alternatives were considered:

- Plan A (supported by the Reagan administration) to supply the contra rebels with arms and humanitarian aid,
- Plan H (supported by the Democratic leadership) to supply only humanitarian aid, and
- Plan N (initially unsupported) to supply no aid of any kind.

Straffin furnishes an interesting simplification of the strategic situation existing at the time.[10] He divides the congressmen into four voting blocs: conservative Republicans, moderate Republicans, moderate Democrats, and liberal Democrats. Each bloc includes 124 voters except the moderate Republicans, who number only 62. All the conservative Republicans preferred A to N to H, while all the moderate Democrats preferred H to A to N, and all the liberal Democrats preferred N to H to A. The moderate Republicans, in contrast, preferred A to H to N.

The first vote was between A and H, with H winning by sixty-two votes. The second vote, between H and N, was won by N, also by sixty-two votes. Solid Democratic support for H was responsible for the first result while, in the second, liberal Democrats combined with conservative Republicans to defeat H. With congressmen on both sides of the aisle voting sincerely, the contras received no aid at all. Might more strategic voting behavior have led to a different outcome?

Insincerity on the final vote is futile since, at that point, only two alternatives remain, and there is no way to game a two-candidate election. So if H defeats A on the first vote, N will again beat H in the second, 248 votes to 186. But if A defeats H in the first, A will also defeat N in the second, by 310 votes to 124. Had the moderate Democrats voted insincerely for A in the first election, the outcome would have been A, which they preferred to N. They cannot tease the outcome H from any agenda that pits A against H in the opening round, since H wins the final round only if opposed by A.

Observe next that A defeats N by 310 votes to 124, N defeats H by 248 votes to 186, and H defeats A also by 248 votes to 186, when all votes are sincere. Hence the agenda that pits A against N in the opening round leads to outcome H when sincerity rules, whereas the one that pits H against N in the opening round leads eventually to A. For any outcome, there is at least one agenda that leads to it with sincere voting. On the other hand, the possibility of insincere voting behavior can never be ignored, because no democratic scoring system can ever be immune to insincerity.

A democratic scoring system is one in which every ballot has an equal chance of being counted. No system can guarantee that every ballot will be counted, but the better ones minimize and equalize the chances that any one ballot will be ignored. The least democratic vote-scoring system imaginable is a "dictatorial" one, in which a single privileged voter's (possibly golden) ballot counts for sure, while none of the others count unless they agree with his. The distinguished voter is then in a position to dictate the outcome of every election.

Allan Gibbard, in 1973, and Mark Satterthwaite, in 1975, both proved mathematically—and independently of each other—that whenever there are three or more candidates to choose from, the only vote-scoring system that is immune to strategic insincerity is the dictatorial one, in which only the dictator's ballot really counts. All other vote-scoring systems are vulnerable to strategic insincerity. Not surprisingly, this discouraging result has come to be known as the Gibbard-Satterthwaite theorem. It means that, like it or not, every citizen of a democratic society is perforce involved, from time to time, in a highly strategic many-sided competition.

The first attempt to formulate a theory of many-sided competition was made by von Neumann and Morgenstern in 1944. By focusing on the behavior of coalitions, they produced what is now known as the cooperative theory of (many-player) games. The coalitions they found most interesting—when their collaboration began in 1939—were the labor unions then gaining both members and influence. Given a chance to revisit the subject today, they would probably be more concerned with lobbying groups and political action committees than with today's often corrupt and increasingly impotent unions.

Von Neumann and Morgenstern began with a two-stage analysis of many-player games. After attributing a value to each potential coalition of players, in a manner to be explained shortly, they sought to explain how those values might be used to calculate what they took to be a solution of the game. In retrospect, their use of the term "solution" appears overoptimistic, as subsequent generations of scholars have largely rejected it in favor of an array of situation-specific alternatives. Of these, the core solution, the Shapley value solution, the bargaining set solution, and the so-called nucleolus are among the most frequently invoked.

If there are three players in a game, there are $2 \times 2 \times 2 = 2^3 = 8$ potential coalitions, including the empty coalition and the coalition of the whole. If there are five players in a game, there are $2^5 = 32$ potential coalitions. With seven players, there are $2^7 = 128$ coalitions, and so on. Every victim of "the new math" will recognize the symbol $\emptyset$, denoting the empty set or (by extension) the empty coalition. The coalition of the whole is of even greater significance and will be denoted by the uppercase Greek letter omega (Ω).

To find the value of a potential coalition K, von Neumann and Morgenstern asked what K could win in a two-player zero-sum game against the coalition $\sim K$ (pronounced "not K") of all players excluded from K. The answer, of course, is VAL(M), where M is the matrix of the two-person zero-sum game in which the members of K seek to maximize their combined winnings, while the members of $\sim K$ seek to minimize that quantity. Hence VAL(K) = VAL(M).

Since large coalitions can ordinarily win more than small ones, VAL(Ω) would seem to represent the largest possible "pot" of potential winnings available to the several players. Von Neumann and Morgenstern began by considering all possible divisions of a pot of that size and sought—by process of elimination—to identify those upon which "rational players" might agree. No such player need agree to a division that pays her less than she can win on her own, without joining any coalition at all. So, for ease of reference, von Neumann and

Morgenstern coined the term "imputation" to describe any division of a pot of size VAL(Ω) in which every player receives at least as much as she could win on her own. To them it seemed axiomatic that a set of mutually acceptable imputations would constitute a solution of the game in question.

As no individual player need agree to a division that allows her less than she can win on her own, no coalition K of players need agree to a division in which the members' pooled winnings fail to equal or exceed VAL(K). Any such division, said von Neumann and Morgenstern, is "dominated" via the coalition K. A typical problem in the cooperative theory of games asks for the solution of the three-player game in which:

$$\text{VAL}(\emptyset) = \$0,$$
$$\text{VAL}(1) = \text{VAL}(2) = \text{VAL}(3) = \$0,$$
$$\text{VAL}(12) = \text{VAL}(13) = \text{VAL}(23) = \$1,$$
$$\text{VAL}(\Omega) = \text{VAL}(123) = \$P,$$

where $\$P$ represents a pot size greater than \$1. Cooperative game theory represents a bold and highly abstract effort to deduce the solutions of games from this and only this information, treating any other available data concerning the contest and its contestants as irrelevant.

If P equals \$1.50, there is a single undominated imputation. If P is larger, there are infinitely many such imputations. If P is smaller, there are none at all. The set of all undominated imputations is often called the "core" of the game and is regarded by many as its natural solution. This is true whether $P = \$1.53$, in which case all undominated imputations closely resemble the symmetric one that allows each player \$0.51, or whether $P = \$100.00$, in which case each of the three (highly asymmetric) imputations that award \$98.98 to one player (thereby leaving only \$1.02 for the others to share) is undominated. Because no one has explained the means by which von Neumann and Morgenstern proposed to solve the foregoing game when P lies between \$1.00 and \$1.50 better than they themselves did, that explanation will not be repeated here.[11]

Von Neumann and Morgenstern viewed elections as "simple games" in which the value of every coalition K is either 0 or 1.[12] If VAL(K) = 1, then K is called a winning coalition. If VAL(K) = 0, then

K is said to be a losing coalition. Should a player be added to a winning coalition *K*, or removed from a losing coalition *L*, *K* remains a winning coalition and *L* a losing one. Winning coalitions are called minimal if the loss of any one member turns them into losing coalitions, and losing coalitions are called maximal if the gain of any additional player turns them into winning coalitions. Winning coalitions seem unlikely to grow beyond minimal winning size, since that would reduce each member's share of the spoils.

Martin Shubik and Lloyd Shapley suggested in 1954 that a remarkable evaluation formula previously discovered by Shapley could be combined with the theory of simple games to create a plausible index of voting power in a legislative body.[13] To illustrate the use of their method, they considered the Security Council of the United Nations. As then constituted, it included five permanent members—each with veto power—and six rotating members. To pass, a resolution had not only to be approved by seven of the eleven members but also to avoid all five potential vetoes. Ignoring the possibility of abstention by a permanent member, Shapley and Shubik interpreted this to mean that a minimal winning coalition had to include all five of the permanent council members along with two others. On that assumption, they were able to calculate that the six rotating members together held only about 1.3 percent of the voting power in the council, the other 98.7 percent being evenly divided among the five permanent members.

In 1965, the Security Council was expanded to include ten rotating members, with a total of nine members needed to pass a resolution. Interpreting that to mean that a minimal winning coalition must now include all five permanent members along with four others, Shapley and Shubik suggested that the rotating members of the council now share 1.9 percent of the voting power, instead of 1.3 percent, the remaining 98.1 percent being again shared equally among the five permanent members. Two different proposals are currently under consideration for the further expansion of the Security Council from fifteen to twenty-four members, possibly to include four new permanent members without veto power. Depending on which, if either, is adopted, the original permanent members might retain as little as 97 percent of the council's voting power.

For all that has been said and written about von Neumann and Morgenstern's remarkable theory of many-player games, it was not by following in their footsteps that John Nash shared, in 1994, the Nobel Prize in Economics with John Harsanyi and Reinhard Selten. Instead, the Nobel committee honored the three for striking out in a quite different direction. It honored them for developing a rival theory, pertaining to a form of many-sided competition that von Neumann and Morgenstern—though well aware of its existence—deemed unworthy of their attention. The Nash theory (a.k.a. noncooperative game theory) concerns a form of competition in which the players are either unwilling or unable to create effective coalitions and/or conclude binding agreements.

Nash's so-called noncooperative theory applies to situations in which all forms of cooperation are as strictly prohibited as bid rigging at auctions, price-fixing in commerce, and jury tampering in courts of law. According to Sylvia Nasar, whose prizewinning biography of Nash was begun soon after he received the Nobel Prize, the decision to honor the developers of the noncooperative theory—rather than those of the older and more mathematically sophisticated cooperative theory—was based on the committee's finding that the former has (mainly through its effect on economic science) been of greater value to mankind.[14]

Backward induction, as it applies to many-player tree games with perfect information, is an important constituent of noncooperative game theory. Each of the strategies it assigns to the several players in a game of interest turns out—more or less automatically—to be optimal against the rest. Thus if all players are commanded by some "higher power" to employ the strategies assigned them by backward induction, none may hope to benefit by unilateral disobedience. Bilateral disobedience can still be rewarding, but unilateral disobedience cannot. Accordingly, an assignment of strategies to the players in a game is said to constitute a "Nash equilibrium" assignment if each component strategy is optimal against the others. Once the players have gravitated to such an equilibrium, no one individual stands to gain by disturbing it. A particular game may exhibit one, many, or no Nash equilibriums.

Nash sought—successfully, as it turned out—to emulate von Neumann by demonstrating that with or without perfect information, all many-player tree games possess "equilibrium solutions" in terms of mixed strategies. Yet in the larger quest to develop as successful a theory of many-player games as von Neumann had for two-player zero-sum games, he failed abysmally. Nash's equilibriums are typically harder to find than von Neumann's saddle points, and are often of little consequence when found. The game of prisoner's dilemma (PD) dramatically illustrates the latter point. The only Nash equilibrium for PD consists of the strategies directing each prisoner to squeal on the other. Yet, as is well known, the decision to "rat out" a partner in crime can be terminally ill-advised.

The tragedy of the commons works much the same way, because the only Nash equilibrium assignment of strategies for playing it leads each farmer to participate in the overgrazing of a common pasture.

The dynamics of various populations, of fish, of fowl, and of fur-bearing fauna, have been studied in considerable detail. Sometimes they can be approximated by surprisingly simple equations relating the instantaneous rate of growth of a given population to its current size. Left to themselves, such populations tend to exhibit large positive growth rates when small, and large negative growth rates when excessively large. The intermediate population size for which the growth rate is exactly zero is often described as the carrying capacity of the surrounding environment.

Judicious harvesting policies make it possible to extract maximum sustainable yields from such populations. Should a particular population consist of a commercially valuable species of fish—such as the once-plentiful New England cod—the size of the annual catch will depend on three factors: the size of the population itself, the size of the fishing fleet, and the fleet's proficiency in locating fish. The latter has increased significantly since the middle of the twentieth century, due to the development of electronic fish-finding technology and to the size, speed, and general seaworthiness of commercial fishing vessels.

Each maritime nation must decide how large a fishing fleet to maintain. Some nations resolve such issues by decree, while others control fleet size through the issuance of licenses. The equations of population dynamics, along with the size of the rival fleet, permit the construction of a system of curves exhibiting the effect of fleet size on sustainable catch size. The topmost curve in figure 6.2 illustrates the relationship that would exist between one's own fleet size and one's own sustainable catch if there were no rival fleet.

The lower curves show the effects on one's own sustainable catch size of successive expansions of the rival fleet, from small to medium to large to extra large and beyond. For present purposes, the composition of the rival fleet is irrelevant. It makes no difference whether all its members fly the flag of a single nation, or whether no two flags are alike. The larger the combined rival fleet, the lower lies the curve depicting a nation's own prospects. Each such curve rises to a peak at which one's own sustainable catch size, and therefore one's fleet size, are simultaneously optimal. To increase one's own fleet size beyond that point, or to reduce it below that point, is to reduce the size of one's own sustainable catch.

Because successive peaks do not align one above another, the decision to send a given number of ships to sea should be predicated on an estimate of the size of the combined rival fleet. It is therefore comforting to know that a fleet that is 5 percent too large or too small reduces the size of a nation's own sustainable catch by a mere fraction of 1 percent in the standard fish-harvesting model. Imprecise estimates of rival fleet size are more than adequate for present purposes.

Next observe that the two-nation version of the global fishery management game can be turned into a board game. The game board is depicted in figure 6.3. Let the nations involved be called X and Y, and let x and y denote the sizes of their respective fishing fleets. The rules of the game provide that the referee shall initiate play by placing a coin at a starting point S of his or her choosing and that the players shall thereafter take turns moving it. The player representing X may

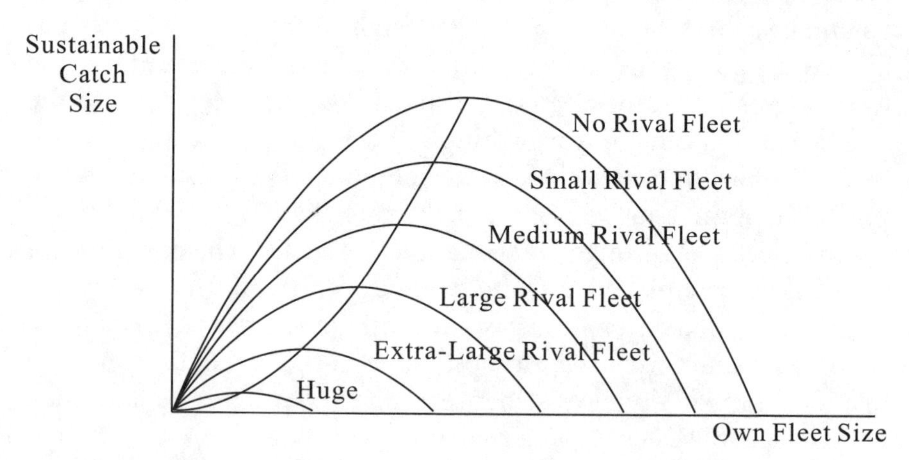

Figure 6.2. A nation's sustainable catch size depends on the size of its own fishing fleet, as well as that of the (combined) rival fleet.

move the coin horizontally, thereby changing the horizontal coordinate x of the (center of the) coin, while the player representing Y may move it vertically, thereby changing the vertical coordinate y of the (center of the) coin. Each player must act independently of the other and may neither propose nor attempt to enforce any agreement of any kind between them.

Each move will alter the rates at which the two nations are currently harvesting fish, and the players must wait to respond, after each move, for the respective catch sizes to settle down to their new sustainable levels. The referee will signal the player waiting to move when he or she may proceed. All this is important to players rewarded on an accrual basis, at ever-changing rates proportional to the catch size of the fleet under management. The funny thing is that no matter where the referee places the coin initially, it will likely come to rest—after a series

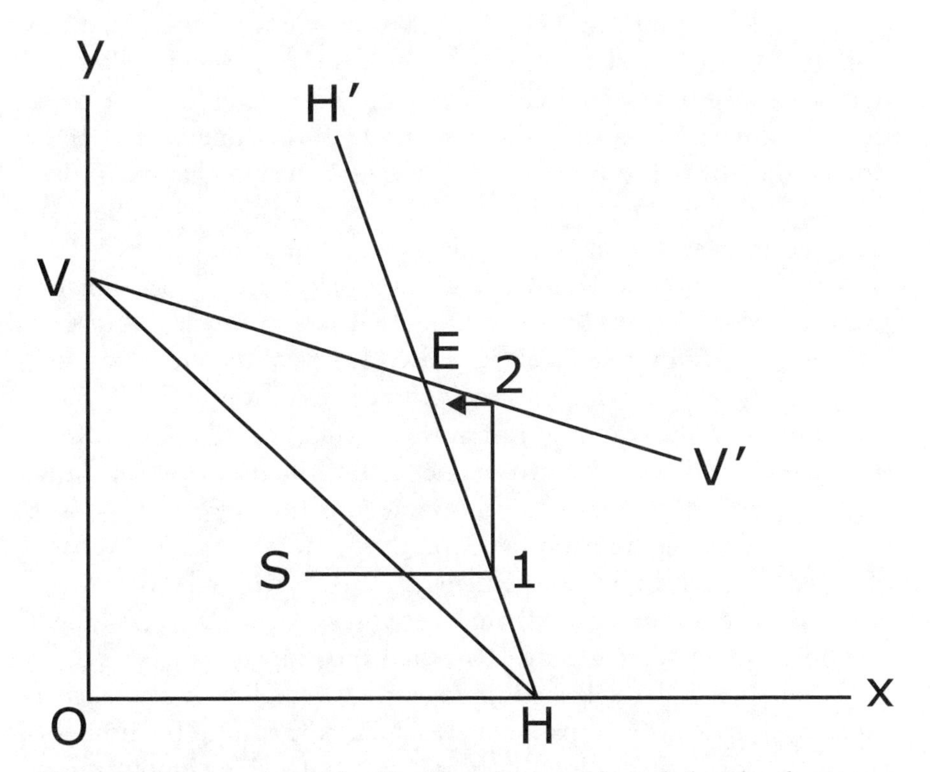

Figure 6.3. The route by which self-serving players might push the token from an arbitrary starting point S *to the vicinity of a point* E *of Nash equilibrium.*

of moves by the players representing X and Y—in the immediate vicinity of point *E* in figure 6.3. That is because *E* alone has the property that neither self-interested player has any reason to move the coin away from *E*. Hence, in its travels about the game board, the coin seems unlikely to come to rest before arriving in the immediate vicinity of *E*. This is not particularly good news.

To see why, let F^* and C^* be the coordinates of the highest point on the highest curve in figure 6.2. C^* then represents the size of the largest sustainable combined catch, while F^* represents the size of the combined fleet required to land it. So, for as long as the coin remains at rest anywhere on the line $x + y = F^*$ in figure 6.3, the two nations will continue to share the largest possible catch C^*. The intercepts of that line with the coordinate axes have been labeled *V* and *H*, respectively.

The line segment *VH* is the site of all efficient locations (x, y), since they and they alone permit the two nations to share the maximum sustainable catch. The line segments HH' and VV' in figure 6.3 are called destination curves because they are the natural destinations of horizontal and vertical moves of the coin, respectively. They are that because, among the points on a given horizontal line, the point that maximizes X's catch is the one at which that line crosses HH'. And among the points on a given vertical line, the one that maximizes nation Y's catch is the one at which that line crosses VV'. So, unless the coin is already at rest on HH', nation X can immediately increase the size of its own catch (typically at Y's expense) by moving the coin horizontally to or toward HH'. And unless the coin is already at rest on VV', nation Y can immediately increase the size of its catch (typically at X's expense) by moving the coin vertically to or toward VV'.

Only when the coin is at rest at the intersection *E* of the destination curves HH' and VV' can neither nation increase the size of its catch by moving the coin in an allowable direction. In other words, *E* is the only Nash equilibrium position in the two-competitor version of the common-fishery management game. By placing a kernel of seed corn (or the point of a pencil) anywhere on figure 6.3, and pushing it first horizontally to or toward HH', then vertically to or toward VV', and so on, the reader will soon discover that virtually any sequence of allowable self-serving adjustments will bring the coin to the immediate vicinity of *E*, suggesting that *E* is a point of stable equilibrium in the competition in question.

The alternately horizontal and vertical movements of the coin whereby the players representing X and Y fine-tune the sizes *x* and *y* of their fishing fleets are known as Cournot adjustments, after the French mathematician Antoine-Augustin Cournot who first investigated them in 1838. The path from *S* to 1 to 2 depicts the first few of the adjustments that would lead (rather quickly, in the present case) from the starting point *S* to the immediate vicinity of *E*. Such sequences of "successive adjustments" are intended to mimic the process by which enlightened competitors would most naturally seek to discover "that employment of the capital [especially human capital] at their disposal" which yields the most satisfactory result. Like the method of "fictitious play" described in Chapter 5, and like many of the machine-learning techniques described in Chapter 1, the discovery process ordinarily leads to a Nash equilibrium point *E*.

Because point *E* in figure 6.3 lies visibly to the north and east of the line segment *VH*, the sum $x^E + y^E$ of the coordinates x^E and y^E of *E* exceeds *F**, the size of the optimal fishing fleet. Hence the competitive fleet sizes x^E and y^E to which the Cournot adjustment process (or any of its more popular variants) leads are too large, in the sense that they cause the target population to be overfished. The deployment (x^E, y^E) has little in common with the ones (x, y) situated on *VH*, which yield the largest possible sustainable catch C^*. Competition dispatches too many boats in pursuit of too few fish. As in Straffin's miniature free-agent draft, and as in the case of the English villagers, the competitors' "study of their own advantage" leads to a socially undesirable outcome—an undeniable tragedy of the commons.

The pattern on display in figure 6.3 is typical of many-player games. One or more points *E* of Nash equilibrium stand well apart from an entire (continuous) spectrum of socially optimal outcomes, such as the ones on *VH*. More or less natural learning processes lead self-interested competitors to the immediate vicinity of a Nash equilibrium position like *E*, located far from the nearest socially optimal outcome. Contrary to the wisdom dispensed by the yellow economic press, the "study of his own advantage" is an unreliable guide to "that employment [of one's own resources] which is most advantageous to the society."

An allocation is often called Pareto optimal if any alternate allocation that deals more generously with one individual must necessarily deal less generously with another.[15] In the present case, only the allocations (x, y) that lie on *VH* are Pareto optimal, since they alone share the property that $x + y = F^*$. All other allocations, including x^E and y^E, send either too few or too many boats to sea and, in consequence, fail to land the largest possible sustainable catch. Only in the rarest of circumstances does any self-interested competitive process lead to a socially (a.k.a. Pareto) optimal outcome.

Like fictitious play, and like a host of machine-learning techniques, Cournot adjustments are intended to imitate real-world "learning in games." An entire book has been written on that subject, in which fictitious play and the Cournot adjustment process are adapted to different competitive environments.[16] Appropriate references to the advanced technical literature are included at the end of each chapter. Although most of the techniques discussed are as simpleminded as the ones described above, a final chapter on "sophisticated learning" raises the bar significantly. Continued progress in this direction is to be expected.

It should be mentioned in passing that laboratory experiments provide less than overwhelming support for the thesis that "rational competitors" naturally gravitate toward Nash equilibrium conditions. On the contrary, numerous experiments suggest that a substantial fraction of human subjects can overcome the impulse to seek personal advantage long enough to reach socially optimal outcomes. A series of experiments conducted by D. H. Stern, pitting two or more business-school students at a time against one another in a game not unlike the foregoing fisheries management game, were among the first to yield such results. Stern found, as have others, that cooperative behavior leading to socially optimal outcomes is almost as frequent as socially wasteful Nash equilibrium behavior.[17]

The three-nation version of the fishery management game is harder to visualize, in part because—as in three-dimensional tic-tac-toe—the associated "game board" must occupy a portion of three-dimensional space. Destination curves must be replaced by destination surfaces, and so on. But computer versions of the game are easily implemented in which any number of "nations" may vie for advantage in the harvesting of fish. The main conclusions are unchanged: agreements designed to secure the health of the underlying fish population invite cheating. The socially (that is, Pareto) optimal positions in the higher-dimensional playing space again form a continuum located at some distance from the (typically unique) Nash equilibrium. Hence, by a series of self-serving adjustments, the international fishing fleet soon comes to exceed the magnitude F^* that corresponds to the maximal sustainable catch size C^*. Small wonder that the Pew Oceans Commission report of 2003, titled "America's Living Oceans," found those oceans to be "in crisis." And, later that same year, the journal *Nature* found that up to 90 percent of the stocks of the ocean's major predators—including Atlantic cod and bluefin tuna—have been wiped out.

Auto dealers compete first to attract prospective buyers to their showrooms and then to sell them cars. Upon entering the premises, a potential customer is soon approached by a salesperson whose job is to learn what sort of car—if any—he or she is prepared to buy, before "introducing" the prospect to as many such cars as possible before pressing to "close a deal." Dealers with large inventories enjoy an obvious advantage over dealers with small ones in this important phase of the competition. Size matters. On the other hand, dealers tend to purchase the cars on their lots with borrowed money, and are obliged to service the concomitant loans. Thus, each dealer must balance inventory size against carrying cost.

If all but one of several identical auto dealers were obliged to keep N cars on their lots at all times, there would be an optimal number N^* of cars for the remaining dealer to carry in inventory. If N were very small, N^* would naturally exceed N. But if N were very large, N^* would doubtless be smaller than N. Hence any N for which N^* is about equal to N would seem to constitute a happy medium, as well as an approximate Nash equilibrium inventory level for each competing dealer, in the sense that that inventory level would be roughly optimal for each against the rest.

Case explains at some length how to compute approximate Nash equilibrium inventory sizes for collections of nonidentical dealers, and how to incorporate other strategic variables—such as advertising budgets—into the analysis.[18] The resulting Nash equilibriums exhibit a welcome tendency toward "robustness" in the sense that the prescribed inventory levels, advertising intensities, and so on tend to perform well even if the opposing dealers do not perform exactly as expected. Many games of commerce may be shown to possess comparably robust equilibrium strategies.

Modern traffic planners make extensive use of techniques for computing Nash equilibrium points, including the methods of fictitious play and Cournot adjustment. Before adding a freeway entrance, widening a city street, or designating a street one-way, they seek to predict the effect of the proposed change on existing traffic patterns. To that end, they assume that a short period of adjustment will lead to a new equilibrium in which—as in the old—no single driver can shorten his or her travel time by switching to a faster route. This Nash-inspired method has gained acceptance by appearing to predict congestion sites and "link travel times" with tolerable accuracy and by being applicable to the design of "intelligent transportation systems."

Intelligent transportation systems are those in which information concerning anticipated route congestion is communicated to individual drivers soon enough to enable many to circumvent potential delays. The most familiar examples are the traffic helicopters operated by local radio stations during morning and evening drive times. But far more sophisticated systems—offering far more detailed and driver-specific information—are under development. The U.S. Department of Transportation has funded extensive research on the design and operation of such systems and has issued guidelines concerning them.[19]

In 1992, a team from the University of Michigan carried out a pilot experiment intended to demonstrate the feasibility of such systems. They began by compiling a massive database of travel requirements for the city of Troy, Michigan. Troy is a bedroom community of approximately eighty thousand residents located some twenty miles north of downtown Detroit. Its main traffic network includes about two hundred intersections, connected by five hundred streets and roads. From survey data collected by a consortium of nearby city governments, the team was able to estimate not only the number of vehicles likely to utilize the road network during peak-load periods but their points of origin, destinations, and times of departure. Altogether, some 16,500 vehicles could be expected to enter the network during the busiest twenty-four minutes of the morning drive and to depart within thirty-six minutes of entry.

As the fraction of vehicles equipped with a guidance system known as SAVaNT increased, both guided and unguided vehicles experienced improved travel times. At low levels of market penetration, the SAVaNT-equipped vehicles reached their destinations 8–10 percent faster than they otherwise would have, while the rest of the vehicles realized lesser gains, due to improved utilization of the network as a whole. At higher penetration levels—above 50 percent—the benefits were shared almost equally between guided and unguided vehicles, again due to superior network utilization.

Some years later, a different team compared the SAVaNT results with those from a newer system known as Alliance, which assigns individual drivers to routes intended to minimize not their own individual travel times but the average travel time experienced by all users of the network. Somewhat to the team's surprise, Alliance barely reduced average travel times at all. Though intended to reduce individual travel times—if need be at the expense of other drivers—the route assignments produced by SAVaNT turned out to utilize the network almost as efficiently as those produced by the collectively motivated Alliance. There was no significant tragedy of the commons on the highways and byways of Troy, Michigan.

Such findings are unusual. The more frequent result—at least in studies performed to date—is that Nash equilibrium outcomes (often loosely described as competitive outcomes) are socially quite inefficient. Such is the case in international fishery management, where self-serving behavior has driven one commercially valuable fish population after another to the brink of extinction, and it is arguably the case in forestry and agriculture, where valuable crop, grazing, and timber land is lost every year to overuse by self-serving corporations. The result is a significant tragedy of the commons, in the form of overproduction, especially of commodities.

Contrary to the teachings of Adam Smith, "the study of his own advantage" need not lead an entrepreneur "to prefer that employment which is most advantageous to the society." Although it seems to work more or less that way in and around Troy, Michigan—and on other crowded road networks around the world—it obviously doesn't work that way for the management of commonly held natural resources. A hundred years ago, the eastern white-tailed deer population was well on its way to extinction before the several states began to issue deer-hunting licenses and to prosecute unlicensed hunters. The same applies to moose, elks, antelope, swans, wild geese, and the more palatable varieties of wild duck. Effective regulation came too late to save the passenger pigeon, which once blackened the skies of the American Midwest, and a similar fate probably awaits elephants, rhinos, and other big-game species, including the great apes.

In an ongoing series of articles about medical care, the *New York Times* columnist Paul Krugman points out that although Americans pay more for care than the citizens of any other nation, they get less for their money.[20] He attributes the difference, in part, to the fact that administrative costs eat up about 15 percent of the premiums paid to private health insurers, but only about 4 percent of the budgets of public health insurance programs. Public insurance plans cost less than private ones because they don't squander resources in an effort to screen out high-risk clients or charge them higher fees. U.S. doctors are obliged to employ extra office personnel just to deal with the insurance companies. A clever assistant can cut down the insurer's rejection rate from 30 percent to perhaps 15. Small wonder that the United States, which relies more heavily on private insurance than any other nation, bears heavier administrative costs.

Krugman deems it perverse but true that the American system, "which insures only 85 percent of the population, costs more than we would pay for a system that covered everybody." Why, he wonders, doesn't competition make the private sector more efficient than the public sector? Isn't that what competition is supposed to do? To be sure, he concedes, there is no one reason why "we put up with such an expensive, counterproductive health care system." Yet few of the other reasons are as much to blame as the "ideological blinders" current leaders seem to wear. As he puts it, "Decades of indoctrination in the virtues of competition and the evils of big government have left many Americans [including American leaders] unable to comprehend the idea that sometimes competition is the problem, not the solution."

Many-player games and many-sided competition are everywhere. The policy process must contend with them every day. Although science has begun to understand the subject, mysteries remain. Whereas two-player zero-sum competition seems well understood, at least conceptually, more complex forms of competition continue to confuse, as the learned debates still surrounding the prisoner's dilemma clearly demonstrate. Subsequent chapters will deal with specific types of competition, especially competition in the wild and economic competition.

CHAPTER 7

Competition in the Wild

Manfred Milinski is an experimental psychologist whose experiments often involve the small scaleless fish known—in recognition of the sharp spines between their heads and dorsal fins—as sticklebacks. He has observed, among other things, that if a single stickleback is fed water fleas at both ends of its tank, it will spend most of its time at the end where food is more abundant. It will, however, check from time to time to see if opportunities at the other end of the tank have improved.

With six fish in the tank, and an equal number of water fleas per minute entering at each end, Milinski found the number of fish at each end of the tank to be about equal. He further observed that any division into feeding parties of equal size represents a Nash equilibrium solution of the six-player game in which each fish strives to eat as well as it can, since no one of the fish can then obtain more food by switching ends. When twice as many water fleas are fed into one end of the tank as the other, four of the fish will typically congregate at the food-rich end of the tank, leaving only two to feed at the food-poor end. Any such division into feeding parties of two and four fish is again a Nash equilibrium solution of the modified six-player game, since—as before—no single competitor can profit by switching ends. Even though the brain of a stickleback is somewhat smaller than a kernel of corn, they are able to solve this simple game. In this they are doubtless aided by the fact that the game has no fewer than fifteen distinct Nash equilibrium solutions.

If the fish were named Abel, Baker, Charlie, Doug, Elvis, and Fabio, there would be five equilibriums in which Abel occupied the food-poor end of the tank along with one other fish, four in which Baker occupied that end of the tank with a fish other than Abel, three in which Charlie occupied that end of the tank with a fish other than Abel or Baker, and so on, to a total of fifteen. The game departs from equilibrium whenever a fish changes ends unilaterally, and it moves from one equilibrium to another anytime a fish from one end of the tank changes places with one from the other end.

The solutions are quite different if one of the sticklebacks is larger and more aggressive than the rest. In that case, the five smaller fish may be forced to concede the food-rich end of the tank to him and to share the food-poor end among themselves. By so doing, they receive only 40 percent of the nourishment they would upon equal division. But if the five can cooperate to drive the bully away from the food-rich end of the tank, so that he alone occupies the food-poor end of the tank, none of the six need give up more than 4 percent of their pro rata shares.

To find out how adept his fish were at finding new equilibriums, Milinski altered the rates at which water fleas were fed into the left and right ends of the tank. At first, the flow rates at the two ends were equal, and the fish about equally divided. Then he increased the flow rate at one end of the tank to double that at the other, and watched as four of the fish gravitated to the richer end. Finally, he interchanged the flow rates at the two ends and watched as four of the fish materialized at the newly enriched end of the tank. Throughout the experiment, the number of fish in the left half of the tank was recorded at regular intervals.

The results—averaged over eleven runs—appear in figure 7.1.

Milinski's elegant experiment illustrates a number of important points regarding competition in nature. Such competition seldom involves only two competitors and is rarely zero-sum. It may include numerous Nash equilibriums, and may or may not make it easy for the contestants to find them. Finally, the investment of time and energy required for the discovery of an equilibrium—represented in this case by the distance between the two ends of the tank—may be large or small. If searching is overly expensive, equilibrium solutions need not be worth the effort of discovery.

In an unreported experiment, Milinski increased the flow rate at the abundant end of the tank to five times that at the other end, and observed that the number of fish dining at the more abundant end

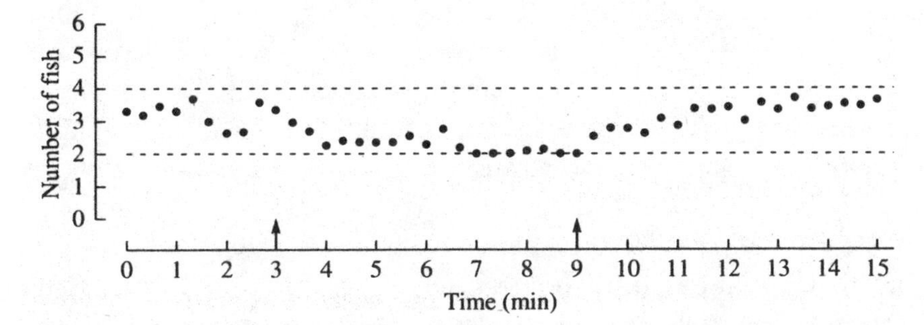

Figure 7.1. The number of fish foraging at the west end of Manfred Milinski's tank varied over time. The east end was the more abundant source of food between three and nine minutes, and the less abundant source at all other times.

tended to approximate five, even as individual fish came and went in search of ever greener pastures. But he never put eighteen fish into a large circular tank with feeding stations at one, three, five, seven, nine, and eleven o'clock, to see if three fish would indeed dine at each one, or doubled the flow rate at stations one, five, and nine to see if four of the fish would dine at the more abundant stations, while just two remained at the others. Nor did he put different species of fish in the tank together, to see if one species would dominate the other, especially in times of famine. As a result, the extent to which an experimenter can control the number of fish dining at each of several feeding stations in a large circular tank by adjusting the food flow rates at each station is only partially known.

Modern television enables even the most sedentary citizen to watch bucks, boars, bulls, bears, cocks, rams, stallions, and other amorous males do battle over estrous females. Spectacular though such competition can be, it rarely ends in death or serious injury because, as doubt concerning the identity of the eventual victor gradually evaporates, the projected loser is likely to allow discretion to override valor. Often, he does so without physical contact. After a few minutes of posturing and threatening, one of the parties feigns loss of interest and departs. If and when the battle is actually joined, bucks, bulls, and rams are genetically programmed to butt heads, cocks to claw at each other, and stallions to hamstring their opponents by severing (with their teeth) the exposed tendons of their opponents' hind legs.

Combat between male (buck) deer is particularly stylized, escalating through a series of predictable stages. What begins with a shouting match, serving both to attract estrous females and to challenge other bucks to defend their claims thereon, escalates first to a parallel walk, in which two males promenade back and forth several times, taking each other's measure at close range, and then to a shoving match with antlers interlocked. Only if neither party concedes does this initial test of strength and stamina further escalate into the sort of all-out combat that continues until one of the contestants is knocked to the ground and gored—usually to death—by the victor's antlers.

Each stage provides a separate opportunity for the rivals to avoid a fight to the finish. The buck with the more impressive mating call is likely to be the larger and fitter of the two and may persuade his less stentorian rival to abandon the conflict sight unseen. Failing that, the parallel walk affords each rival an opportunity to assess at firsthand the other's size, strength, and apparent fitness. The shoving match, should it come to that, offers a third opportunity for the underdog to end the engagement without risking serious injury. Each stage permits the antagonists to gain additional information concerning the adversary at some additional risk. The parallel walk is more dangerous than the shouting match, and the shoving match is more dangerous still. Injuries can occur at any but the shouting stage.

The danger in the parallel walk lies in the fact that by lagging a step or two behind his opponent, a contestant can obtain unobstructed access to the latter's unguarded flank. The temptation to turn and drive his antlers into that undefended target must be all but irresistible. Yet bucks in the wild typically forgo this momentary advantage by speeding up and pulling abreast of their fleetingly vulnerable opponents. Many naturalists have expressed amazement at this admirable restraint.[1] These majestic beasts seem to behave as if bound by genetically programmed rules of fair play. In Europe, gamekeepers have long enforced the rule—which obviously benefits the population—by shooting violators. But what prehistoric force could have formulated such a rule in the first place? Before we attempt to answer that intriguing question, it seems worthwhile to examine a few additional examples of limited warfare in the wild.

In many species, such as the bear and the orangutan, males and females have little contact outside mating season. In others, a single dominant male travels with a more or less permanent group of adult females, together with their young. Wild stallions act this way, maintaining their authority by forcing adolescent males to leave the herd. Gorillas in the wild do much the same thing. Only the fittest of the banished survive long enough to challenge for the leadership of an established herd, or to assemble a new one from strays. Stray females being relatively uncommon in the wild, young stallions and gorillas usually have to lure potential harem members away from existing herds. Sometimes the purpose can be achieved by verbal invitation alone—screeching from hilltops—and sometimes it requires the challenger to do battle with an established leader. Depending on the outcome, the challenger may acquire some, all, or none of the contested herd.

A pride of lions may contain several adult males, the most dominant of which—known as the alpha male—deems it his exclusive right to breed with resident females. Only when the alpha leaves the pride to consort with an estrous female is there likely to be an opportunity (should a second female become estrous) for some lower-ranking male to breed. Should such a usurper attempt to breed in the presence of the alpha male, he would immediately be attacked by the latter and (upon defeat) driven from the pride. A banished adolescent may hope in the fullness of time to acquire a pride of his own, either by taking over an existing pride or by enticing restive females to join him in building a new one. Upon defeat by a younger male, an aging alpha must abandon his pride and is likelier to perish in solitude than to acquire—or even join—another.

At one time, it was believed that such male dominance and aggression—toward females, juveniles, and weaker males—explained just about everything that happens in the animal kingdom. Only when trained observers (mainly primatologists) began to conduct multiyear field studies did they learn that subtler explanations are required for much of what goes on. Shirley Strum, among the first to observe a troop of baboons over a period of years, describes her unexpected findings in her book *Almost Human*.[2] She found baboons to be particularly worthy of study because they, humans, and patas monkeys are the only primates that currently live at ground level rather than in trees.

The earliest primates (prosimians) entered the fossil record some sixty million years ago, soon after the dinosaurs disappeared. The first ones had long snouts for smelling and eyes on the sides of their heads, as well as apelike hands for climbing trees. But because life in the treetops rewards the acuteness of vision needed for grasping branches more richly than an acute sense of smell, flat-faced relatives with large round eyes facing forward—as they must to produce the overlapping fields of vision needed for depth perception—soon appeared. Later, as the areas of the brain related to vision and manual dexterity grew in size, primates developed—as Strum puts it—"an unequaled capacity for exploring and dismantling the world." As prosimians, monkeys, the lesser apes, the great apes, the early hominids, the earliest humans, and finally modern humans evolved in turn, a new and undoubtedly unique way of experiencing the world evolved with them.

Beginning in 1972, and continuing for more than fifteen years, Strum studied a troop of about sixty baboons known as "the Pumphouse Gang." It was one of several such troops living on the forty-five-thousand-acre Kekopey Ranch, located some seventy miles northwest of Nairobi, Kenya, in Africa's Great Rift Valley. Shortly before her arrival, the group included six adult males, seventeen adult females, and thirty-seven juveniles. A seventh adult male joined the troop at or about the same time Strum did. The adult females ranged in age from about six to thirty-four, while the adult males varied from ten to almost thirty. Whereas females seldom if ever left their natal troops, males were expelled as they approached breeding age. Wandering males may remain with a new troop for as little as a month or as much as ten years. But their stays are always short by comparison to a female lifetime of up to forty years.

At least one sex has to leave the natal group to prevent inbreeding. Among gorillas, some males and virtually all females depart as they approach sexual maturity. Among baboons, chimps, and bonobos—long described as pygmy chimps, but now considered a separate species—males do the bulk of the traveling. Before 1960, when Jane Goodall began her (now famous and still ongoing) observation of the chimps at Gombe, on the shores of Lake Tanganyika, no one had watched any group of primates long enough or closely enough to document family trees and record the comings and goings of individual group members. Since then, such studies have been conducted among all four great ape species.

By 1978, Strum had become an accomplished baboon watcher. She could identify every member of the Pumphouse Gang by sight and knew which pairs were related by blood or friendship. She had even learned to predict what a given Pumphouse baboon would do in most situations. Yet she was troubled by the fact that she was seeing things she had not expected to see, and was not seeing the things she had expected. Although, for instance, an adult male baboon seems an almost ideal fighting machine, she rarely witnessed altercations. Nor did she see any evidence that male dominance and aggression could explain reproductive success. Indeed, she could discern no evidence that a "male dominance hierarchy" even existed within the Pumphouse Gang. The females plainly had one, but the males didn't seem to. Friendships, kinships, and guile seemed to explain more of the group's behavior than aggression and dominance.

Needing group protection to survive at ground level—as early humans must surely have needed it—baboons have learned to invest both time and effort in the alliances and friendships group living requires. In particular, male baboons have developed elaborate strategies for gaining acceptance in new troops. The process requires weeks of effort and is often abandoned while still incomplete. Baboons, says Strum, have learned to be "nice" to one another because group membership is as critical to their survival as the air they breathe and the food they eat. If baboons have learned these lessons, she asks, must the earliest humans not have learned them as well?

Primate survival strategies combine body plans with behavioral patterns to achieve success. The small groups in which gorillas live—seldom if ever exceeding twenty individuals—seem a useful adaptation for creatures of their enormous size. Larger groups would be unable to find enough food in a single locale. In contrast, the larger groups in which baboons dwell seem necessary for survival on the food-rich savannas they share with lions, leopards, hyenas, and more. Baboons, incidentally, are a remarkably successful species. Until recently, they seem to have outnumbered humans in Africa. Allegedly more intelligent gorillas and chimps are, by comparison, currently threatened with extinction.

A somewhat subtler adaptation concerns testicle size. Although a male gorilla weighs about four times as much as a male chimpanzee, a chimp's testicles are about four times as large as a gorilla's. Large testicles contribute to breeding success among chimps because female chimps are extremely promiscuous and because large testicles tend to produce more sperm. Since an estrous female chimp may copulate with numerous males in a single estrous cycle, the swain with the largest testicles is the one most likely to achieve paternity. Because male gorillas tend to dominate their females, testicle size is of little value to a gorilla and might even represent a handicap. Goodall points out, in this regard, that chimps at Gombe often have their testicles bitten while hunting the colobus monkeys whose meat contributes most of the protein in their diet.[3]

John Maynard Smith and George R. Price were the first to explain, in Darwinian terms, how overly aggressive modes of behavior become rare in a population—without disappearing entirely—while more pacific ones grow more common.[4] This they did in terms of a symmetric two-player but non-zero-sum wilderness version of the dating game in which the randomized Nash equilibrium strategy for each player is to concede with 90 percent probability and to contest the issue otherwise. The reward structure for the game is as follows: If one player concedes while the other contests, the latter will realize the breeding opportunity at issue. If both concede, the prize is equally likely to go to an interested third party as to either contestant. But if both contest the issue, both are equally likely to sustain a serious injury, the consequences of which far outweigh the gain or loss of a single breeding opportunity. That structure is incorporated in the reward matrix shown in figure 7.2, which displays only the rewards accruing to ME. The rewards accruing to THEE follow by interchanging the 0 and 1. As long as the probability that a randomly chosen opponent shall concede the issue remains constant at 90 percent, it doesn't matter whether ME contests a given issue, since both aggression and concession are then rewarded equally (0.3) on average. But as soon as the probability that such an opponent shall concede the current issue falls below 90 percent, concession offers a better average reward than aggression, and vice versa.

Otherwise put, the convention whereby one concedes with 90 percent probability and contests the issue otherwise is to some extent self-reinforcing. Should the population be invaded by a number of more aggressive individuals, the probability that a randomly chosen opponent shall concede would fall below 90 percent, and concession would begin to offer a better average reward than aggression. Conversely, should there be an invasion of passive individuals, the probability that

		THEE		
		concede	contest	90% mix
ME	concede	1/3	0	0.3
	contest	1	−6	0.3

Figure 7.2. The reward matrix for ME in a game between two equally prime bucks in the wild seeking to realize a single breeding opportunity. The matrix for THEE is obtained by interchanging the 0 and 1.

a randomly chosen opponent shall concede would rise above 90 percent, rendering aggression more profitable on average than concession. Not every Nash equilibrium exhibits any such self-reinforcing property. The ones that do are said to be composed of evolutionarily stable strategies, or ESSs. ESSs have become a cornerstone of modern theoretical biology and underlie the current understanding of Darwinian evolution.

Maynard Smith's idea, upon which he expanded at length in his book *Evolution and the Theory of Games,* is that a particular individual wanders about in the wild, encountering other members of his own or a rival species, and having to decide each time how to respond to the challenges they represent.[5]

By no means should the reader conclude that breeding opportunities are the only prizes deemed worth fighting for in the animal kingdom. An adolescent bear with a fresh-caught salmon between his teeth would have to fight to retain possession if the local alpha male were to express an interest in acquiring his catch. Likewise, a gang of hyenas preparing to dine on a fresh-killed zebra or wildebeest would have to fight to retain possession if a hungry lion were to arrive on the scene. And so on. Animals in the wild—like military leaders in the field—can only afford combat when the expected benefit exceeds the expected cost.

If, in a particular animal's experience, the probability that a randomly chosen opponent in the game of figure 7.2 will quickly concede has exceeded 90 percent, then he should respond aggressively. But if in his estimate the probability of a prompt concession is less than 90 percent, then he himself should concede. Only if his experience suggests that the probability of an easy concession is exactly 90 percent should he entertain doubt concerning the proper course of action. Almost any theory of learning in games would then predict that deviations within the population from a 90 percent concession rate will prove self-correcting.

The critical 90 percent concession rate is an immediate consequence of the particular entries in the reward matrix. As the natural unit of reward in the wilderness dating game is a single breeding opportunity, the 1 in the reward matrix represents exactly one such opportunity, while the 0 represents the absence of such opportunity, and the 0.3 represents a three-in-ten chance of realizing such an opportunity. Accordingly, the −6 at the bottom of the middle column represents something far more consequential than the gain or loss of a single breeding opportunity. If that −6 were replaced by a −2, the critical probability would shrink to 75 percent. Or, should that −6 be replaced by a −13, the critical probability would climb above 95 percent. And so on. The optimal concession probability is highly sensitive to the quantity at the bottom of the middle column of the reward matrix, which reflects the severity of the injury one is likely to suffer if combat ensues.

While heavily armed species, such as elk, wolves, and baboons, rarely engage in actual combat, lightly armed species can afford to be more aggressive. Although doves in the wild fight often, they are seldom seriously injured because as soon as one appears to gain the upper hand, the other flies away. When doves are confined together in a small cage, however, the weaker of the two is slowly pecked to death.[6] Doves' biblical reputation as birds of peace is largely undeserved. Even the most imposing species of wild animals, such as lions, elephants, and gorillas, do have to fight occasionally, lest they be taken advantage of by a more aggressive subspecies. Dian Fossey, who died studying (and trying to protect) gorillas in the wild, found through autopsies that the corpses of most adult males exhibited signs of serious injury—such as broken tooth fragments lodged in skulls—inflicted during early adulthood.[7]

Goodall and Strum report similar findings among chimps and baboons. Though males of both species prefer ritual to actual physical combat, few if any great apes reach adulthood without acquiring a few battle scars. Fossey, in particular, was impressed by the ability of adult gorillas to recover from serious wounds. Yet the decision to accept or avoid combat is by no means the only one in nature that is best left to some random device. As a soccer player has to decide whether to aim for the left- or right-hand corner of the goal, so rabbits fleeing from wolves must decide how often to dodge right and dodge left. And while it seems unlikely that a fleeing rabbit would pause to flip an unbiased coin before each change of direction, one is entitled to doubt that any bias in favor of, say, right-hand turns on the part of the local rabbit population would long go undetected by local wolves.

The vigilance games played by small foraging animals furnish yet another example of competition among wild animals. Such creatures must divide their attention between foraging and scanning for predators. The small American finches known as juncos, some of which are more alert than others, have been closely observed for generations. One whose neighbors devote much time to sentry duty can spend most of its own time eating. But one whose neighbors are full-time eaters needs to be more vigilant. It is not difficult to imagine that a stable balance within the flock, appropriate to its surroundings, soon emerges.

It matters not to a strategist whether all members of a flock spend 10 percent of their time on sentry duty or half the members spend 20 percent. From a strategic point of view, the result will be the same. A biologist, on the other hand, will consider the difference important. In the former case, the population is homogeneous; in the latter, it is split into two types. There is, of course, no a priori reason why a single flock cannot split into three, eight, or twenty-seven types, each with its own propensity to scan for predators. Only the overall population average will affect the fitness of a randomly chosen member of the flock.

Shoppers play a similar sort of vigilance game. Though advised by consumer advocates to clip coupons, shop on Wednesdays, and (above all) compare prices, relatively few actually follow such advice on anything like a consistent basis. Yet as long as there remains a critical mass of comparative shoppers, suppliers must keep their prices competitive to retain the business of this discerning minority. Should that minority become too small, undue price variation would creep back into the market, increasing the incentive to join the group. Or, should the minority grow too large, individual members can relax without penalty. Again, a self-sustaining balance seems likely to emerge. Whereas dating games involve only a single species, vigilance games typically involve at least two.

Interactions between competing species have been analyzed for many years. After World War I, the Italian biologist Umberto D'Ancona noticed that—according to the available statistics—an unusual number of predatory fish had been caught in the Adriatic during the war. Wondering if the increase could be attributed to the presence of two hostile navies, he raised the question with his prospective father-in-law, the senator and mathematician extraordinaire Vito Volterra. That sage responded by formulating a pair of differential equations from which he was able to extract a surprisingly unambiguous answer for D'Ancona's question. Because the same equations had previously come (for different reasons) to the attention of the American scientist Alfred J. Lotka, they are now known as the Volterra-Lotka equations.

A typical solution consists of a pair of curves, not unlike the ones in figure 7.3, showing the rise and fall in the predator (dashed line) and prey (solid line) populations over time. Notice that both curves peak five times in eighteen years with—as might be expected—predator peaks lagging prey peaks by at least six months.

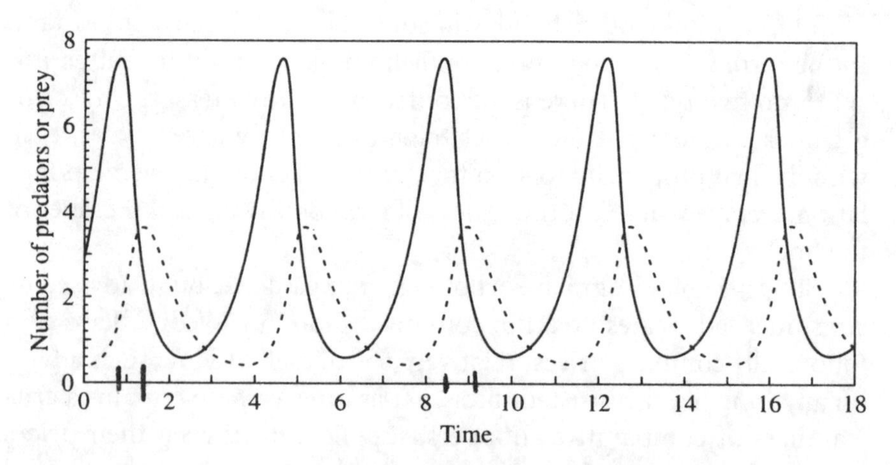

Figure 7.3. The sizes of the predator and prey populations in the Volterra-Lotka model of predator-prey interactions. The predator population varies somewhat less than does the prey population, and peaks somewhat later.

The answer to D'Ancona's question is contained in what is now known as Volterra's principle, according to which increased or decreased fishing pressure has opposite effects on predators and prey. Since a scarcity of prey reduces the rate at which predators can reproduce, while a dearth of predators increases the rate at which prey can reproduce, prey populations tend (somewhat surprisingly) to rise during times of increased harvesting pressure, whereas predator populations tend to decline. Conversely, when harvesting activity is curtailed, predator populations tend to increase as prey become scarcer, exactly as D'Ancona observed.

Volterra's principle has found confirmation in some catastrophic failures of insect pest control.[8] Pesticides tend to act not only on the target species of caterpillars, weevils, aphids, or whatever but also on their natural enemies. Indeed the effect on the enemies can exceed that on the target species because—being higher up on the food chain—the enemies tend to ingest the poison in concentrated form. Moreover, since the pests' gestation periods are likely to be shorter than those of the predators, they are often better able to adapt to the new conditions. Small wonder that insecticide campaigns so often backfire, permitting pests to prosper at the expense of their natural enemies.

Biologists were not immediately convinced by Volterra's explanation, finding it too simple and too far-reaching. But he himself found mathematical ecology so fascinating that he devoted the final years of his celebrated career to the subject. His work attracted the attention of a trio of gifted Russians, who proceeded to demonstrate that the ecological warfare between a single species of predators and a single species of their prey can have only four possible outcomes: The environment may be so inhospitable to the prey population that (1) it is unable to sustain even the smallest conceivable community of predators. This occurs when the predators are unable to capture enough prey to sustain themselves. With greater hunting success, they could (2) reach a stable critical mass. Typically, the mass so reached turns out to be stable in the sense that if perturbed by some random event, the sizes of both the predator and the prey populations will revert (after a few desultory oscillations) to their original (pre-perturbation) values. Were they still more efficient hunters, the predators could (3) launch the two populations into a never-ending series of oscillations, such as the ones depicted in figure 7.3. In that case, the popula-

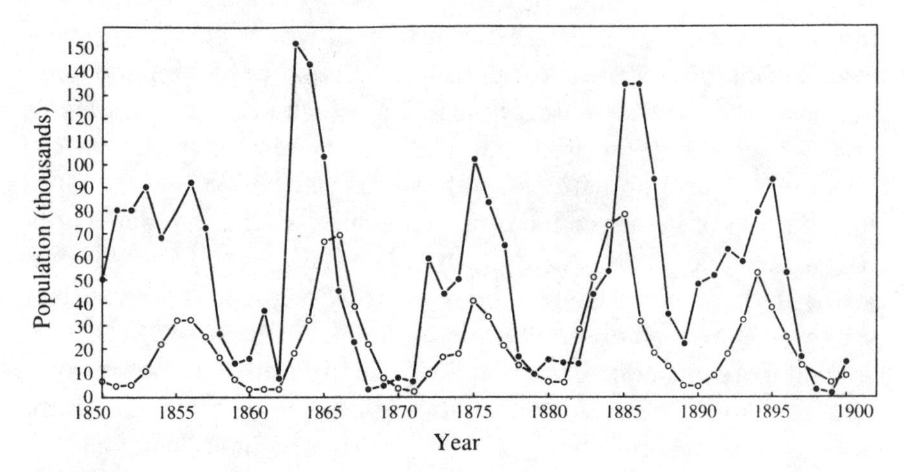

Figure 7.4. The most famous graph in population biology depicts the periodic rise and fall for over a century of the Canadian lynx and snowshoe rabbit populations in the regions surrounding Hudson's Bay.

tions might become so small at the low points of their oscillations that some trifling event, such as a minor epidemic or an unusually hot, cold, wet, dry, or windy season, could (4) render either or both extinct.

Fish in the Adriatic are not the only documented example of fluctuating populations in the competition for survival. Records kept by the Hudson's Bay Company for more than two hundred years show—as does figure 7.4, which has been called the best-known graph in ecology—that the numbers of lynx and snowshoe rabbit caught for their furs by Canadian trappers peak more or less regularly at ten-year intervals, before collapsing to very low levels. Since lynx are known to prey on rabbit, Volterra's equations would seem to explain the variations in the Hudson's Bay Company's fur acquisition experience, along with that of the Adriatic fishermen.

For all the (qualitative) resemblance between figures 7.3 and 7.4, some scholars remain unconvinced that Volterra's explanation is the correct one. They point out, among other things, that most predation has little effect on the size of the prey population. Indeed, because the overwhelming majority of the animals brought down by predators are well beyond reproductive age—consisting mainly of sick, injured, or superannuated individuals who become separated from their herds—

most predators can be thought of as scavengers lacking the patience to wait for their victims to die. This and other objections to Volterra's explanation of oscillatory population sizes in interdependent species are summarized in some detail by Karl Sigmund.[9]

The fluctuations in figure 7.4 are not the result of any detectable evolution in the lynx population or its ability to convert captured rabbit into newborn lynx. On the contrary, they are a natural feature of the interactions of two genetically stable populations. One can imagine that a predator population might evolve through all three of the stages described above, before being driven into extinction by some trifling natural disaster. But it is difficult to imagine anything of the sort happening in mere decades to species that breed only once a year.

If there is any one lesson to be learned from the study of competition in the wild, it is that "there are many ways to skin a cat," no few of which are low-cost, subtle, and (relatively) nonviolent. From the chest-beating and branch-shaking that gorillas display preceding actual combat, through the parallel-walk sequences performed by rutting deer, to the elaborate mating dances performed by certain bird species, animals have evolved a host of ways to attract mates without resorting to potentially fatal combat. Indeed, species unable to evolve such nonviolent alternatives seem the exception in nature, rather than the rule.

To many readers, the most interesting aspect of competition in the wild is the resemblance it bears to economic competition. The emergence, divergence, and eventual extinction of species have often been compared to the rise, stagnation, and eventual fall of giant corporations, as well as economic institutions like feudalism, demand-deposit banking, and the factory system. These similarities were no less obvious to Darwin himself than to the social Darwinists such as Herbert Spencer and William Graham Sumner who built upon the Darwinian foundation. Indeed, although Darwin himself scrupulously avoided the subject in his own writings—thinking his central theses likely to prove controversial enough as they apply to the plant and animal kingdoms—he is said to have been rereading Malthus's *Essay on the Principle of Population as It Affects the Future Improvement of Society* during the time he began committing his own ideas concerning the origin of species to paper.[10] The rest of this book will be devoted to the subject of economic competition.

CHAPTER 8

Auctions

Legend has it that the first producing oil well was drilled by Colonel Edwin Drake, near Titusville, Pennsylvania, in 1859. Others concede that similar wells may already have been in operation near Baku, on the Caspian Sea, for as many as ten years.[1] There seems little doubt, however, that the first offshore oil well was drilled near Summerland, California, in 1887, by a man named H. L. Williams. Noticing that the best wells in the area were the ones nearest the water's edge, he decided to sink his own well from the end of a wharf extending three hundred feet out to sea. When it proved to be an exceptional producer, imitators abounded. Soon, the longest connecting wharf reached more than twelve hundred feet out into the Pacific. The first stand-alone rig, unconnected to any shore, was drilled at Caddo Lake, Louisiana, in 1911.[2]

During the 1920s and 1930s, wells were drilled in the lakes, swamps, and bayous of southern Louisiana. But there was no great pressure to operate far from land, or in deep water, since Depression-era demand remained sluggish, and promising dry-land well sites remained. Moreover, nobody seemed to know who owned the right to drill in remote offshore locations. All that changed in the aftermath of World War II as demand strengthened, wartime technologies proved adaptable, and Congress passed the U.S. Submerged Lands Act of 1953. It established that—under ordinary circumstances—the federal government controls mineral rights located three or more miles from shore, while the states control rights within that limit. Subsequent legislation authorized (indeed encouraged) the secretary of the interior to lease federal lands to prospective developers. As offshore production increased in the gulf, so it increased worldwide.

The first drilling platform built out of sight of land was erected by the Kerr-McGee Corporation in the Gulf of Mexico in 1947. By 1949, forty-four exploratory wells had identified eleven distinct oil fields beneath gulf waters. From a negligible total in 1947, worldwide production of offshore oil grew to about 14 percent of the total in 1974, and to about a third today. Likewise, by the mid-1990s, offshore production of natural gas had come to represent 20–25 percent of the worldwide total. It seems unlikely that world demand for offshore oil will soon diminish.

It would be a mistake to assume, despite the foregoing litany of technical achievement, that progress in the offshore oil and gas industry was either easy or uninterrupted. A host of engineers and scientists have made rewarding careers in the oil industry by helping to surmount—one by one—the challenges of underwater exploration, offshore communications, weather forecasting, tidal and current prediction, and horizontal drilling, to name but a few. Such structures as Shell Oil's production platform Bullwinkle, capable of surviving North Sea storms while anchored in 1,350 feet of water, and such vessels as the self-positioning drill ship *Eureka*, which can hold its position over a deepwater well site in any combination of twenty-foot seas and forty-mile-per-hour winds, are the results of lengthy evolutionary processes. Like Rome, such installations were not built in a day.

It would be equally wrong to assume that hardware requirements were the sole impediment to expanding offshore production. The entire offshore enterprise almost came to a screeching halt during the early 1960s, when the firms involved began to suspect that—despite burgeoning production—they were losing money offshore. Closer scrutiny of the figures revealed that most if not all of the firms involved were in fact making money from their offshore operations. They just weren't making as much of it as they did from their bread-and-butter operations on dry land. As a result, corporate boardrooms became battlegrounds in which production departments eager to increase offshore production went head-to-head with accounting departments conscious only of meager rates of return.

At first, nobody knew quite why the offshore returns on investment were so low. The usual suspects, including the difficulty of interpreting deepwater seismographic data, were soon exonerated. Every research lab in the industry had a few good people working on the problem, and most of them quickly narrowed their focus to the sealed-bid auction process by which firms acquired the rights to drill on specific segments of the ocean floor. The companies could have been making money hand over fist if they hadn't paid so much for drilling rights. Subtle dangers inherent in the process of bidding competitively for properties of highly uncertain value seemed to have lured industry brass into paying too much for what they received. The most famous instance involved the Destin Dome, an enormously promising site off the coast of the Florida Panhandle for which BP paid $200 million in 1968. Three holes were drilled there, all of which came up dry.

Most of the investigators worked in teams, and most of the teams began with a literature search, which revealed only that the science of the day offered little guidance on auction strategy. The first technical paper on the subject had been written only a few years earlier, in 1956, by one Lawrence Friedman.[3] Neither it nor any of the others then in print appeared to shed much light on the problems besetting the oil companies. As a result, the teams in question quickly concluded that they were on their own and would need to develop methods new to the annals of the decision sciences.

No one will ever know the full extent of the discoveries made during the flurry of activity that followed. Most of the firms involved chose to regard anything and everything to do with their own auction-strategy selection process as a closely guarded secret, rather than background information suitable for publication in the open scientific literature. A few, on the other hand, concluded that they could only benefit from having the perils of overbidding clearly understood throughout the industry, and they actively sought to publicize their newly acquired understanding of those perils.

Perhaps the most insightful—and surely the most famous—of the resulting publications was the work of three employees of the Atlantic Richfield Company (ARCO), who discovered something they described as a "winner's curse" lurking within the sealed-bid auction process, dooming the buyer of an object at auction to find it less valuable than anticipated.[4] Moreover, and perhaps more important, they discovered a way to demonstrate the power of the curse experimentally, and thereby to assess the vulnerability of rival auction strategies to it.

To familiarize audiences with the subtleties of the sealed-bid auction process, Edward Capen, Robert Clapp, and William Campbell made a game of it—a game only slightly more complicated than finger matching or rock, scissors, paper, in which almost anyone can participate. Instead of a yearling racehorse, an art object, or the right to drill for oil in a remote portion of the ocean floor, the threesome auctioned off jars of nickels. Before the auction, prospective bidders were allowed to look at the jar, to shake it gently, and to compare it with a standard $2 roll of nickels. But they could not open the jar or otherwise measure the contents. Thus each participant had to guess the value of the nickels in the jar—much as an oil company must guess the quantity of recoverable hydrocarbons underlying a specific piece of the ocean floor—before deciding how large a bid to make and recording his or her decision on the postcard provided. Capen and Clapp revealed, in a follow-up paper written a few years later, that the trio had failed on only one occasion to sell the jar of nickels for more than it was worth.[5] What better way to demonstrate the reality (and potency) of the winner's curse?

Several variants of the game have been tried over the years. In one of the more informative, respondents are asked to divulge their estimates of the value of the jar of nickels, as well as the amounts they wish to bid. A prize of $5 or $10 is typically offered for the best estimate, to make sure that respondents are as diligent with their estimates as with their bids. The results of playing this version of the game reveal that, whereas the *average* of the estimates received from a particular audience ordinarily approximates the value of the contents of the jar rather closely, the individual estimates—and hence the bids received—vary all over the place.

Such a jar—in fact containing 341 nickels—was offered at a party to the person who most nearly guessed its value. Thirty-six estimates were received, the best being an overestimate by a single nickel. All the guesses are displayed in figure 8.1. From it, the root cause of the winner's curse is readily apparent. For if Capen and Clapp had tried to auction off that jar of nickels to that audience, and if all in attendance had been prepared to bid just 50 percent of what they privately considered the jar to be worth, the 341 nickels (worth $17.05) would have brought $22.50. Even that is a conservative estimate, since most inexperienced players of auction games are prepared to bid more than half of what they think a liquid asset is worth.

One can infer the outcome of lots of other transactional guessing games from the contents of the table in figure 8.1. If, for instance, the jar were given to the woman who thought it contained 173 nickels, and the one who guessed that it contained 632 nickels were the designated buyer, chances are the two could agree on a price. But if the roles were reversed, it seems unlikely that a deal could be concluded. Then again, if five of the respondents were chosen at random to participate in a sealed-bid auction for the jar, the chances are better than fifty-fifty that one of the highest five estimators would be among those included. So if all in attendance were prepared to bid just 60 percent of what they believe the jar to be worth, it should bring at least $19.50, though worth only $17.05. The winner's curse can thus afflict the victor at even a poorly attended auction, provided that all present are moderately serious about acquiring the prize. And so on. There is no apparent limit on the variety of guessing games that can be built around data sets like the one in figure 8.1.

Clearly, sealed-bid auction bidding is a two-stage process. After

900	600	500	356	256	200
859	550	480	350	250	173
691	525	480	342	250	150
655	523	415	310	240	145
650	505	400	300	200	99
632	500	381	284	200	42

Figure 8.1. A total of thirty-six guesses was received in a contest to guess the number of nickels in a jar actually containing 341 nickels. Underestimates slightly outnumbered overestimates.

deciding what the object on offer is worth, one must decide what fraction of that amount one is prepared to bid. Capen and Clapp found, as have others since, that no clear consensus exists among novice bidders as to what constitutes an appropriate bidding fraction. A few respondents have exhibited a bid fraction of 0 percent by refusing to bid at all, even after taking part in the competition to estimate the value of the jar of nickels. Another few have exhibited fractions in excess of 100 percent by offering more than they claimed to think the nickels worth, in their zeal to obtain them. But a solid majority of the reported bid fractions have fallen in the interval between 20 and 90 percent. A coherent theory of auction bidding would presumably identify a significantly narrower range of advisable bidding fractions, if not a unique optimal choice for each of several (presumably identical) bidders.

Such an optimum, should one exist, would constitute a Nash equilibrium for the game in which rival bidders perforce engage, since it would be optimal for each to use against the others. Numerous authors, beginning with Michael Rothkopf and Robert Wilson during the 1960s, have computed Nash equilibriums for various technical formulations of the basic sealed-bid auction.[6] But even those efforts have failed to identify a single bid fraction for use in any and all circumstances. How could they? At the very least, the efficacy of a particular bid fraction must reflect the difficulty of the associated estimation problem. When estimation is easy, one may have confidence in one's own value estimate and may bid a large fraction of it without fear of overpaying. When estimates are hazy at best, bidders must guard against the possibility that their own best estimates are excessively high by selecting significantly smaller (more conservative) bid fractions.

Estimation is easy when independent estimates of the same unknown quantity tend to agree. Otherwise, estimation is difficult. Figure 8.2 depicts—as ticks on a line—two different populations of independent estimates. The ones on the upper line are tightly clustered, suggesting substantial agreement as to the magnitude of the unknown quantity. The ones on the lower line are more widely dispersed, suggesting significant disagreement. But reasonable people can't always agree as to which of two sets of ticks on a line is the more tightly clustered, or identify the points about which they cluster. Standards of comparison are needed.

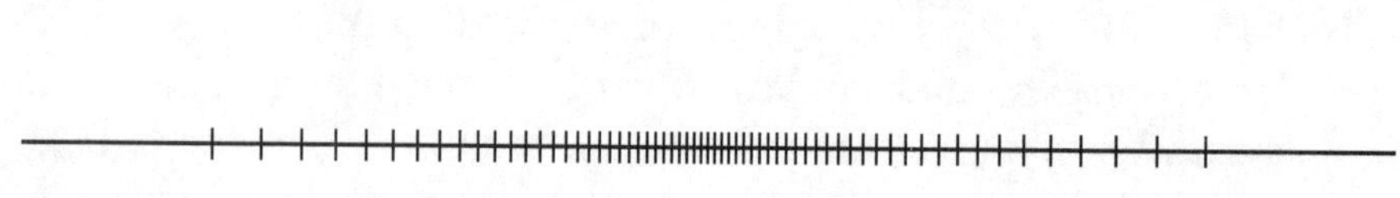

Figure 8.2. A population of ticks on a line may cluster more or less tightly together.

There is an obvious way to identify the point about which a population of ticks on a line clusters: Measure their distances from, say, the left edge of the page, and calculate the arithmetic mean (a.k.a. ordinary average) μ—the lowercase Greek letter mu—of those distances. The "center of clustering" is then the point at distance μ from the edge of the page. Some would argue that it is prudent to discard the leftmost and rightmost ticks (dropping a small but equal number of each) before averaging—to obtain a "trimmed mean" of the given measurements—on the grounds that the most extreme values tend to be the least reliable. Opinions vary on this, with different practitioners advocating different practices. Yet the simple untrimmed mean (a.k.a. ordinary average) μ remains the most frequently employed "measure of central location."

Once a center of clustering has been identified, there is an almost equally obvious way to measure the degree (tightness) of clustering about it. Describe the distances from the center of clustering to the individual ticks as "deviations," and calculate their average. The smaller the average, the tighter the cluster. For technical reasons, this second average is traditionally computed in a manner that may strike the reader as strange: After the individual deviations have been squared, they are averaged together in the usual way, and the square root extracted.[7] The result is known as the root-mean-square (RMS) of the several deviations and is denoted by σ, the lowercase Greek letter sigma. Together, the mean μ and standard deviation σ of a population of estimates reveal a good deal about those estimates. In particular, they offer a quick way of identifying a Gaussian bell curve with the population, as described in Chapter 4.

Capen, Clapp, and Campbell argue that by comparing ARCO's

carefully archived pre-auction estimates of the values of drilling sites that ultimately turned out—after several years in production—to be of roughly equal value, they could confirm that ARCO's estimates had indeed clustered about the (approximately) common value of those sites. They just hadn't clustered very tightly. On the contrary, the degree of clustering—as measured by the standard deviation σ—in what the three regarded as high-risk offshore bidding situations is somewhat looser than that found among the foregoing nickel-jar estimates. Even in medium-risk situations, the clustering was only slightly tighter. Among other things, these findings confirm the relevance of nickel-jar experiments.

Useful though they are for testing particular auction strategies, such experiments are a feeble tool for discovering prudent, effective (perhaps even Nash equilibrium) bidding strategies. For that, one needs a model—perhaps resembling the auto-agency models of Chapter 6—which lends itself to both hand and machine computation. Capen, Clapp, and Campbell bridged the gap between data analysis and model building by observing that the *logarithms* of independent estimates of the value of a particular offshore drilling site—rather than the estimates themselves—follow an exceedingly familiar statistical pattern. They tend to cluster about the logarithm of the "true value" of the site, and to resemble a random sample of values drawn from a Gaussian normal population.

Given that the logarithms of independent estimates of the value of an offshore drilling site—or a jar of nickels—behave like Gaussian normal populations clustering about the logarithm of the true value of the property, with a standard deviation predictable from experience bidding for similar properties, and knowing the bid fractions for all participants, one can compute the probability that a particular bidder will "win" the property in a sealed-bid auction, along with each individual bidder's "expected profit" from participation. The actual profit is of course not predictable, due to the effects of random estimation errors on the outcome of the contest. A search for advantageous bid fractions can then be conducted, more or less as was done in behalf of the auto dealers of Chapter 6, who needed to identify advantageous advertising and inventory policies. The Nash equilibrium bid fractions will presumably be numbered among the most advantageous of all.

Consider first a symmetric "common value" auction, in which the object at auction is of equal, though unknown, value to all participants, presumed equally adept at estimating that unknown value. If all but one of the bidders are presumed to employ a common bid fraction f, the remaining bidder will have an optimal bid fraction f^*, which may or may not equal f. Indeed, if f^* is plotted against f, it is found that f^* exceeds f when f is small but is exceeded by f when f is large. It is advantageous to outbid timorous opponents but foolhardy to outbid fearless ones. Moreover, there is ordinarily a single intermediate value of f for which $f = f^*$. Call it f^N, short for "f-Nash." The situation in which all participants employ the bid fraction f^N constitutes a Nash equilibrium, since none can then increase his own expected profit by switching to a different fraction.

Asymmetric auctions, though more difficult to analyze, typically have computable Nash equilibriums as well. This is as true for ordinary "English" outcry auctions—in which the auctioneer gradually increases the asking price until only one bidder remains in the competition—as it is for the sealed-bid variety. It is also true for "Dutch" outcry auctions, in which the auctioneer gradually reduces the asking price from an initially prohibitive level until it reaches one agreeable to someone in the audience. The winner's curse seems to be a factor in every variant of the auction process, which has been tamed but not eliminated by research in the field.

Different oil companies developed different approaches to the auction-bidding problem, including one designed to exploit the fact that drilling sites are auctioned in batches rather than one at a time.[8] This means that a firm should strive to maximize its expected profit from a succession of auctions, rather than a single one, and that instances of overpayment can be offset by windfall acquisitions, meaning those that turn out to be worth significantly more than any potential bidder would guess. This approach mitigates the winner's curse to some extent, and it enables firms to bid somewhat more aggressively than they otherwise might.

Among its many signal achievements, auction theory has provided the fledgling discipline of experimental economics with an excellent opportunity to test the part of game theory having to do with Nash equilibrium solutions. Given that many-player game theory has become an accepted tool of economic analysis during the last thirty years, and that Nash equilibrium theory is the most frequently invoked branch of many-player game theory, all such opportunities are to be embraced with open arms. Before we examine the test results, it seems appropriate to digress briefly on the subject of experimental economics.

Until the late 1940s and early 1950s, it was widely believed that (as such arbiters of informed opinion as the *Encyclopaedia Britannica* continued until quite recently to repeat) "there is no laboratory in which economists can test their hypotheses."[9] But amid the burst of scientific activity that followed World War II, a few daring thinkers began to wonder if the credibility of "the dismal science" might not be enhanced by the creation of one or more such laboratories. Although their initial results ignited no great burst of enthusiasm in a (largely academic) discipline that had survived without the experimental method for a century and a half, these intrepid iconoclasts persevered. In 1986 they formed the Economic Science Association, with the avowed intent of turning economics into a genuine science. By 1998, the *Journal of Experimental Economics* had begun publication, and several dozen laboratories devoted in whole or in part to the subject were active around the world. In 2002, Vernon Smith and Daniel Kahneman shared the Nobel Prize in Economics for their contributions to the rapidly maturing field. Auction theory was but one of many areas of interest to the new discipline.

The first task for experimenters interested in auctions was to establish the reality of the winner's curse phenomenon. Because Capen, Clapp, and Campbell had not conducted their demonstrations under laboratory conditions, and because professional economists distrust evidence that anything as "market oriented" as a sealed-bid auction could result in an ill-advised transfer of ownership, a need was felt for additional confirmation. Furthermore, because early experiments had involved "single-shot" auctions, in which just one jar of nickels was auctioned off to a given audience, it was widely assumed that a little practice would enable even novice bidders to avoid the curse.

Empirical evidence for the existence of a winner's curse has since been adduced in auctions for book publication rights, in professional baseball's free-agency market, in corporate takeover battles, and in real-estate auctions.[10] It is seldom easy to attribute inflated transaction prices unequivocally to the winner's curse, due to the changing times and underlying conditions in which auctions invariably take place. But the fact remains that overpayments are observed in virtually every experimental auction market, and the risky nature of the auction process itself furnishes an entirely plausible explanation.

The first to test the winner's curse hypotheses under laboratory conditions were Max H. Bazerman and William F. Samuelson, in 1983, using MBA students at Boston University as subjects.[11] By auctioning off in succession four jars containing, respectively, 800 pennies, 160 nickels, 200 large paper clips (worth $0.04 each), and 400 small paper clips (worth $0.02 each) for an average price of $10.01—resulting in an average loss of $2.01 for the "lucky" winners—they quite firmly established both the reality and the persistence of the winner's curse as it applies to novice bidders. Subsequent experiments by John H. Kagel, Dan Levin, and others demonstrated that more experienced bidders are better able to avoid the curse but are by no means immune to it.[12] Specifically, in auctions including only three to four bidders, moderately experienced participants tended to make money—albeit less than Nash equilibrium theory would predict—while at auctions including six or seven bidders, they continued to incur actual losses.

Nash equilibrium theory turns out to be a rough predictor at best of subject behavior in auction games played under laboratory conditions. Nowhere is this weak correlation more clearly illustrated than in the matter of sellers' revenues at different types of auctions. Nash equilibrium theory predicts quite unequivocally that ordinary English outcry auctions should generate higher selling prices than sealed-bid auctions, in part because the winner's curse is less potent at the former than at the latter, and in part because the bidders at English auctions can profit from the information conveyed by the calling out of rival bids. Yet Levin, Kagel, and Jean-François Richard found that at least in the laboratory, English auctions produce less rather than more revenue for the seller.[13]

Laboratory experiments are also well adapted to predicting the effects of proposed rule changes. One such change involves the so-called second-price sealed-bid auction rule, whereby the prize still goes to the highest bidder, who is only required to pay the amount of the second-highest bid. This rule change has been proposed—and employed from time to time—as a method of mitigating the effects of the winner's curse in sealed-bid auctions, in part by causing the results to more closely resemble those of an ordinary English auction, at which the auctioneer ceases to solicit higher bids as soon as the second most extravagant bidder drops out of the contest.[14]

Nash equilibrium theory predicts that the pre-auction release of relevant public information—meaning information that could help interested parties assess the value of the object at auction—should increase the seller's revenue by encouraging higher bids. Yet in the laboratory, such information seems to depress bids by enhancing bidders' ability to avoid the winner's curse at both first- and second-price auctions. Likewise, increasing the number of bidders at a second-price auction should in theory cause individual bidders to decrease their bids, in deference to the winner's curse. Yet experiments show no such adjustment of bid fractions in response to additional bidders. Although experienced bidders do seem able to avoid the winner's curse in previously encountered situations, they don't seem to understand the phenomenon well enough to apply the lessons learned in unfamiliar situations. Experimental methods confirm some predictions of Nash equilibrium theory and contradict others. Efforts to resolve issues upon which the two conflict are still in their infancy.

In recent years, the number of dollars changing hands at auctions has grown by leaps and bounds. The increase is due, in large part, to the massive radio frequency (a.k.a. spectrum) auctions conducted on six continents since 1981. The first one held in the United States began on July 25, 1994, in the Blue Room of the Omni Shoreham Hotel in Washington, D.C., and lasted for five days. Bidding was conducted electronically, on-site. Since then, such auctions have raised more than $40 billion for the U.S. Treasury, while similar auctions in Europe have raised almost $100 billion. The British spectrum auction of 2000 alone, taking place at the height of the so-called tech bubble, brought in a reported £22.5 billion ($34 billion).

These auctions are complicated to conduct, because the right to broadcast at a given frequency in a given geographical area is difficult to put a price on. A specific license may be worth a great deal to a firm that already owns the right to use the same frequency in surrounding areas, yet it might be worth nothing at all to a firm that already controls another frequency in the license area. This is especially true in the United States, where FCC regulations deny any firm access to more than 45 MHz in any one region, to leave room for at least five firms to compete in every geographical market. Much careful thought was given to the design of the first U.S. spectrum auction, which has since become the de facto standard against which competing designs are judged.

A typical spectrum auction involves a dozen or more closely related licenses and a pool of eligible bidders. The eligibility rules are often complex, since bidders may bid only on those licenses for which they have already purchased, for a nominal fee, the right to bid. This practice helps to ensure that every bidder is in earnest about acquiring the licenses in which it has expressed interest. The bidding proceeds by rounds, until one passes without a single new bid being received. At that point the auction is over, and each license is awarded to the highest bidder. After each intermediate round, the highest bid and bidder for each license are announced to all in attendance, along with minimum bid increments for the coming round and any changes in eligibility. Eligibilities change from round to round because of the so-called activity rule, according to which each bidder must bid on a stated fraction of the licenses for which she is eligible to bid in order to retain her current level of eligibility.

The activity rule, together with the minimum bid increments announced at the beginning of every round, have proven highly effective at bringing such auctions to a timely conclusion. They have not always done as well at preventing collusion. Paul Milgrom describes the German spectrum auction of 1999, at which ten licenses were on offer and a 10 percent bid increment was in force.[15] Mannesmann opened the bidding by jump-bidding the price of five of the licenses to DM 20 million and the other five to DM 18.18 million. This bid was a clear invitation to some other large firm to claim the remaining five licenses by bidding DM 20 million (or 110.01 percent of DM 18.18 million) for each one. T-Mobile did, and no further bids were received. Milgrom cites this as an example of poor auction design and suggests that the German government could have earned substantially more by a more judicious choice of auction rules. It is hard to imagine that the expertise required to design and implement the complex spectrum auctions of the 1990s would have existed without stimulation—both financial and intellectual—from the offshore oil lease auctions of the 1960s and 1970s.

Spectrum auctions and oil lease auctions are by no means the only sorts of auctions that reward thoughtful design. Matching auctions—

intended to pair students and colleges, workers and firms, marriageable men and women, and so on—are another area in which such efforts have proven worthwhile. It would be hard to find a better example than the National Resident Matching Program (NRMP), through which some twenty thousand new doctors a year find their first jobs as residents in American hospitals.

The program has been in operation since the early 1950s and is distinguished from its rivals and predecessors by its longevity. Alvin Roth has argued at some length that the matching protocols that survive are distinguished from those soon abandoned by the ability to produce *stable* matchings.[16] A matching is deemed unstable if it leaves apart one or more pairs that would mutually prefer to be together. In his opinion, the matching procedure adopted in 1952 by the NRMP has endured largely because it tends to produce stable matchings.

The NRMP protocol was revised in 1998 to take account of changes in the demographics of new physicians. Whereas medical-school students were almost exclusively male in 1952, they are now almost half female, and in many cases married. Thus the population of new physicians includes numerous husband-wife teams who need to be assigned to residencies in the same geographical area, if not in the same hospital. Moreover, the growth of specialization mandates that many new physicians now require more than one residency. The revised protocol is designed to produce stable matchings that accommodate these contemporary requirements.

Among the technical documents digested and subsequently ignored by the oil company teams formed during the 1960s to investigate the low rates of return (allegedly) being earned offshore was one published in 1961 by William Vickrey, a professor of economics at Columbia University. Vickrey was later to share (with James Mirrlees in 1996) the Nobel Prize in the economic sciences, in part for his work on auction theory. The paper in question inaugurated the study of what have come to be known as "private value" auctions, in which each bidder knows exactly what the property at auction is worth to himself but can only guess its value to rival bidders. Vickrey began the practice of analyzing such auctions on the assumption that the unknown valuations constitute a sample drawn at random from a known probability distribution, such as the Gaussian bell curve peaking at a specific point.

Even though few if any real-world auctions actually work this way, the technical literature concerning private value auctions dwarfs that concerning the common value auctions discussed above, where the prize is worth more or less the same amount to all concerned, though nobody knows what amount that is at auction time. The reason seems to be tractability: it has proven possible to deduce all manner of significant-sounding theorems about private value auctions, including some that link private value auction markets with the sorts of markets economists used to study before auctions became fashionable.

Perhaps typical of the sweeping statements made concerning private value auctions is the so-called revenue equivalence theorem, which asserts that auction design is of no consequence to the seller at such an auction, since any sensible auction mechanism generates the same expected sales revenue. Other celebrated theorems concern the contrast between risk-averse and risk-neutral bidders, and the identity of the Nash equilibrium strategies in specific auction games. Yet others insist that such auctions are "efficient" in the sense that—under ordinary circumstances—the bidder who imputes the highest value to the prize will always win it.

The original statute directing the FCC to conduct spectrum auctions stipulated that such auctions should commence no later than July 1994, that they should encourage minority ownership, and that they should promote "efficient and intensive use" of the radio spectrum. The meaning of the term "efficient" was not spelled out but was eventually construed to imply—in the words of then vice president Al Gore—"putting licenses in the hands of those who value them most."[17]

Perhaps the final word on auction design was written by Paul Klemperer, who, along with Ken Binmore, designed the English spectrum auction of the year 2000. After underscoring the importance of auction design with the observation that the third-generation mobile phone auctions of 2000–2001 generated as little as $20 per capita in Switzerland and as much as $600 per capita in England, he went on to declare that the fine points (including efficiency) addressed by what he terms "advanced auction design theory" are far less important than the more mundane and traditional considerations, such as attracting a large number of bidders and discouraging collusion among them in order to elicit a favorable price.[18] In short, a little auction theory goes a long way.

CHAPTER 9

Competition in Financial Markets

There are many kinds of competition in financial markets, none more extensively chronicled than that between managed mutual funds. Money pours from the rest, into the funds that perform best. There is also competition between brokers for commissions, and between investment counselors to attract clients—in part by counseling them to minimize brokers' commissions. Then there is competition between the so-called hedge funds for the patronage of already wealthy clients prepared to bear extraordinary risk in the hope of earning supernormal returns. There is also competition for circulation among *Barron's*, the *Financial Times*, and *The Wall Street Journal*; *Forbes*, *Fortune*, and *Money* magazines; and the Bloomberg Channel and CNBC, to name but a few. Last but not least, there is the grand competition among individual investors to better their own (financial) conditions.

A vast store of reading material is available to those involved. While much of it touts transparent get-rich-quick schemes, the best stress the importance of basic understanding and common sense. Andrew Tobias's *The Only Investment Guide You'll Ever Need* comes highly recommended. *Beating the Street*, by Peter Lynch and John Rothchild, appeared soon after Lynch led Fidelity's Magellan Fund through a threefold growth spurt in just five years. The authors insist that skilled investors can increase their net worth at above-average rates and suggest plausible ways of going about it. *The Great Mutual Fund Trap*, by Gregory Baer and Gary Gensler, presents charts and tables showing that low-cost index funds routinely outperform managed mutual funds and should be preferred by most investors.

Burton Malkiel's *Random Walk Down Wall Street* makes much the same point, after hammering home the more fundamental lesson that stock price movements are best compared to the proverbially unpredictable motions known to mathematicians as random walks. *Buffettology*, written in part by Warren Buffett's ex-daughter-in-law Mary Buffett, explains her understanding of the master's methods, right down to the worksheets he allegedly uses to evaluate the companies in which he becomes interested. The methods described are highly reminiscent of those advocated by his former teachers Benjamin Graham and David L. Dodd in *Security Analysis: Principles and Technique*, the first edition of which appeared in 1934.

Although the more recent of these books introduce scientific theories and evidence, they are mainly concerned with the art of investing. The science of investing, though born at the tail end of the nineteenth century, was without practical consequence until the latter part of the twentieth century. Then, after smoldering for perhaps twenty post–World War II years in remote corners of academia, it burst into Wall Street flame with a suddenness that surprised even its most ardent practitioners. Overnight, "rocket scientists" became star performers, able to command six-figure salaries and more. The history of the ensuing revolution in money management has been told many times, never better than by Peter Bernstein in his eminently readable memoir *Capital Ideas*.[1]

The science of investing was the brainchild of one Louis Bachelier, who wrote his doctoral dissertation on the subject in 1900, based on his observations of the Paris Bourse, or stock exchange. Although he worked under the direction of Henri Poincaré—arguably the most eminent mathematician in the world at the time—the mathematical community took little notice of his findings. In part because finance was then considered a somewhat unsavory (if not necessarily dishonest) occupation, rather than a suitable subject for scientific inquiry, his groundbreaking dissertation on the evolution of asset prices in financial markets did not impress the panel of judges before whom he was obliged to defend it. They awarded it only a respectable "*mention honorable*," rather than the "*très honorable*" required to launch an august mathematical career in early-twentieth-century France. As a result,

Bachelier divided more than twenty years (with time out for military service during World War I) between teaching high school and serving as an adjunct lecturer at a series of second-tier universities, before securing a permanent chair at provincial Besançon, near the Swiss border, in 1926.

Bachelier was a distinct outsider in the cliquish, elitist world that was turn-of-the-century French academic life, having begun his studies at the plebeian University of Paris—open at that time to any high-school graduate—rather than at one of the *grandes écoles* he might have attended if his (moderately wealthy) parents had lived but a few years longer. His choice of an unfashionable dissertation topic only distanced him further from the French mathematical mainstream. As a result, his dissertation lay unnoticed for more than half a century, until Leonard J. Savage chanced upon a copy gathering dust on a library shelf and began bringing it to the attention of various scholarly acquaintances. Among those acquaintances was the future Nobel laureate Paul Samuelson, who, as luck would have it, had recently embarked on his own study of financial markets. It was Samuelson who made Bachelier's work known to the growing number of economists and statisticians who were only then beginning to study such markets in earnest.

Between the time Bachelier wrote and the time his work was rediscovered, the world experienced two world wars, a withering stock market crash, and a great depression. The latter pair stimulated a good deal of soul-searching—and some fruitful investigation—into the nature of economic life in general and financial markets in particular. Among the more productive investigators on the financial side was one Alfred Cowles, scion of the family that owned the *Chicago Tribune*. Stricken with tuberculosis soon after his 1913 graduation from Yale, Cowles spent ten years in a sanatorium outside Colorado Springs. The cure must have been successful, for he died in 1985, at the age of ninety-three.

While still in Colorado, Cowles began to help manage the family finances. To that end, he perused a variety of newsletters offering guidance to individual investors. In time he found the flow of advice too voluminous to fathom and resolved to confine his attention to those advisers who consistently offered the most reliable advice. He then set out to decide, as scientifically as he could, which advisers those were. In 1928, he purchased subscriptions to twenty-four of the most widely circulated newsletters and monitored their performance for four eventful years—years that included the height of the 1920s bull market, the crash of 1929, and the economic slump of 1931–32.

To his surprise, Cowles discovered that exactly none of his twenty-four publications had managed to foretell either the crash of 1929 or the steady market decline that followed. He began to wonder if there was any way to predict stock market prices. In an effort to find out, he contacted a number of leading authorities on economics and statistics, including professors Harold T. Davis at Indiana University and Irving Fisher at Yale. The latter turned out to be an old friend of Cowles's father and the president-elect of the then newly formed Econometric Society.

While Davis helped Cowles acquire the latest in Hollerith punch-card computing equipment from IBM, Fisher encouraged him first to fund the publication of the society's journal *Econometrica* to the tune of $12,000 annually (well over $100,000 in 2007 dollars) and later to establish the Cowles Commission for Research in Economics at Colorado Springs. The commission moved to the University of Chicago in 1939, under the directorship of the future Nobel laureate James Tobin, and later to Yale, where it continues to prosper.

Cowles summarized his own early research on the accuracy of market forecasts in an article in the July 1933 issue of *Econometrica* titled "Can Stock Market Forecasters Forecast?" After a painstaking analysis of the twenty-four publications that had started him on his quest—an analysis that would have been virtually impossible to complete without his state-of-the-art IBM equipment—along with data from three other sources, he found it "doubtful." In each test, he noted that the market as a whole had outperformed the forecasters as a group. Moreover, of the few forecasters who did manage to beat the Dow Jones Industrial Average (DJIA), none had done so by a margin too large to be accounted for by blind luck. More money could have been made with a portfolio selected at random from the stocks listed on the New York Stock Exchange (throwing darts at the financial page) than by heeding so-called expert advice. In 1944, he published a yet more extensive study that reached essentially the same conclusion.

In 1938, under Cowles's direction, and inspired by Fisher's ideas on the proper construction of market indexes, the commission began to publish an index of stock market prices that was far more comprehensive than the DJIA. With only punch-card computers to work with, it was infeasible to publish such an index more than once a month. But with the advent of modern electronic computers during the 1960s, the daily (even hourly) compilation of such an index became commercially viable. Soon thereafter, the Cowles Commission index became the Standard & Poor's 500. The DJIA and the S&P 500 tell roughly the same story, as indicated in figure 9.1. Dow Jones Inc. can't afford to let them drift apart, because the world would soon lose interest in the DJIA if it began to differ significantly from the more authoritative S&P 500.

John Burr Williams was another important Depression-era figure. He entered Harvard graduate school in 1932, in search of someone (anyone) who could explain to his satisfaction what had caused the nation's apparently robust economy to self-destruct. Although he failed—like Diogenes with his lamp, searching for an honest man—to find the object of his quest, he stayed long enough to devise what remains the accepted method of calculating the "present value" of an asset expected to yield a predictable stream of revenue for years to come. He explained his method, and forged it into what he called the dividend discount model of stock evaluation, in his Ph.D. thesis of 1938. Published that same year by Harvard University Press as *The Theory of Investment Value*, his thesis became an instant classic. By that time, at

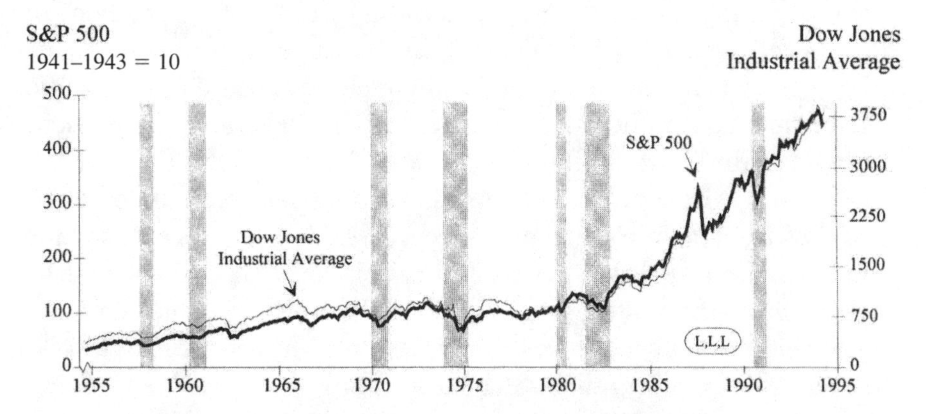

Figure 9.1. The Dow Jones average of thirty industrials (DJIA) moves as it must, more or less in unison with the more comprehensives S&P 500 index.

least two commercial publishers had rejected the manuscript on the grounds that it contained (gasp) algebraic symbols, and Williams had agreed to defray part of Harvard's printing cost.

Williams began by reflecting on the "time value of money." Suppose you owe $1,000.00 to each of three people. The first debt is due in three months, the second in six, and the last in a year. Suppose, too, that you have access to a risk-free savings account that pays interest at an annual percentage rate (APR) of 3 percent, continuously compounded. How much would you need to put in that account today in order to discharge the foregoing obligations in a timely manner? Because $992.64 will grow into $1,000.00 in three months, $985.33 will grow into $1,000.00 in six months, and $970.87 will grow into $1,000.00 in a year, $2,948.84 should suffice. Moreover, assuming your creditors have access to similar accounts, they should now be willing to accept $992.64, $985.33, and $970.87, respectively, for your three (gilt-edged) IOUs, since each of those amounts will appreciate into $1,000.00 by the respective settlement dates. In theory, each should be equally willing either to go on holding your IOU or to allow you to redeem it today at the indicated discount.

The sum $2,948.84 is known the world over as the present value of your three IOUs. No one has since proposed a more appropriate—or even a significantly different—method of ascertaining that value. Williams proposed to compute the present value of a share of stock in much the same way, by adding up the appropriately discounted present values of all future dividends and adjusting the total downward to account for the uncertainties surrounding the eventual payment of those dividends. The greater part of his book explains how to decide what a given firm's dividends will be in the long run and how much to adjust the total for uncertainty. Correcting for uncertainty is an essential part of the computation.

For those who found Williams's advice too ambiguous or time-consuming, Benjamin Graham offered a collection of rules for deciding when a share of stock is priced appreciably above or below what he called its "intrinsic value." Armed with such knowledge, one can make money either by buying shares that are currently underpriced or by selling (short if necessary) shares that are currently overpriced, on the theory that competitive markets seldom allow shares to remain mispriced for long. Warren Buffett, Graham's foremost disciple, has amassed an immense fortune by practicing and refining what Graham preached.

Graham was careful to note that his methods relied on the market only to restore mispriced shares to their correct level relative to other shares. He specifically disowned the currently popular notion that market prices are correct in any absolute sense. As he put it in an interview conducted at his home in Aix-en-Provence a few days before his death, "I don't see how you can say that the prices made in Wall Street are the right prices in any intelligent definition of what right prices would be."[2] In doing so, he was clarifying the conclusion reached in the textbook he wrote with Columbia professor David Dodd to the effect that "the real accomplishment of the many thousand analysts now studying not so many thousand companies is the establishment of proper *relative prices* in today's market for most of the leading issues and a great many secondary ones."[3] Graham was nothing if not careful to avoid overstating his own case.

Benoit Mandelbrot, who wrote the chapter on Bachelier in *The New Palgrave: A Dictionary of Economics*, describes Bachelier as the founder of the "coin tossing" theory of market prices.[4] That theory asserts that share prices evolve more or less as if someone tossed a fair coin each day to decide whether the price of a given issue should go up or down by a specific dollar amount. If that really happened, the price in question would perform what is often described as a "random walk." So long and intensely have such walks been studied by mathematicians that little about them remains unknown. The simplest ones take place on long, straight sidewalks where someone tosses a coin at regular intervals and moves one paving stone to the east or west according to whether the coin comes up heads or tails.

Although it is impossible to predict where the walker will actually be after a few dozen tosses, it is easy to calculate the probability with which he or she shall occupy a particular stretch of pavement. Specifically, the probability that the walker shall stand on or to the right of paving stone A is equal to the area below an appropriately constructed Gaussian bell curve and above the stretch of pavement in question, as indicated in figure 9.2. After N tosses, the top of the bell will still be located directly above the walker's original position, at an altitude inversely proportional to the square root of N. That means that the top of the bell must descend at a steadily decreasing rate of speed. As it does so, the width of the bell must increase, so that the area beneath it may remain constant.

Figure 9.2 exhibits three successive stages in the process by which

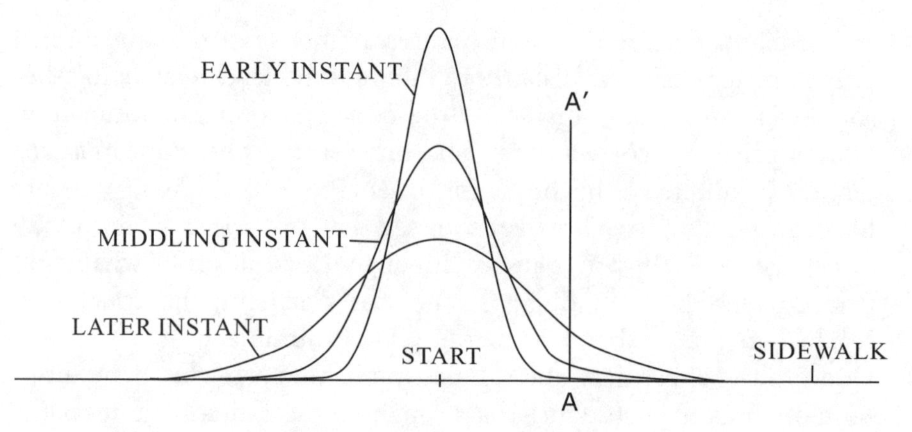

Figure 9.2. The probability that a random walker will stand more than n *paces to the right of his or her starting point after taking* N *random paces can be computed from a Gaussian normal bell curve that widens and flattens with the passage of time.*

the walker's probable location diffuses, like heat through an iron rod, away from a point of original concentration. At the earliest of the three instants depicted, the walker's probable whereabouts are still so concentrated near the starting point that he or she has virtually no chance of standing to the right of *A*. This is reflected in the fact that the area confined between the tallest of the three bell curves and the sidewalk, to the right of the line segment *AA′*, barely exceeds zero. But as time passes, that area grows progressively larger, indicating that the walker is more and more likely to stand at least that far to the right of the starting point. It is the intimate (albeit approximate) relationship between random walks and Gaussian bell curves that makes such walks mathematically tractable. If the coin were biased in favor of eastward movements, the top of the bell would move to the right at a constant rate of speed while descending at the previously described (diminishing) rate.

The modern version of the coin-tossing theory of random price movements differs from Bachelier's in two important respects. Both modifications were proposed by Paul Samuelson during the 1950s. He suggested, first, that the logarithm of a particular share price, rather than the price itself, should perform a random walk and, second, that the coin employed should be slightly biased in favor of price increases, represented as eastward movements. These modifications are consistent with the observed fact that share prices do tend to rise over extended periods (say, a decade or more) and that share prices never become negative, as some inevitably would in Bachelier's original formulation. The fact that the coins employed are biased in favor of price increases causes the stock market to resemble a casino whose owners are content to operate at a slight loss, slowly channeling money to the players instead of taking it from them. As a result, all investors could conceivably emerge with a profit, though it seldom seems to work that way in practice.

Although Mandelbrot does not personally subscribe to the coin-tossing theory of share price evolution, he understands it as well as anyone. In his 2004 book on that and related subjects, he credits four men, in addition to Bachelier, with the main steps in its development.[5] Conceding that many others have contributed to the theory, and that he may have overlooked some deserving individuals, he opines that Harry Markowitz, William Sharpe, and the team of Fischer Black and Myron Scholes made the boldest strokes in completing the canvas begun by Bachelier. Three of the four received Nobel prizes for their achievements, while Black surely would have had he but lived a few years longer.

Markowitz's most important contribution to the science of investment was to shift attention from individual issues to combinations (a.k.a. portfolios) thereof. Treating as observable the appreciation rate of a particular share price, as well as the tendency of the monthly rate of appreciation to deviate from the long-term average rate, he noted that many different combinations of publicly traded assets must exhibit the same appreciation rate. Moreover, of the many that do exhibit a given rate, the portfolios that appreciate most steadily (in the least volatile manner) seem the most desirable. Last, the higher one sets one's target rate of appreciation, the more unsteadiness (volatility) one must be prepared to tolerate.

To discover which of the portfolios that promise to appreciate at a given rate seem likely to do so most steadily, Markowitz was obliged to consider the "covariance matrix" of the eligible assets. Such a matrix consists of equally many rows and columns of numbers between $+1$ and -1, as shown in figure 9.3. If i and j are any two numbers between 1 and 5 inclusive, the number in row i and column j is the covariance of the prices of assets i and j, which measures the tendency of those two asset prices to rise and fall in unison.

$$\begin{pmatrix} 1 & .2 & 0 & .7 & .3 \\ .2 & 1 & .4 & .6 & 0 \\ 0 & .4 & 1 & -.3 & .8 \\ .7 & .6 & -.3 & 1 & .9 \\ .3 & 0 & .8 & .9 & 1 \end{pmatrix}$$

Figure 9.3. A covariance matrix is its own mirror image, when reflected about the diagonal consisting entirely of 1s.

If the prices of assets *i* and *j* were bound to move in perfect unison—whatever that might mean—the covariance would be +1. If their movements were equal and opposite, their covariance would be −1. In other cases, the covariance will lie somewhere between +1 and −1. When two prices move independently of each other, as if generated by two different random walkers, using two different coins, their covariance is 0.

The negative entries in the covariance matrix are of particular interest, since they represent opportunities to reduce portfolio volatility. If, for instance, two negatively correlated asset prices exhibit the same appreciation rate, they can be combined in a portfolio that also appreciates at the same rate but does so with less volatility than either of the two ingredient assets. If two negatively correlated assets have different appreciation rates, then any combination of the two will have an intermediate rate of appreciation, and at least a few such combinations will exhibit less volatility than either of the ingredient issues.

Markowitz proposed to construct, for each achievable appreciation rate, the least volatile portfolio capable of sustaining that rate of appreciation. The result would be a list of "efficient" portfolios, from which a final selection could be made. The efficient portfolios with low volatilities would perforce exhibit low appreciation rates, while the ones with high appreciation rates would exhibit high volatilities. By selecting an efficient portfolio, the chooser could rest assured that no less volatile alternative with an equal or larger appreciation rate had been overlooked.

Markowitz soon discovered that to construct the least volatile portfolio with a given appreciation rate, one needed to solve what is known as a quadratic programming problem involving a covariance matrix with numerous rows and columns. Because quadratic programming problems are an order of magnitude more difficult to solve than the linear programming problems mathematicians were only then learning to solve on their still rather primitive electronic computers, he found it infeasible to carry out his scheme with more than a short list of eligible assets. Even today, with all the improvements that have been made to computers since the 1950s, the cost of finding the efficient portfolios constructable from the list of stocks included in the S&P 500 index, or the Russell 2000 Index, remains prohibitive.

The breakthroughs required to convert Markowitz's idea into a practical method of portfolio selection were made by William Sharpe, working initially under Markowitz's guidance while a graduate student in economics at UCLA. Acting on Markowitz's suggestions that (1) because the entries in the relevant covariance matrices are predominantly positive, there is a strong tendency for the prices of all financial assets to rise and fall together, and (2) an appropriate index fund might be included among the assets eligible for inclusion in the portfolio to be constructed, Sharpe built a program to solve the quadratic programs created by discarding all correlations save those between individual assets and the included index (which might or might not be the S&P 500) of market prices.

In a 1962 address to the Econometric Society, Sharpe pointed out that the quadratic programs so obtained are far easier to solve than the ones formulated by Markowitz, which discard no correlation, and that the resulting efficient portfolios compare very favorably with the ones produced by Markowitz's original (more laborious) technique. He further noted that if investors are free to borrow or lend at the risk-free rate of interest, there is a single best portfolio of which every investor should hold some amount while borrowing or lending as needed (at the risk-free rate of interest, of course) to produce the desired combination of volatility (risk) and appreciation (reward). By 1964, Sharpe was prepared to argue that the previously unidentified "best portfolio" was none other than the market itself, which may (since 1971) be held in the form of a publicly traded index fund. The upshot of all these developments has been the illustrious capital asset pricing model (CAPM)—pronounced "cap-EM" by those in the know—in use ever since at financial institutions the world over.

Myron Scholes and Fischer Black, the remaining two of Mandelbrot's five key developers of the coin-tossing theory of asset price evolution, discovered a remarkable formula for the price of a European call option. Call options are contracts that entitle the buyers to purchase designated assets (typically shares of stock) from the sellers at stated exercise prices on (or perhaps before) given expiration dates. If the market price exceeds the exercise price on the expiration date, the buyer can turn an instantaneous profit by exercising his or her option to buy the designated asset at the lower (exercise) price and sell it forthwith at the higher (market) price. If the market price fails to exceed the exercise price, the buyer simply allows the contract to expire unexercised.

Because most options are never exercised, they are relatively inexpensive to own. That goes for "put" options as well as "calls." Puts obligate the sellers to purchase designated assets from the buyers at stated exercise prices on (or perhaps before) given expiration dates, if asked to do so. Should an exercise price exceed the corresponding market price on (or perhaps before) the relevant expiration date, the holder can again turn a quick profit by buying low and selling high.

Options offer an effective way of dealing with risk. If a man fears that his shares in XYZ Corp. are about to take a dive, he can purchase a put entitling him to sell some or all of those shares at or near today's price at any time during the next ninety days. If one firm owes another a million euros, payable sixty days hence, and expects euros to appreciate significantly against the dollar, it can purchase calls entitling it to buy the required number of euros at that time, at or near the current price. And so on.

Unlike European options, which can only be exercised on their expiration dates, American options can be exercised on *or before* those dates. Black and Scholes obtained their formula by solving—with a timely assist from Robert Merton—a certain mathematical problem. The version of that problem corresponding to European options is substantially easier to solve than the one corresponding to American options. That is why no variant of the Black-Scholes formula applicable to American options has yet been found. As matters stand, the version of the problem corresponding to American options is only approximately solvable, and that with the aid of a computer.

In April 1973, the Chicago Board of Options Exchange (CBOE)

opened its doors. On its first day, with 284 members on the floor making markets, just 911 call option contracts on shares of sixteen corporations were traded. A year later, more than 20,000 contracts on the stock of thirty-two corporations were being traded. No fewer than 567 members had paid $40,000 (up from $10,000 the previous year) for the privilege of making markets on the floor. Three years later, volume had risen to more than 100,000 contracts. The Black-Scholes formula, published the month after the CBOE commenced operation, proved to be exactly the tool traders on the floor needed to evaluate the contracts they were dealing in. Given just five numbers—the current price U of the underlying asset, the exercise price E, the time T until the option expires, the risk-free rate of interest R, and the volatility σ of the random walk performed by the logarithm of U—the formula permits the prompt evaluation of a put or call (European) option.

The use of the formula soon became so prevalent that, as Fischer Black once explained, "market prices are usually close to formula values even in situations where there should be a large difference: situations, for example, where a cash takeover is likely to end [prematurely] the life of the option."[6] That being the case, options traders who ignore the Black-Scholes model do so at their own peril: they can lose money even when their own instincts are more reliable than the formula, because widespread use of the formula can prevent the market from behaving as it "properly should." Within a year of the formula's publication, Texas Instruments began to market a pocket calculator preprogrammed to calculate the Black-Scholes value of an option at the push of a few buttons.

The success of the Black-Scholes formula has greatly encouraged the growth of the market for derivative securities. Derivatives, as they are commonly known, are two-party contracts containing multiple clauses, each of which specifies either an option or a "forward obligation." Such obligations commit one party to deliver, and the other to purchase, a particular bundle of assets at a designated time, place, and price. Futures are just forwards sold in an organized market, such as the Chicago Board of Trade, where the risk of nonpayment is minimal. Coffee, sugar, wheat, corn, oil, precious metals, and currencies are among the commodities most often traded on futures markets. The number of ways in which options and forwards can be combined to reduce perceived risk is unlimited. Indeed, there was once a flourishing market in "designer derivatives," customized to combat the risks confronting a particular (typically corporate) customer. Today, however, corporate risk managers are more inclined to make do with off-the-shelf products.

The name "derivative" refers to the fact that any value the market may impute to such contracts derives exclusively from the values of the underlying assets. The market for derivatives is currently immense—far larger, by most measures, than the markets for the underlying assets. Derivatives are typically evaluated clause by clause, in what is known as a Black-Scholes environment. The essential characteristics of such an environment are the random walk performed by the logarithm of the market price of the underlying asset, the absence of transaction costs, and the known, constant, risk-free rate of return at which investors may choose either to borrow or to lend. It is doubtful that such markets could have grown to anything like their current size without the contribution of Black and Scholes.

Not included among Mandelbrot's five main developers of the coin-tossing theory of financial price evolution are some highly influential individuals. Robert Merton gave Black and Scholes the arbitrage argument that allowed them to justify their famous formula and moved Black to concede on at least one occasion that "it should probably be called the Black-Merton-Scholes formula."[7] Moreover Merton's development of continuous time finance vastly accelerated the development of financial economics as a branch of economic theory, even as it encouraged the practical application of asset pricing and options pricing theory.

Merton also completed one of the more telling empirical studies ever devised to illustrate the difficulty of predicting stock market behavior. In a paper published in 1981, he pointed out that $1,000 invested in the stock market on January 1, 1927, and shifted back and forth between stocks and short-term interest-bearing bonds at the beginning of each month in such a way as to reap the greater of the two returns available during that month from those two sources would have grown to $5,362,212 by December 31, 1978.[8] Had those 624 stocks-or-bonds decisions been faultlessly made, the initial investment would have appreciated at an average annual rate of 34.71 percent for fifty-two consecutive years. That is the rate of return a money manager with perfect market foresight extending only thirty days into the future could have obtained. It far exceeds the 10 or 11 percent per annum that mortal money managers struggle to achieve.

During the 1960s, Eugene Fama drew attention to the notion of an efficient market and devised ways to test the so-called efficient market hypothesis. In its purest form, that celebrated hypothesis contradicts the wisdom of Benjamin Graham—who, as mentioned earlier, trusted free markets only to produce correct relative prices—by asserting that private citizens can't beat the financial markets using publicly available information because such markets "know too much." They acquire all relevant information more quickly than private citizens can hope to acquire it, and they employ it more surely and rapidly to eradicate any gaps between asset prices and "intrinsic asset values."

Concluding that the unknowability of intrinsic asset values renders the efficient market hypothesis untestable in its purest form, Fama produced three dilute forms of the hypothesis. His "weak form" asserts that there is no way to increase one's net worth more rapidly, by trading in a given market, than the associated market index while using only information concerning the history of prices in that market. The "semi-strong form" adds that using information gleaned from annual reports, new-product evaluations, stock split announcements, and the like won't help, either. Finally, his "strong form" asserts that there is no way you can achieve above-average growth without recourse to insider information. He found evidence to support all three forms of the hypothesis, although most interested parties found his case for the stronger forms to be less persuasive than the one he advanced in support of the weak form.

Paul Samuelson's early work on warrant pricing was informed by Bachelier's achievements and prompted his observations that (1) the logarithms of asset prices seem more likely than the prices themselves to perform coin-tossing random walks and (2) the coins involved should be slightly biased in favor of price increases. Moreover, it was Merton's involvement in that work (initially as Samuelson's graduate assistant) that launched his own mercurial career in financial economics. It remains something of a mystery, especially to those familiar with Samuelson's work on warrant pricing, that he failed to discover the Black-Scholes formula on his own.[9] For warrants are nothing more than long-lived call options issued by corporations.

They entitle the bearer to buy or sell shares of stock in the issuing corporation, and they expire after years rather than months.

In 1965, Samuelson wrote a very influential paper titled "Proof That Properly Anticipated Prices Fluctuate Randomly." In it he assumed the yet-to-be-named efficient market hypothesis to be true in its pure form—the one Fama would later conclude is experimentally untestable. He assumed, in other words, that markets process information so effectively that at any given time, current market prices represent the best possible estimates—given only publicly available information—of the corresponding asset values. On that assumption, he demonstrated that market prices should indeed fluctuate randomly, just as Bachelier had observed them to do on the Paris Bourse more than half a century earlier. The mere fact that Samuelson—whose stature within the economics profession was and still is second to none—had assumed the efficient market hypothesis to be true in its purest form made it acceptable for younger economists to assume the same thing. For at least a generation, virtually all of them did.

The significance of the efficient market hypothesis is obvious. If the market is truly unpredictable, it is foolish to go on trying to predict it. Rules like Graham's for picking stocks destined to appreciate more rapidly than the market average are doomed to fail. Searchers' time would be better spent collecting hen's teeth, looking for the philosopher's stone, or designing perpetual motion machines. Market-neutral tools, like Black-Scholes and CAPM—which leave the user no reason to care which market prices go up and which go down—are the best one can hope for. Many of those who take the efficient market hypothesis most seriously, such as Burton Malkiel, advocate index funds.[10] Although they can't possibly appreciate faster than the S&P 500 index (or whichever index they are designed to mimic), they can and do benefit their shareholders by charging lower management fees.

Vernon Smith, who shared the 2002 Nobel Prize in Economics with Daniel Kahneman, and who is widely regarded as the father of experimental economics, has recently turned his attention to the formation of market bubbles, such as the ones that afflicted Japan's Nikkei index during the late 1980s and the NASDAQ index until March 2000. In a series of experiments dating back to the 1980s, he and his coworkers have developed a standard format for market bubble experiments.

Each of several subjects is seated at a computer terminal and endowed with a sum of money, in addition to a portfolio of shares in a fictitious commercial enterprise. Subjects are encouraged to communicate by computer, but in no other way. Each share pays a dividend at the end of each five-minute trading period, and each experiment consists (for ease of comparison) of fifteen trading periods. In some experiments, the dividend is a fixed $0.24 per trading period, while in others it is chosen at random after each period from the following list: $0.00, $0.08, $0.28, $0.60, so that the average (or expected) dividend is again $0.24.

The subjects are encouraged to exchange shares for cash, and vice versa, if ever and whenever they deem it advantageous to do so. In experiments with nonrandom dividends, the buy-and-hold value of each share starts out at 15 × $0.24 = $3.60 and decreases by $0.24 at the end of each trading period. In experiments with random dividends, the *expected value* of the unpaid dividends is still $3.60 at the outset and still decreases by $0.24 at the end of each trading period. During each such period, the amount of money in circulation is equal to the sum of the initial monetary endowments, augmented by the sum of dividends already paid, if any.

Were there a rule prohibiting the sale and purchase of shares, subjects could earn nothing more than the dividends to which, as shareholders, they are entitled. But in the absence of such a rule, they may hope to increase their earnings at the expense of their trading partners. Subjects are engaged in a sort of two-way auction in which there lurks a significant winner's curse, since there is nothing to prevent an individual subject from paying too much per share. As in the one-way auction experiments of Capen, Clapp, and Campbell, Smith and his coworkers found subjects fell prey to the winner's curse with distressing regularity. When the prevailing share price exceeds the actual or expected value of the unpaid dividends on a single share for several consecutive trading periods, a market bubble may be said to have formed.

Smith and his coworkers have identified a number of factors that seem to encourage bubble formation. In a paper published in 2001, he and two coworkers found the outcome to be affected by (1) whether or not dividends are paid at the end of the period in which they are earned, (2) whether or not market participants have up-to-date information concerning the state of supply and demand, and (3) how

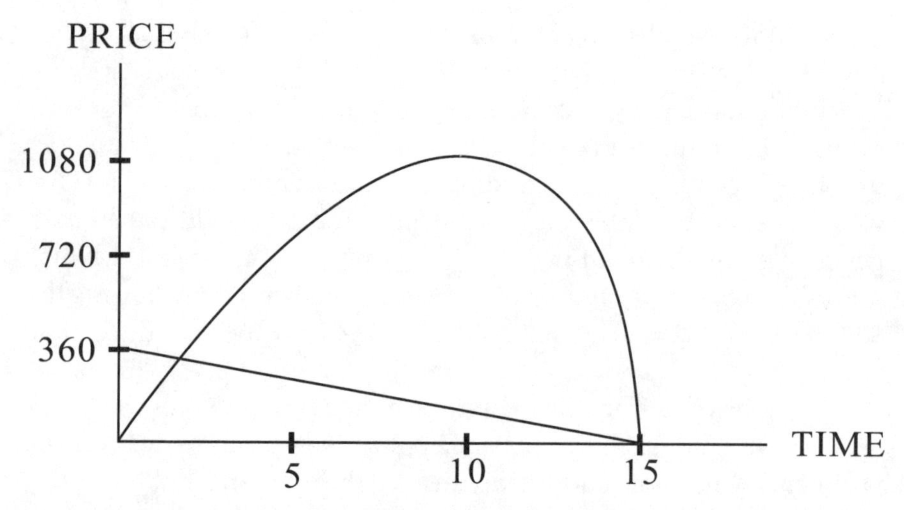

Figure 9.4. In markets starved for cash, the "going price" of a security seldom exceeds the value of its outstanding dividends. In other conditions a bubble may form, enabling shareholders to dispose of their shares for far more than their intrinsic worth.

much cash is in circulation relative to the number of shares.[11] The results were roughly as indicated in figure 9.4.

In three of the six experiments represented, the share prices formed bubbles by soaring, as does the curve in the figure, far above the straight line representing the steadily declining value of unpaid dividends. In each of the three bubble-plagued experiments, there was an abundance of cash in circulation from the beginning, augmented periodically by the payment of dividends, and poor information about conditions of supply and demand. In the bubble-free experiments, the opposite conditions prevailed. By far the most significant factor was the amount of cash in circulation, as compared with the value of the dividends to be paid. More than a few subjects seemed compelled to convert any available cash into income-bearing assets, with seeming disregard for the price. Perhaps they were subscribers to the "greater fool theory," whereby it is assumed that no matter what assets are selling for today, a greater fool will emerge to pay more for them tomorrow. Clearly, the pure form of the efficient market hypothesis is false as it applies to Smith's experimental markets. Why then believe it true of actual markets?

The most famous efficient market denier is, of course, Robert

Shiller, the Yale economist who both coined the phrase and wrote the book *Irrational Exuberance*, which remained on the *New York Times* list of bestsellers for most of 2000.[12] He offers a number of reasons for disbelieving the famous hypothesis, the most telling of which dated back to a paper published in 1982, near the beginning of Shiller's career. In it he pointed out that the pure form of the efficient market hypothesis is not quite as untestable as Fama thought. For although there seems to be no way of deciding whether it is true at present, John Burr Williams's dividend discount model provides an obvious—if laborious—way of deciding whether it has been true in the past.

Shiller began by extrapolating a number of historical time series back into the past, as far as New York Stock Exchange records would permit. His first series carries the S&P 500 index back to 1872, roughly as it would have been had the value of the information contained in such a series been fully appreciated in that bygone era. His second and third series reconstruct the dividends paid and earnings announced by the corporations whose share prices are involved in the index. His fourth and last series reconstructs the consumer price index, more or less as it would have been had the Labor Department begun investigating such things in the nineteenth century. Without such an index, Shiller would have been unable to correct the first three series for inflation, as must obviously be done in a study covering more than a century of financial activity.

To calculate on January 1, 2001, the present value of a single share in a (necessarily fictitious) S&P 500 index fund as of January 1, 1901, one just needs to know the magnitude of each dividend paid by each firm involved in the index (including those founded later in the century) during the twentieth century, along with the present value—call it V—of that share on January 1, 2001. Everything else one needs to perform the calculation is present in Shiller's various time series. Only V is impossible to determine without knowledge of the dividends to be paid during the twenty-first century and beyond. The simplest way to estimate V is to assume that dividends will grow in the future as they have grown in the past, at the same average rate at which they grew throughout the twentieth century. The smooth curve in figure 9.5 was calculated on that assumption, in a manner suggested by Williams's dividend discount model.

The curve so obtained shows what a share in an S&P 500 index fund was worth, in the present value sense, on any date between 1872 and 2001. Had he included curves calculated in exactly the same way, on the assumption that the estimate for *V* used in calculating the curve in the figure was 50 percent too high or too low, the three curves would appear to merge to the left of about 1950, meaning that very little doubt surrounds the present value of a share in an S&P 500 index fund at any date prior to 1950, since any plausible estimate of *V* leads to the same conclusion concerning those bygone present values. The other curve in the figure—the erratic one—depicts the actual S&P 500 index, extrapolated back by Shiller to 1872.

Although the more erratic curve represents the historic market price of a single share of an S&P 500 index fund, it bears scant resemblance to the curves generated by the dividend discount model using plausible estimates of *V*. Nor does it closely resemble any of the other curves one might construct by incorporating less plausible estimates of *V* into the dividend discount model. Although the model can be made (by employing ridiculously high estimates of *V*) to generate

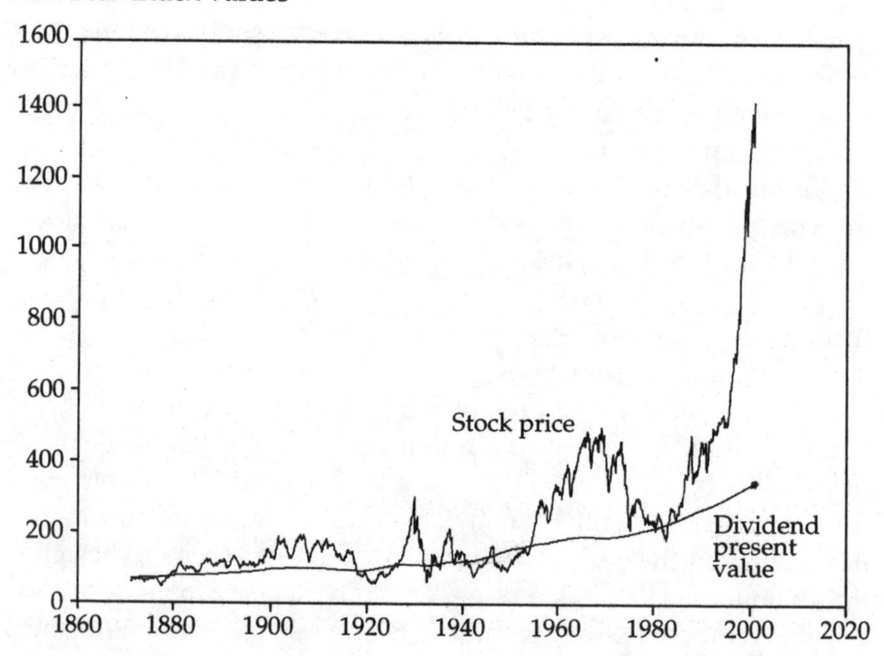

Figure 9.5. While the present value of the stream of dividends to which the holder of a share of stock is entitled tends to grow rather slowly and steadily, the share's market value tends to rise and fall more erratically.

curves that rise to the heights attained by the index during the late 1990s, it cannot be made to generate curves exhibiting anything like the erratic ups and downs of the historic S&P 500 index. Although these and other facts assembled by Shiller are exceedingly difficult (not to say impossible) to reconcile with any conceivable version of the efficient market hypothesis, many economists continue to defend that appealing hypothesis.

Mandelbrot rejects not only the efficient market hypothesis but also the coin-tossing theory of asset price evolution, on the grounds that the latter understates the risk borne by those who hold speculative assets. It does so by oversimplifying the nature of the random walks their prices (or more properly, the logarithms of their prices) perform. To illustrate the distinctions he has in mind, he offers the depictions of four price evolutions shown in figure 9.6 and asks the reader to distinguish between the two real ones and the two fakes.

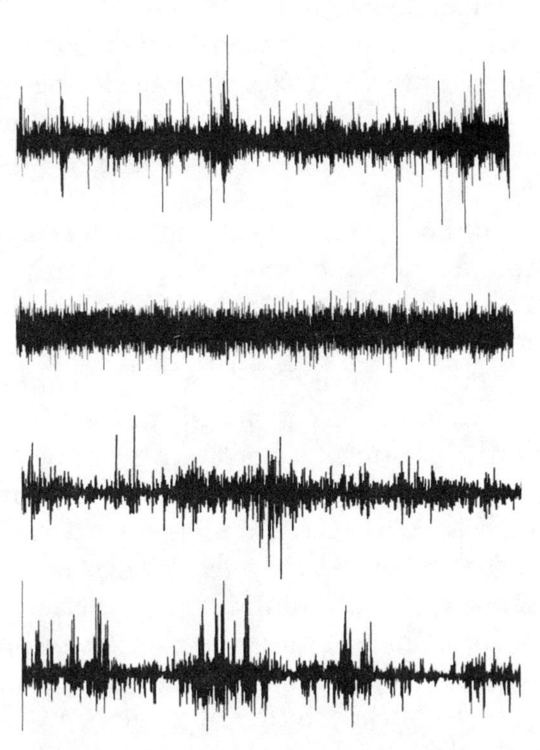

Figure 9.6. Benoit Mandelbrot challenges his readers to pick out the two genuine stock price histories from the above four.

The second, which shows prices varying more or less uniformly over time, stands out from the rest. It was generated by a coin-tossing random walk in which the price changes are drawn at random from a Gaussian normal population. The largest changes barely exceed the average change.

Contrast that with the first and third price series, which are authentic. The first depicts the changes in the IBM stock price between 1959 and 1996, while the third chronicles the dollars-to-deutsche-marks exchange rate over the same period. In these and most other empirical price series, changes are highly erratic, the large ones being numerous, clustered, and far larger than the average change. The fourth series is a fictitious one, generated by Mandelbrot's latest model of "how financial markets work." It appears to the naked eye to contain about as many large changes (both positive and negative) as the two real ones, as well as a similar degree of clustering. Mandelbrot describes the model by which it was generated as "fractional Brownian motion in multifractal time." It is significantly more complicated than ordinary Brownian motion and the coin-tossing random walks that mimic it so closely. While the new model cannot yet be used to pick stocks, trade derivative securities, or evaluate options, he seems hopeful that further research will in time make such applications possible.

The difference between a coin-tossing random walk and one of Mandelbrot's more complicated models is particularly apparent when calculating the "value at risk" while holding a moderately risky asset, such as a position in the dollar market for euro futures. If one were to assume that the prices of such futures perform coin-tossing random walks, one might conclude that the chance of losing as much as 12 percent of one's investment is not larger than 5 percent. But if one assumes that the prices evolve according to one of Mandelbrot's more complicated models, the chances of losing 12 percent or more (possibly much more) are many times larger. A single bank can suddenly lose more than its entire worth, forcing it to default on its obligations to other banks, which in turn must do likewise, precipitating an international crisis. It is part of Mandelbrot's message that modern financial theory—which tends to treat all random evolutions as coin-tossing random walks—vastly underestimates the likelihood of such catastrophes.

Mandelbrot quotes Lawrence Summers—former secretary of the U.S. Treasury and former president of Harvard University—to the effect that the global financial system endured no fewer than six genuine crises during the generally prosperous 1990s: Mexico in 1995; Thailand, Indonesia, and South Korea in 1997–98; Russia in 1998; and Brazil in 1998–99. He might also have mentioned the narrowly averted bankruptcy of Long-Term Capital Management, which threatened during the summer of 1998 to bring down several of the larger banks in Germany, Switzerland, and the United States. Mandelbrot sees no reason to suppose that the risk of such disasters will abate until the world ceases to entrust its financial well-being to shamans—his characterization of central bankers—instead of scientists.

By no means certain that science is ready to outperform the shamans, Mandelbrot echoes the opinion once expressed by Wassily Leontief—a longtime Harvard economist and 1973 Nobel Prize winner—to the effect that "in no field of empirical enquiry has so massive and sophisticated a statistical machinery been used with such indifferent results." He concludes only that science has plenty left to learn about the risk, reward, and potential ruin inherent in global financial markets and that the most important lessons will remain unlearned unless and until slow, painstaking scientific research is given the resources to learn them.

With little fanfare, a different method of managing money has been lurking in the background for several years, offering an alternative to CAPM- and Black-Scholes-related techniques. Known as the Kelly rule, it traces its origin to the early 1950s, when scientists and engineers at Bell Labs were still absorbing Claude Shannon's new "information theory." Beginning with the observation that no channel can have an unlimited capacity to transmit information, Shannon's theory goes on to specify the maximum rate (measurable in bits of information per second) at which a given "noisy" channel can transmit information accurately, and to demonstrate that properly encoded messages can at least approximate that ideal rate, known as the channel capacity. Without explaining how a message must be encoded to exploit the full potential of a given channel, Shannon's theory demonstrates that—with mathematical certainty—codes with that capability must exist. For many years thereafter, the best available codes exploited little more than half of a typical channel's capacity.

John L. Kelly Jr., regarded by many who knew him as the second smartest man (after Shannon himself) at Bell Labs, found a novel interpretation of Shannon's theory.[13] Kelly envisioned an inveterate horseplayer endowed with a private channel of communication over which he or she receives a steady flow of track-related information unavailable to rival bettors. Although the information received need not be 100 percent accurate, it must be reliable enough to convey a distinct advantage. On that assumption Kelly was able to demonstrate that the sustainable rate at which the horseplayer's bankroll may appreciate cannot exceed the rate—again measured in bits of information per second—at which he or she receives accurate information unavailable to the betting public.

The key decision to be made concerning each successive race concerns the fraction—call it f—of the horseplayer's current bankroll to be placed at risk. William Poundstone, whose recent book *Fortune's Formula: The Untold Story of the Scientific Betting System That Beat the Casinos and Wall Street* relates the tangled history of Kelly's formula in amusing detail, expresses the required decision rule in the form $f = \text{edge} \div \text{odds}$.[14] Here "edge" refers to the statistically expected gain from the contemplated bet, while "odds" betokens the multiple of the amount of that bet which the bettor may hope to collect upon winning. Poundstone's book profiles the surprising cast of characters who have used Kelly's remarkable betting formula—in casinos, at racetracks, and on Wall Street—to parlay nominal amounts of money into veritable fortunes, as quickly as possible, and without risk of going broke, by means of a mathematically informed sequence of wagers.

For technical reasons, Kelly's formula is often described as the "geometric mean" formula for selecting bets. It has a variety of demonstrated virtues, including the fact that it maximizes the long-term average rate at which bettors' bankrolls appreciate, that it causes such bankrolls to reach preselected target levels as rapidly as can be expected, and that it enables them to do so without risk of ruin. Stanford mathematician Thomas Cover has compared the Kelly rule's array of desirable properties to those of the mathematical constant $\pi = 3.14159265\ldots$, which may be defined as the circumference (in feet) of a circle one foot in diameter and which keeps turning up in branches of mathematics having nothing to do with circles. "When something keeps turning up like that," he suggested to Poundstone, "it usually means it's fundamental."

The next chapter in the saga of Kelly's rule concerns Edward O. Thorp, the first man to devise a winning strategy for casino blackjack. Having become interested during a Christmas visit to Las Vegas in 1958, Thorp began to suspect that the game as then played could be beaten simply by betting more heavily when the deck is stacked in the player's favor than at other times. Upon arrival at MIT the following summer, to assume an entry-level position on the mathematics faculty, he realized that the school's spanking-new IBM 707 computer could be exactly the tool he needed to confirm his suspicion. So, after teaching himself FORTRAN—the then new thing in computer programming—he set out to discover a card-counting strategy capable of turning a profit at the tables in Las Vegas. In time, he succeeded.

Given that he hoped to make a name for himself among mathematicians, Thorp concluded that the most effective use he could make of his discovery would be to publish an account of it in the most prestigious journal he could find—the *Proceedings of the National Academy of Sciences*. But to have his paper published there, he would need the sponsorship of a current member of the academy who would be willing to submit something on so unusual a topic to its flagship journal. Accordingly, Thorp approached Shannon, who had recently joined the MIT faculty on a permanent basis and who was the only mathematically inclined member thereof who was also a member of the academy. Shannon was quick to acknowledge that Thorp had indeed enhanced mankind's understanding of blackjack and offered to submit a lightly edited version of his paper to the journal. Shannon also advised Thorp to read Kelly's paper.

Seeing the relevance immediately, Thorp hastened to incorporate a form of Kelly's rule into the revised version of his strategy that appeared in book form in 1962.[15] He also made at least a dozen trips to Nevada during the 1960s and seems to have cleared—after learning to detect cheating dealers and avoid the other main hazards of casino gambling—about $25,000 using his system. He later estimated that he might have earned as much as $300,000 a year, using his system, had he been able to play full-time without interference from management. But that, of course, was never to be. Casino managers soon came to regard card counting as a form of cheating, and to deal accordingly with those suspected of engaging in the practice. More than one of those suspects, upon refusing a polite invitation to vacate the premises voluntarily, was physically beaten. Then, too, both Thorp and Shannon soon realized that financial markets represent the biggest casinos of all—casinos in which $300,000 was (even then) a mere pittance.

For Shannon, stock market participation involved the investment of personal funds and the convening of regular meetings at MIT on the science of investment. On the rare occasions when he himself addressed those meetings, they had to be moved to one of the largest lecture halls on campus. Poundstone managed to obtain access to enough of Shannon's personal records to confirm that (starting with next to nothing) his portfolio had grown to more than half a million dollars as of 1986, when Shannon reached seventy years of age.

By his own admission, Shannon had gone through a learning period during which he bought and sold so frequently that transaction costs had eaten up the lion's share of any gains. But having absorbed that lesson, he was subsequently able to increase his net worth at a rate of about 28 percent per annum. And although his records are incomplete in that they omit mention of stocks once owned and later discarded—a considerable omission for accounting purposes—Poundstone is able to certify that the estimate is not grossly exaggerated.

For Thorp, on the other hand, the record is far more complete. That is because he quickly yielded to requests from family and friends to begin investing their money along with his own. His main vehicle of investment was Princeton-Newport Partners (PNP), a hedge fund that maintained offices in both Princeton, New Jersey, and Newport, California, between 1969 and 1988. Figure 9.7 depicts the fund's performance during the nineteen years of its existence.

It is not only the fact that $1.00 invested at the outset grew into $14.79 during the nineteen years of the fund's existence (for an APR in excess of 15 percent during an era that saw the S&P 500 index appreciate at a mere 8.8 percent) that arouses the jealousy of fund managers everywhere. Nor is it the fact that PNP had to be earning at least 20 percent on the dollar to pay its investors 15 percent. It is actually the steadiness of the growth that most impresses industry professionals, especially those charged with attracting new investors to their funds. Nothing sells like rapid, steady success. Consistent appreciation, and a seemingly limitless ability to identify favorable bets lurking in remote corners of the financial markets, have been the most remarkable features of Thorp's lengthy career in money management. The edge ÷ odds formula is worthless without an adequate supply of edges to exploit.

Thorp is by no means the only money manager to have used the Kelly rule successfully. According to Poundstone, Kenneth Griffin's Citadel Investment Group, James Simons's Medallion Fund, and D. E. Shaw & Co. have done so as well. Baltimore's legendary William Miller, manager of the Legg Mason Value Trust, is another convert, having written in his 2003 annual report that "the Kelly criterion is integral to the way we manage money." This is a significant statement because Miller's fund is the only SEC-regulated mutual fund (Thorp's various funds and the others mentioned above being all-but-unregulated hedge funds) ever to outperform the S&P 500 index for ten consecutive calendar years. In fact, it eventually did so for fifteen calender years, before failing in 2006.

Financial economists are almost universally hostile to the use of the Kelly rule for portfolio management. Samuelson probably spoke

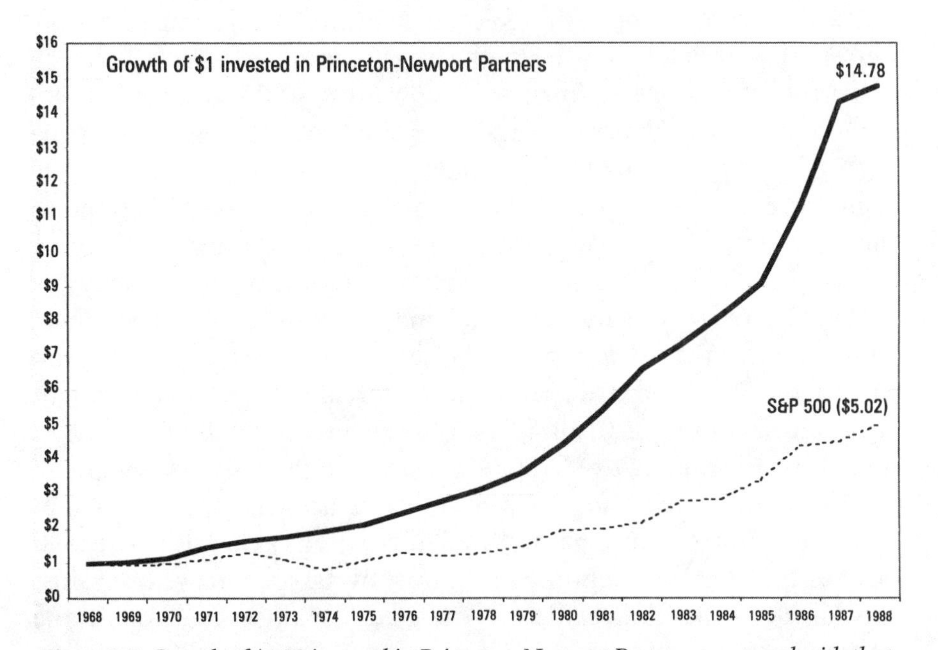

Figure 9.7. Growth of $1.00 invested in Princeton-Newport Partners compared with that of $1.00 invested in the S&P 500, over the nineteen-year history of the partnership.

for most of them when he wrote, in 1969, that "our analysis enables us to dispel a fallacy." The fallacy he wished to dispel was the "notion that, if one is investing for many periods, the proper behavior is to maximize the geometric mean of the return [on investment] rather than the arithmetic mean. I believe this to be incorrect."[16] Two years later, he conceded that "acting to maximize the geometric mean at every step will, if the period is 'sufficiently long,' 'almost certainly' result in higher terminal wealth . . . than any other decision rule."[17]

He went on to caution, however, that this concession should not be interpreted as an assertion that any one decision rule is right for all investors, under any and all circumstances. On the contrary, different decision rules are required to achieve different objectives in the world of investments, as in any other. The obvious implication is that there are numerous investment situations in which it is wiser to maximize Samuelson's arithmetic mean than Kelly's geometric one. But Samuelson neglects to identify the situations he has in mind.

Samuelson was probably thinking that a bankroll entrusted to the Kelly rule tends to appreciate at a highly irregular rate. Sharp peaks are typically separated by wide (and often deep) valleys, as in figure 9.8. According to Poundstone, it depicts the waxing and waning of a Kelly bettor's bankroll during a lengthy sequence of wagers at a single player-friendly wheel of fortune. The promise of the Kelly rule is reflected in the fact that the peaks grow progressively higher with the passage of time. No matter how many peaks there have been, or how high they have been, there is always a higher one to come. No target is out of reach. The sky's the limit for the bettor with unlimited staying power. But the drawback so unacceptable to Samuelson is equally apparent: There are long interludes during which the bankroll is significantly smaller than it once was. Should a sudden need for cash arise during one of those dry spells, the bettor may feel obliged to quit while far behind. Samuelson is right to consider this a highly undesirable (if not disqualifying) feature in a gambling/investment scheme.

Samuelson's criticism takes no obvious account of the fact that mankind has been taming unwanted fluctuations at least since Joseph advised Pharaoh to warehouse the bounty of the seven fat years against the famine of the lean years to follow. Radio engineering texts devote entire chapters to the subject. Gamblers soon devised at least two different ways of damping the random fluctuations in the rate at which bankrolls appreciate under the management by the Kelly rule. For one thing, they discovered that "half Kelly"—which directs the bettor to risk only half the Kelly bet on each turn of the wheel—is a highly effective variant of Kelly's original rule, since it reduces the appreciation rate by only 25 percent while drastically reducing the volatility of the returns. For another, they learned to work in teams. By dividing a common bankroll evenly both before and after each trip to the casino floor, and by using a common (fractional Kelly) strategy against a number of different (but physically identical) wheels of fortune while there, they could smooth their Kelly winnings substantially.

By adapting both techniques to investing, Thorp was able to keep the assets of the Princeton-Newport Partners on the enviable growth curve shown in figure 9.7. A few other legendary investors, including George Soros and Warren Buffett, recorded more rapid appreciation during those years. But—again according to Poundstone—neither did so as steadily as Thorp. The standard deviation of PNP's return was only 4 percent, making it less volatile than the market as a whole. While the S&P 500 lost nearly a quarter of its value during the bear market of 1974, and nearly 29 percent on Black Monday (October 19,

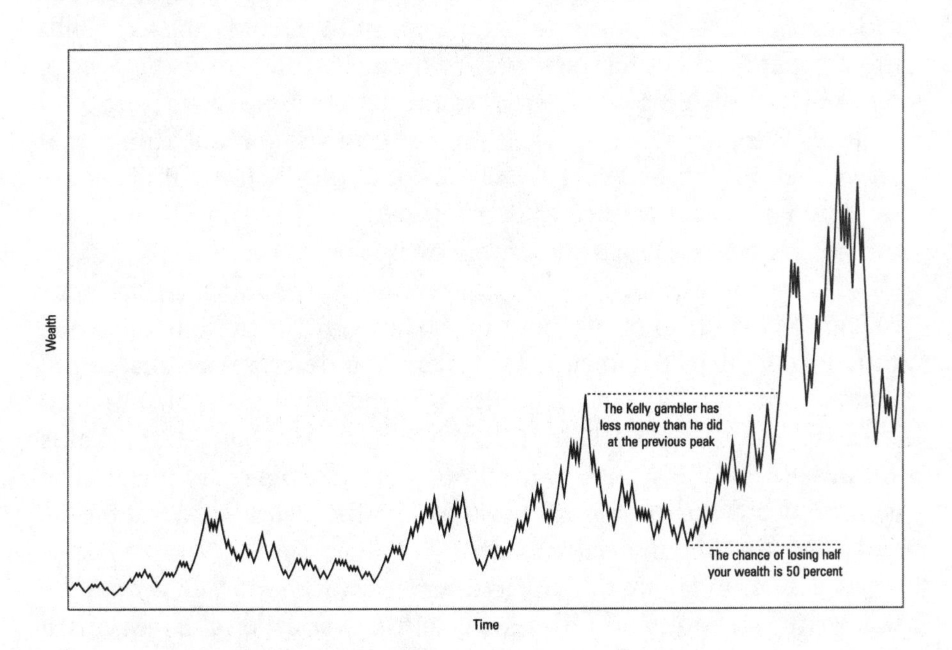

Figure 9.8. The erratic rate at which $1.00 grows when invested according to the unmodified Kelly rule means that the investor's account is usually worth a good deal less than it originally was.

1987) alone, PNP suffered no such reverses. Indeed, PNP broke roughly even for the month of October 1987 and went on to gain an astonishing 34 percent for the year.

The dispute between advocates of the Kelly rule and the (vastly) more numerous champions of the efficient market hypothesis/coin-tossing random walk approach to money management would seem to have all the earmarks of a genuine scientific debate. Both sides formulate and test hypotheses consistent with their underlying theories. But no such debate has ever really gotten under way. Poundstone observes that the efficient market hypothesis/coin tossers have little incentive to participate in such a debate, since they already enjoy the fruits of victory.

They do so because they control the relevant media. Economic ideas are mainly disseminated through the economic literature, which consists primarily of scholarly journals edited by prominent economists. Papers submitted to those journals by non-economists—such as the diverse crew of mathematicians, statisticians, professional investors, and information theorists familiar with the merits of Kelly's rule—have a way of being rejected. Investment manager Jarrod Wilcox related to Poundstone, "At one point I was so daring as to submit a paper [explaining the merits of the Kelly rule] to *The Journal of Finance*. The review said, 'This contradicts everything we've learned in finance.' Well, it really doesn't. But it contradicts so many things that are so well established that the claws come out."[18] Journal pages are too often regarded as turf, to be protected from hostile theory, contrary evidence, and other uninvited intrusions.

As a result, few business schools even bother to include the Kelly rule in the MBA programs they offer to aspiring portfolio managers, a situation the mathematician Nils Hakansson describes as "shameful." Poundstone quotes William Miller, who has enjoyed such unparalleled success using the Kelly rule to manage his Legg Mason Value Trust, as saying, "My guess is most portfolio managers are unaware of it, since it did not arise from the classic work of Markowitz, Sharpe, and others in the financial field."[19]

When Poundstone told the mathematician Thomas Cover that he was writing a book about the Kelly rule, Cover called it a story with everything but an ending. Like many in the Kelly camp, he won't consider the question settled until leading economists recant their erroneous testimony. Wilcox describes the stalemate between the two schools of thought as follows: "You've heard of Kuhn's paradigm shift? That's what's going on here." As he sees it, "Until you get one of the leading lights at MIT or Stanford to endorse it, you're not going to have the paradigm shift."[20] He neglects to explain how that might come to pass.

CHAPTER 10

Orthodox Economic Thought

Americans are daily bombarded with the latest statistics concerning "real" economic growth; "core" currency inflation; "seasonally adjusted" unemployment rates; "bellwether" interest rates; federal budget deficits; foreign trade deficits; tax revenues; corporate profits; stock market indexes; energy prices; consumer, producer, and import price indexes; hourly wage rates; automobile sales; housing starts; median incomes; farm incomes; poverty levels; environmental degradation; productivity; competitiveness; consumer confidence; and more. Readers of the financial press are exposed to more detailed versions, broken down in telling ways by talking heads offering insights into the meaning of it all.

The perceived importance of such statistics is due, in large part, to the influence of John Maynard Keynes. Coming to prominence in the aftermath of World War I by writing voluminously and (often) lucidly about mass unemployment during the era in which it first came to be regarded as a treatable condition, Keynes changed forever the public view of economic policy. Though fully conversant with orthodox economic thought—a subject he had mastered in a mere eighteen months of study at Cambridge University—Keynes was unable to reconcile it with anything like a plausible plan of attack on unemployment. Even with 20 percent of the workforce idle in both Germany and the United States, most mainstream economists opposed government employment programs on the ground that any such interference with the hallowed "market mechanism" must inevitably do more harm than good. Keynes soon came to regard such talk, along with the theory behind it, as arrant nonsense.

Taking the side of those in every European parliament who favored governmental intervention in the form of massive public works—financed, if necessary, by deficit government spending—Keynes railed against the Oxbridge intelligentsia who (like the Hoover administration) proposed to take no action to alleviate the suffering they saw around them. To advertise the benefits of government spending during an economic slowdown, Keynes identified the so-called multiplier effect. It is among the few of his contributions to the subject embraced even by the many who would (if they could) expunge his every thought and deed from the memory of man.

The Penguin Dictionary of Economics defines "the multiplier" to be the increase in national income caused by a onetime infusion of cash into the economy, divided by the amount of that infusion.[1] The resulting ratio varies with the (proper) fraction f of each windfall dollar allocated by a typical citizen to immediate consumption. Keynes described f as society's "marginal propensity to consume." If the government were, for instance, to accept—suddenly and without prior notification—responsibility for damages of a certain sort incurred in the remote past, and were it to elect in the spirit of that decision to honor some tribe's long-ignored claims for such damages, to the tune of $1,000 per claimant, the money would come as a windfall to the surprised recipients. Each will be forced to decide how much of every dollar to spend and how much to save. The unsaved portion will be passed along to other families like themselves, who will spend a like fraction of what they in turn receive. And so on. If each recipient spends the same proper fraction f of his or her windfall, the ultimate boost to the economy resulting from the government's infusion of $1,000 into it may be calculated in the following manner:

Round	**Expenditure**
0	$\$1{,}000$
1	$\$1{,}000 \times f$
2	$(\$1{,}000 \times f) \times f$
3	$(\$1{,}000 \times f \times f) \times f$
⋮	⋮
N	$(\$1{,}000 \times f^{N-1}) \times f$

As a result, the total expenditure during the first N rounds is \$1,000 multiplied by $(1 + f + f^2 + f^3 + \ldots + f^N)$. Having devoted most of his college years to the study of mathematics, Keynes immediately recognized the terms of the sum in parentheses to be those of a "geometric sequence." The indicated sum therefore equals $(1 - f^{N+1}) \div (1 - f)$ for every value of N, and—because f and all its powers are *proper* fractions—the total cannot exceed $1 \div (1 - f)$. Finally, the amount by which the latter exceeds the former must be very small if N is at all large. So, even if the process should halt prematurely, the total will not differ significantly from \$1,000 $\div (1 - f)$.

Taken literally, the foregoing computation would suggest that every dollar infused into the economy of a nation in which everyone saves 10 percent of his or her income—so that $f = 90$ percent—will increase the national income by roughly \$10. Or, if the same dollar were infused into an economy in which everyone saves only 5 percent of his or her income—so that f equals 95 percent—then every dollar injected into the economy would increase national income by roughly \$20. And so on. In reality, such calculations have to be taken with a healthy grain of salt, since there are many in every nation who spend their paychecks before they earn them and a frugal few who save at least half of all they earn.

Keynes meant his argument to show how the response to government spending can dwarf the infusion of funds required to trigger it. Economic stabilization is both possible *and affordable*. Today, all developed nations have economic stabilization policies, formulated in the main by quasi-independent central banks such as the Federal Reserve that consult all manner of data in the process of deciding when to raise or lower the key interest rates under their control. Only when economic growth begins to falter must legislatures step in with economic stimulus packages—the design of which also requires large quantities of data—calculated to forestall recession. In theory, the debt incurred to stimulate a flagging economy can easily be repaid when the threat of recession has passed. In practice, repayment is almost never made. Still, few economists of any persuasion would dispute the wisdom of stabilizing economic growth rates.

In January 1995, the U.S. Senate debated a constitutional amendment requiring the federal government to balance its budget annually. Such an amendment would have defeated all future attempts to stabilize the "real" rate of economic growth. Seeking to explain to his colleagues the gravity of the step they were about to take—one that the House of Representatives had already voted to send to the states for ratification—the late Daniel Patrick Moynihan (D.-N.Y.) took daily to the Senate floor a copy of the graph shown in figure 10.1. After about 1950, as the value of economic stabilization became clear to the Truman and Eisenhower administrations, the nation's GDP—which had frequently grown at rates in excess of 10 percent in the past and had fallen to or below −10 percent on several occasions—all but ceased to deviate from the range between 0 and 5 percent. Gone were the periods of national euphoria that came and went with growth spurts in excess of 10 percent. But gone, too, were the periods of widespread pain and suffering that necessarily accompany severe recessions. It is easy to forget how devastating the loss of a job can be and that, in a workforce of 130 million, the unemployment rate need decrease by only 0.1 percent to put 130,000 people out of work. That's enough breadwinners to populate a city of 300,000.

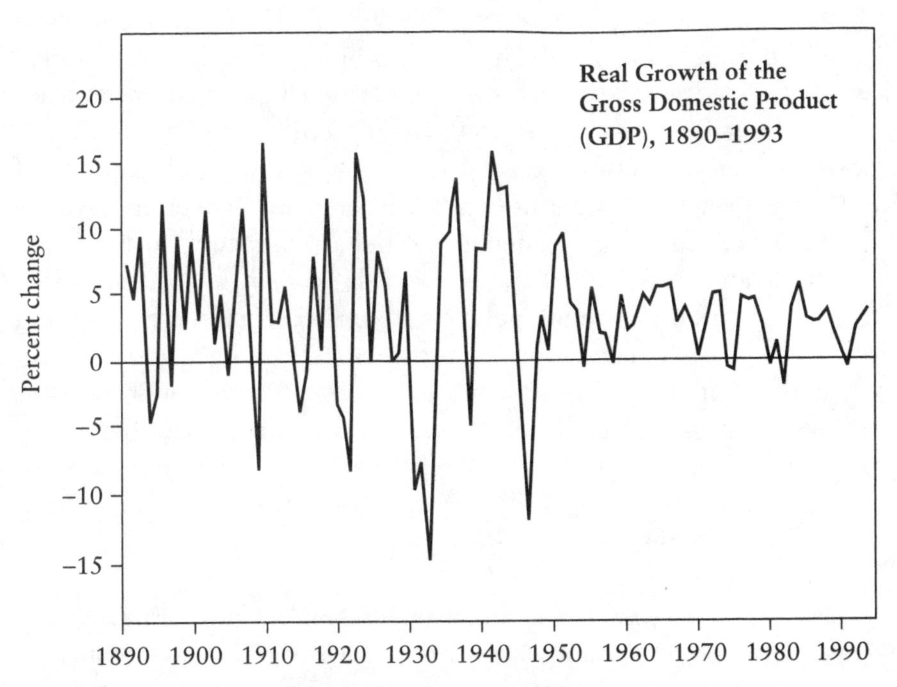

Figure 10.1. The violent fluctuations in the rate of growth of U.S. GDP have largely been brought under control since the early 1950s, when the Truman and Eisenhower administrations began to take Keynesian policy prescriptions seriously.

The availability of accurate, abundant, and up-to-date economic data has revolutionized the public policy process, both here and abroad. Capitols abound with technocrats seeking—with limited success—to craft legislative reforms that will combine all or most of the intended effects with a bare minimum of unforeseen consequences. As late as 1930, with the United States sinking into the depths of the Great Depression, the most up-to-date statistics on national unemployment were the ones collected for the 1920 census.

Not until the end of World War II did the Commerce Department begin—under Keynes's considerable influence—to erect National Income and Product Accounts (NIPAs) on the template fashioned by Simon Kuznets (1901–85), a Russian-born but U.S.-trained economist. Kuznets's book on the subject was published in 1941, and went on to become a practitioner's bible for the postwar years.[2] Having compiled his own National Product Accounts for the years 1934, 1941, and 1946, Kuznets possessed the practical experience needed to guide the Commerce Department in its development of a rather more comprehensive system.

Kuznets also wrote a book on GNP, in which he proposed to standardize the methods by which the Commerce Department obtained its official estimates of that important indicator.[3] Standardization made it possible to compare GNPs from successive years, as must be done to obtain meaningful estimates of the nation's unsteady rate of economic growth. Later, although the differences between the two indicators are of little consequence, the department took to reporting GDP instead of GNP. In 1971, the Swedish Academy of Sciences cited Kuznets's work on national income and product accounting—among other things—when awarding him the Nobel Prize in Economics.

Early champions of mathematical economics were regularly asked where they proposed to obtain the numerical data needed to confirm their theories and implement the resulting policy recommendations. The answer supplied by William Stanley Jevons in 1871 has been quoted many times:

> I answer that my numerical data are more abundant and precise than those possessed by any other science . . . There is not a clerk nor bookkeeper in the country who is not engaged in recording numerical facts for the economist. The private account books, the great

> ledgers of merchants and bankers . . . are full of the kind of numerical data required to render economics an exact mathematical science . . . It is partly the very extent and complexity of the information that deters us from its proper use. But it is chiefly a want of method and completeness in the vast mass of information which prevents our employing it in the scientific investigation of the natural laws of Economics.[4]

Kuznets provided the method missing in Jevons's day, while the Commerce Department dealt with the completeness issue by promulgating rules obliging every active firm to comply with its requests for information.

Among the hungriest consumers of economic data are the giant computer models now in use. A special report by the staff of *The Economist* magazine described the emergence of such models, which are of two main kinds.[5] Macroeconomic models reside mainly in the financial community, where they focus on (relatively minor) fluctuations in economic activity and the design of monetary policies to dampen them. The other species of high-profile models, known as computable general equilibrium (CGE) models, tends to overlook minor fluctuations in order to concentrate on the long-term repercussions of major policy changes, such as international trade negotiations or the proposed elimination of the federal inheritance tax.

The special report begins by asserting that "for each of the big questions facing the world (What do we stand to gain from a global trade deal? By how much has expensive oil retarded growth? What might be the economic costs of an avian flu pandemic?) there is a model that will provide a big numerical answer ($520 billion, 2.5% of world GDP, and $4.4 trillion, respectively)." Such estimates are ubiquitous in today's public policy debates, despite their legendary inaccuracy.

Large-scale economic models came of age during the battles over the North American Free Trade Agreement (NAFTA). While opponents of the 1994 treaty had the catchiest lines—Ross Perot talked of "a giant sucking sound" as U.S. jobs moved south of the border—advocates of free trade furnished the computer-generated numbers that carried the day. The most persuasive computer-generated figures concerned the jobs such a treaty would create in the United States. President George H. W. Bush predicted an immediate gain of 200,000 such jobs, a claim later amplified by the Clinton administration.[6] Twelve years in, according to the special report, "economists have shown little inclination to go back and check" the key figures for accuracy. The notable exception was Timothy Kehoe, an economist at the University of Minnesota, who wrote a paper in 2005 finding fault with most if not all of the publicized predictions.

The Economist also credits "dubious computations" with helping to bring "a belated end" to the Uruguay Round of global trade negotiations, which also ended in 1994. Peter Sutherland, head of the General Agreement on Tariffs and Trade (GATT)—the predecessor of the current World Trade Organization (WTO)—reportedly urged the negotiators to close the deal on the grounds that failure to do so would cost the world as much as $500 billion a year in unrealized "gains from trade." Even confirmed free traders described these claims as "extravagant" and "overblown." Yet they escaped scrutiny because, in the opinion of at least one professional, they emanated from "gigantic" models, opaque even to experts. "Given the appetite of the press and politicians for numerical estimates," he concluded, along with the publicity they bring to researchers, "these models are here to stay."

The appetite for such figures was undiminished at the beginning of the succeeding round of international trade negotiations, launched at Doha, Qatar, in the autumn of 2001. The World Bank was predicting at the time that an ambitious agreement could raise global incomes by as much as $520 billion while lifting as many as 144 million people out of poverty by 2015. "Those figures," the special report observes, found their way into "almost every news report about the Doha round that autumn." Subsequently, as it became apparent that any agreement reached by the Doha negotiators would be less than ambitious, the bank cut its projections. A 2005 publication mentioned a $95 billion increase in global incomes and proposed to release a mere 6.2 million people from poverty.

Macroeconomic models incorporate comparatively little of economic theory. They consist mainly of equations asserting stable and enduring relationships between two or more economic time series, such as average hourly wages and per capita consumption. It is not enough to say that wage increases are ordinarily reflected in increased consumption. Model builders must stipulate—at least implicitly—how much a fifty-cent hourly wage increase is likely to stimulate public and private consumption. Several financial firms now offer website access to state-of-the-art macroeconomic models. Simply by clicking on the indicated icons and filling in a few blanks, visitors can estimate the effects of fiscal and monetary policy reform.

An article describing the Moody's model discusses several such relationships, together with the data from which they were inferred.[7] It reveals, for instance, that total light-vehicle sales are directly related to real net worth per household, and inversely related to the unemployment rate, the three-month Treasury yield, the current rate of personal bankruptcies per household, and the relative value of new and used cars at prevailing prices. It further stipulates that the relationship was inferred from quarterly data extending from the fourth quarter of 1979 to the second quarter of 2003 and furnishes assorted estimates of statistical significance.

Elsewhere in the article, it is pointed out that a certain rather complicated numerical measure of labor compensation per hour is directly proportional to recent changes in the effect of inflation on personal consumption expenditures, and inversely proportional to the amount by which actual employment currently differs from the so-called non-accelerating inflation rate of unemployment (NAIRU).[8] Because the article describes only a small fraction of the equations internal to the Moody's model, the risk that a rival will be able to reconstitute the model from its Web page description is vanishingly small.

Established economic theory does not stipulate that total light-vehicle sales must depend on the yield of three-month Treasury notes, rather than that of short-term CDs, nor insist that such sales be influenced by personal bankruptcies per household. The asserted influences are little more than happenstance, confirmed by statistical inference. Unchallenged theory insists only that every economic variable has the potential to influence every other—either directly or indirectly—although the bulk of the influences are imperceptibly small. This fundamental insight is usually attributed to Léon Walras (1834–1910), a mining engineer by training who held the economics chair in the Faculty of Law at Lausanne between 1870 and 1892. His insight allegedly moved a colleague to quote the poetry of one Francis Thompson: "Thou canst not stir a flower / Without troubling a star."[9]

Walras popularized, at least among economists, the notion of an economy in which the supply of every traded good or service is precisely equal to the demand for it at prevailing prices, while the income at said prices of every firm or household is equal to its cash outlays. Today, such an economy is said to be in general equilibrium, as opposed to those in partial equilibrium, where only a few markets are required to clear and only a few sets of books are required to balance. Walras developed his ideas in three volumes published between 1874 and 1877 describing (among other things) a system of equations capable of being solved in two stages, first for the equilibrium price in gold of every other traded good or service, and then for the quantities destined to change hands at those distinguished prices. Not surprisingly, his achievement spawned little immediate activity.

Only after World War II did the availability of electronic computers stir renewed interest in solving equations as numerous and complex as the ones Walras had in mind. The difference between macroeconomic and CGE models lies mainly in the extent to which the two are required to be consistent with basic economic theory. Generally speaking, the builders of CGE models are significantly more constrained as to the form and content of the equations their models may incorporate. Yet even they enjoy considerable latitude.

Ross McKitrick, an economist at the University of Guelph in Canada, ran two simulations of the Canadian economy's response to a 10 percent increase in the "sales tax" imposed on services.[10] Both runs employed the same CGE model and used the same numerical data. Even though the two differed only in the algebraic form in which they expressed the relevant laws of supply and demand, one simulation predicted that the proposed tax increase would permit government spending to grow by 60 percent, while the other showed a gain of only 14 percent. Clearly the former was the more tempting to tax-and-spend Canadian politicians, but that doesn't mean it was the more accurate estimate. It is a mistake to lose sight of the oldest (and truest) adage in all of computerdom: "Garbage in, garbage out." If the inputs to a program are worthless, the output will be, too. This applies as well to the commands the machine is expected to execute as to the input data.

All this points to an important, though impolitic, question: What exactly does orthodox economic theory have to say about supply, demand, and competitive markets, and how credible are its dicta? Is it safe to incorporate the central tenets of orthodox theory into computer programs likely to affect the fate of nations, or is it possible to identify particular teachings that merit exclusion? Prudence suggests a gradual approach to this, as to any other emotionally charged issue—an approach that begins with the simple and uncontroversial.[11] Because the simplest and least controversial portion of economic theory is that having to do with monopoly, the accepted theory of monopoly conduct will be examined here in unusual detail.

Although Adam Smith published his *Inquiry into the Nature and Causes of the Wealth of Nations* in 1776—the same year Great Britain's American colonies declared their independence—and although he wrote at length about the iniquities of monopoly, another sixty-two years were to pass before anyone explained the process by which monopolists presumably choose their prices. This was accomplished by a then prominent French mathematician named Antoine-Augustin Cournot (1801–77), whose stated purpose was to demonstrate that mathematical techniques can be as useful in economics as they had already become in other branches of science. He began by plotting the monopolist's profit against her price and observing that the resulting curve would ordinarily exhibit a single peak, as in figure 10.2. A secure monopolist will presumably maximize her profit by setting her price equal to the horizontal coordinate (here labeled "Price*") of the highest point on the price-profit curve. Simple as that may seem, it was apparently left unsaid before Cournot. Earlier writers tend to give the impression that secure monopolists are unable to price themselves out of the markets they dominate. Nothing could be further from the truth.

The profit $\Pi(P)$ to be earned by a monopolist offering her wares at P dollars a unit is obtained by subtracting the unit cost C of a single item from P and multiplying the result by the quantity $Q(P)$ the market will accept at that price. Reduced to symbols, the formula reads: $(P - C) \times Q(P) = \Pi(P)$. The horizontal line in the figure—the one labeled "minimally acceptable profit"—is meant to remind us that even a monopolist needs to recover certain "sunk costs," such as monthly rent and utility bills, which remain fixed regardless of how much or how little the monopolized market absorbs. The minimal acceptable profit may be either very low or very high. Should it come to exceed

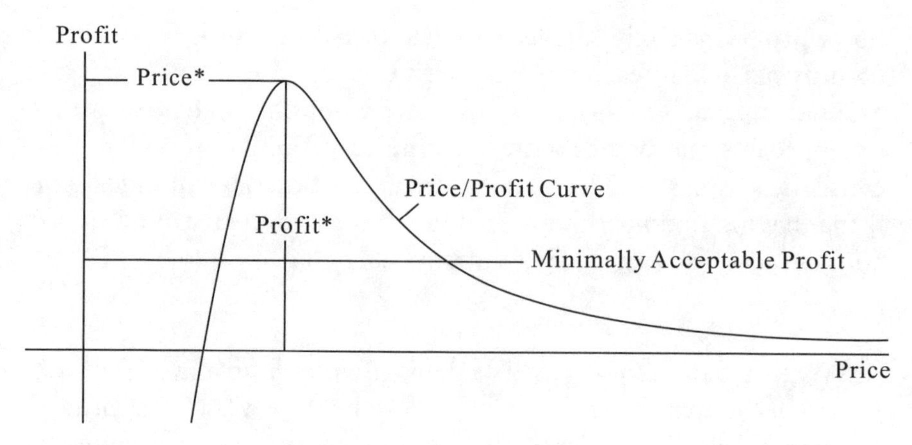

Figure 10.2. A secure monopolist is free to charge the profit-maximizing price.

the maximum obtainable profit, even a monopolist will be forced to abandon the market she dominates. Profitable shops sometimes close their doors because the owner of the building in which they are located decides to increase the rent.

By plotting potential prices P against the quantities $Q(P)$ destined to be purchased at those prices, one obtains the so-called market-demand curve facing the monopolist in question. Although there can be exceptions, such curves typically slope downward to the right, as in figure 10.3, which depicts the (smooth) demand curve used to construct the profit curve in figure 10.2. The actual demand curve is the one that looks like a staircase, the smooth version being no more than a convenient approximation. The fact that a typical demand staircase would have hundreds if not thousands of tiny steps—had anyone the time, energy, and data to draw them all—makes it advisable to think in terms of smooth approximations. The horizontal distances from the vertical axis to the individual "risers" in the staircase are indicated by ticks on the "Price" axis.[12] They correspond to "balk prices" separating the amounts a specific consumer will pay for the item in question from the ones he or she won't.

Cournot's analysis of monopoly prices bears repetition, if only because it is the closest economic theory comes to an undisputable truth. Yet many monopolists have good and sufficient reason to charge more than the profit-maximizing price. Châteaus in Burgundy, for instance, are obliged to do so because they cannot increase the production of their unique and inimitable wines. A fashionable portraitist may inflate his prices merely to gain leisure time. On the other hand, the manufacturer of a particularly "hot" consumer item may choose to sell it for less than the profit-maximizing price to discourage potential competitors. And so on. It is no accident that responsible economic policy analyses are as full of whys and wherefores as any legal brief.

It should never be forgotten that behind every market-demand curve, there always lurks a population of balk prices containing a largest and a smallest member. Such populations may be approximately normal, as discussed in Chapter 4, or they may be far more complex.

Monopolies are not the only markets devoid of competition. The ones in which many sellers service a single buyer are known as monopsonies. As a monopolist offers to sell as much as anyone wants at her price, but nothing at all at any lower price, so a monopsonist offers to buy as much as anyone can deliver at her price, but nothing at all at any higher price. The response of such a market to all potential prices P is traditionally displayed in the form of a supply curve, sloping from lower left to upper right. As with demand curves, real supply curves look like staircases, in which the distances from the vertical axis to the various risers correspond to the "must-have prices" that sepa-

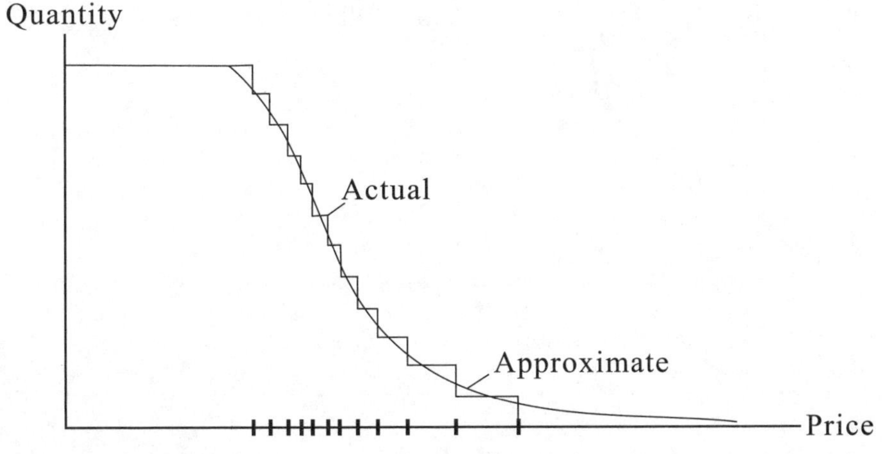

Figure 10.3. Price-profit curves are easily deduced from (smooth approximations of) market demand curves.

rate the prices a particular seller can or will accept from those he or she can or will not. Supply curves, too, are best replaced—for most practical purposes—by smooth upwardly sloping approximations.

If a monopsonist buys at P dollars a unit to resell at M dollars, her profit $\Pi(P)$ can be calculated by multiplying the quantity $Q(P)$ delivered at the market at price P times the difference between M and P. Symbolically: $(M - P) \times Q(P) = \Pi(P)$. The result is another single-peaked price-profit curve, such as the one in figure 10.4, from which the optimal price and optimal profit—again denoted "Price*" and "Profit*"—can be obtained as the horizontal and vertical coordinates of the highest point on the curve.

In yet other markets, buyers and sellers—though numerous—interact only through a middleman who behaves as a monopolist toward the (retail) buyers and as a monopsonist toward the (wholesale) sellers. That is, he offers to buy all the sellers can deliver at a wholesale price p of his choosing and to sell all anyone wants at a retail price P, also of his choosing. The trick a middleman must master is to coordinate p with P in such a way that the quantity of goods flowing in at p is equal to the quantity flowing out at P. A low unit wholesale price p will draw forth only a small supply, which may then be disposed of at a high unit price P.

In contrast, an excessively high wholesale price p may call forth a larger supply than can profitably be disposed of. Between the two extremes must lie a happy medium at which the middleman's profit is maximized. To realize the maximum profit available to him, a middle-

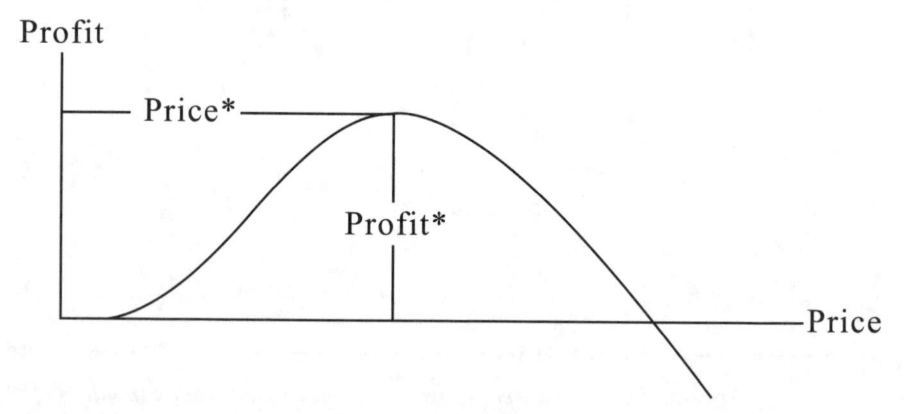

Figure 10.4. A secure monopsonist's profit-maximizing price can likewise be determined from the coordinates of the highest point on a price-profit curve.

man must begin by being able to predict the retail price $P(p)$ needed to equalize the quantities supplied to and demanded of him by the distinct markets in which he buys and sells. His profit Π can then be calculated by multiplying the difference between $P(p)$ and p by the quantity $Q(p)$ both supplied to and demanded of him when his acquisition and disposal prices are p and $P(p)$, respectively. In symbols: $\Pi(p) = (P(p) - p) \times Q(p)$. A possible middleman's price-profit curve appears in figure 10.5. The one shown is again single peaked, although there is no necessity that this be so. If such a price-profit curve should exhibit multiple peaks, ignore all save the highest.

Yet to be addressed is the fact that few firms sell only a single product. Most have one or more complete lines of products. By 1920, the Model T came in open-cab touring car, two-door and four-door closed-cab town car, and light-truck versions. Most brewers now market premium, superpremium, and more affordable "popular-price" brands. Men's shorts come in boxer or brief styles, as well as quarter- and mid-thigh-length versions. Auto rental agencies offer subcompact, compact, midsize, full-size, and (in the larger markets) luxury models. And so on. Sellers ordinarily have not one but several prices to choose, and a need to choose them in a way that minimizes the interference between brands belonging to the same product line. If Buicks didn't cost a lot more than Chevys, and Cadillacs didn't cost a lot more than Pontiacs,

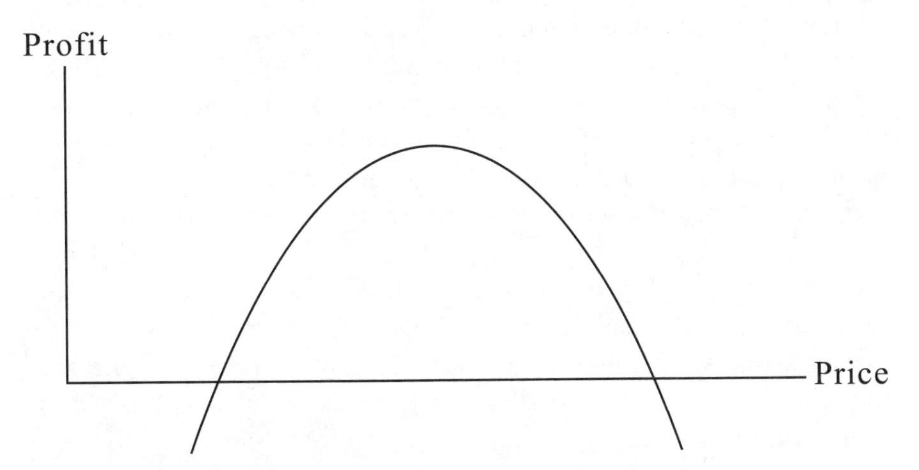

Figure 10.5. A secure middleman, dealing with customers as a monopolist and with suppliers as a monopsonist, can also determine his profit-maximizing price from the coordinates of the highest point on a price-profit curve.

the Buick and Cadillac divisions of General Motors would interfere with the Chevrolet and Pontiac divisions. And so on.

A computer program to calculate a profit-maximizing price list for the XYZ monopoly would need to contain a subprogram for computing the profit to be expected from a given list of prices, which would itself need to contain subprograms for predicting the expected sales volume for each product in each product line corresponding to the given price list. The master program would then search the universe of potential price lists for the one that maximizes total profit. Although computers have, over the years, become remarkably adept at solving maximization problems, it still seems to require as much art as science to solve the practical ones.

In reality, such programs are seldom written, because firms seldom know how to construct the lowest-level subprograms—the ones that decide how much per day or per week of each individual product the market will absorb, on average, during the time a given price list is likely to remain in force. Most firms still rely on shortcuts and rules of thumb invented before the advent of computers. Automobile manufacturers, for instance, typically quote a base price for the stripped-down version of a given model and add to it the price of any extras—such as air-conditioning, power steering, antilock brakes, tinted windows, or deluxe sound—with which a vehicle can be equipped. Not surprisingly, they tend to do it in such a way that a "loaded" subcompact costs about as much as a stripped-down compact, while a loaded compact costs about as much as a stripped-down midsize car, and so on. Brewers do much the same thing, holding the price ratios between their flagship and various ancillary brands roughly constant, so that the everyday (as opposed to the discounted) prices of their entire line of products tend to rise in unison, by a common proportion.

Difficult as it is to construct a computer program capable of calculating a profit-maximizing price list for either a monopolist or a monopsonist, it is an order of magnitude more difficult to do the same thing for firms serving competitive markets. Another important question that seems difficult to approach, save as it applies to a monopolist, has to do with market penetration. Figure 10.6 shows how monopoly profits typically vary with unit costs of production. When one uses primitive techniques, production is expensive, making it impossible to penetrate a market deeply, even if selling at cost. In such circumstances, large profits are unobtainable, and even a profit-maximizing monopolist is best advised to price her wares only slightly above cost. But as technology improves, real unit production costs decrease—meaning that the short vertical line in the figure, marking the lowest currently achievable unit production cost, moves leftward—permitting additional customers to be served. A profit-maximizing monopolist will increase her profit more quickly than her sales volumes by passing most but not quite all of her cost savings on to her customers. While some of the remainder is hers to keep, the rest must be reinvested in improved plant and equipment. Without such reinvestment, not even a royally chartered monopoly could long persist.

Around the middle of the eighteenth century, after millennia of barely perceptible change, costs of production in the shoe industry, the textile industry, the mining industry, the smelting industry, the hardware industry, the glassware industry, the transportation industry, and any number of other industries began to decline so precipitously that even ordinary folk became aware that a revolution was under way. Today, historians call it the Industrial Revolution, and it revolutionized corporate profits as much as anything else.

The reader should by now be convinced—in case he or she needed convincing—that some parts of economic theory may confidently be incorporated into models intended to influence the course of human events. Indeed, if there is anything dubious about the foregoing analyses, it is the assumption that monopolists, monopsonists, and exclusive middlemen are driven to maximize profits. In truth, a good many

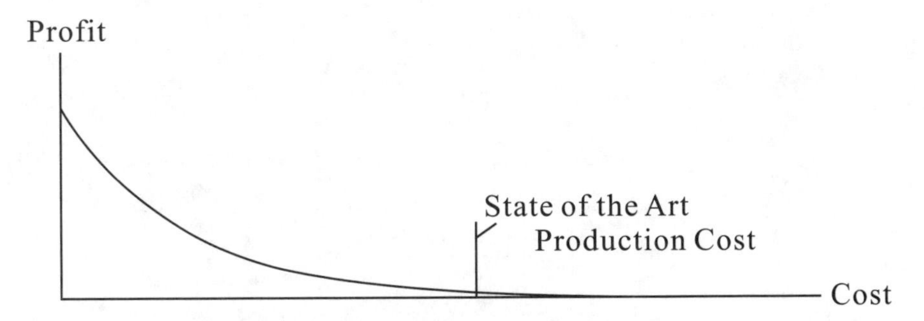

Figure 10.6. When production costs are high, due to primitive technology, a monopolist's profit-maximizing price exceeds her production cost by very little. But, as improved technology reduces cost, profits and markups rise.

of them may be content to earn a comfortable return on investment that is substantially less than they might earn. Such behavior is known as "satisficing," and its occurrence in the animal kingdom is well documented. But it is probably rare among publicly traded corporations, because firms that earn less than they might attract the interest of hostile-takeover specialists. In any case, because monopolies, monopsonies, and unopposed middlemen account for only a small part of modern economic activity, it is important to know whether there is a comparably trustworthy theory of competitive markets that can be incorporated into future generations of economic models. That and related questions will be addressed in Chapter 11.

CHAPTER 11

Economic Competition

What quickly became the orthodox theory of free-market capitalism was formulated, around 1900, by members of the Cambridge school of economic thought. Their unchallenged leader was Alfred Marshall (1842–1924), whose *Principles of Economics* appeared in 1890. The book ran through many editions, the last appearing in the year of his death. Marshall's *Principles* remained the standard introduction to the subject until Paul Samuelson's *Economics* appeared in 1948. That, too, has gone through many editions, though it now has a host of competitors. All describe what has come to be known as the neoclassical model of commercial interaction, the active ingredient of which is a fiction known as perfect competition. In recent years, a few writers have begun to speak of a perfectly competitive model (PCM) differing from its neoclassical progenitor only in the explicit assumption that both capital and labor are free to move across national borders. Prior generations had tacitly assumed them fixed.

The fullest accounts of the neoclassical model are found in books on price theory, value theory, and microeconomics, which purport to articulate the supposedly immutable laws according to which—absent undue government interference—competitive free markets may continue in perpetuity to reconcile the (presumably limitless) wants of individual households (consumers) with the (plainly limited) ability of existing firms (producers) to satisfy those wants. Virtually everyone has experienced the process, and most are vaguely aware of the laws of supply and demand, especially as they apply to the price of gasoline at the pump.

Although the previously encountered Cournot was the first to draw supply and demand curves, it was Marshall who made them famous. A typical pair is shown in figure 11.1, together with a third (darkened) curve called the transaction quantity curve. The latter follows the supply curve up to its point of intersection with the demand curve, and descends with the latter to the right of that point, on the theory that the quantity $Q(P)$ exchanged at a given price P can exceed neither the quantity $S(P)$ suppliers are prepared to supply at that price nor the quantity $D(P)$ buyers demand at that price. In symbols: $Q(P) = \text{MIN}\ [S(P), D(P)]$. A competitive free market will presumably settle on the price P^* at which $S(P) = D(P)$, since dissatisfied buyers can be counted upon to bid up any lower price, while suppliers with unsold goods can be expected to bid down any higher one. An equilibrium of sorts is established as the prevailing price P gravitates toward the "market-clearing" price P^*, so called because it is the only price at which there are neither unsold goods nor unsatisfied demand. There is genuine empirical evidence that competitive free markets do, under suitable conditions, tend to an equilibrium of roughly this sort. Notice next that the point at which the supply and demand curves intersect is—of necessity—the highest one on the transaction quantity curve.

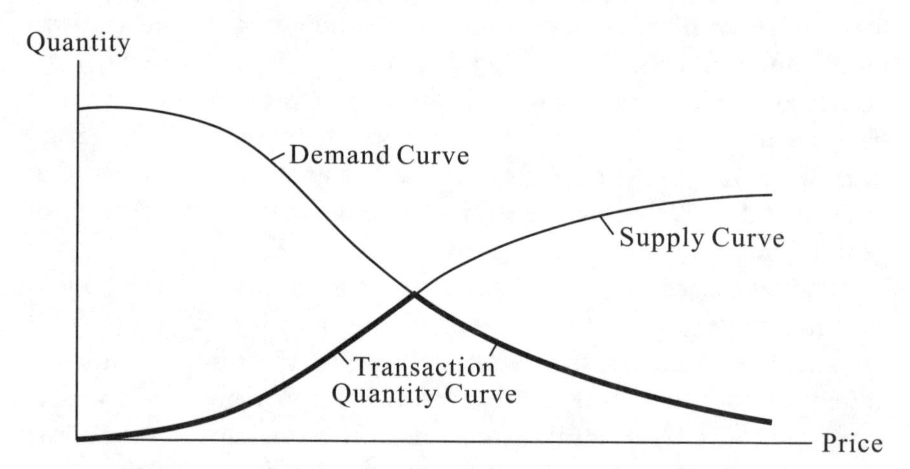

Figure 11.1. The coordinates of the point at which a supply curve crosses a demand curve indicate the market-clearing price and volume. Because supply curves rise and demand curves fall with increasing price, the point of intersection is the highest one mutually acceptable to both buyers and sellers.

As a result, single-product markets exhibit a strong tendency to transfer the largest possible quantity of goods from buyers to sellers. This is an early indication of the following:

> *Efficiency Principle*: Under *perfect competition*, and with no market failures, free markets will squeeze as many useful goods and services out of the available resources as possible.[1]

Usually attributed to Adam Smith, this oft-invoked principle encapsulates much of conventional economic wisdom. It asserts that a purely voluntary price system—ever and always the active ingredient of "competitive" free markets—may be trusted to coax from the farms and factories, harbors and highways, Indians, chiefs, and other essential ingredients of a full-service economy *maximal output at minimal prices*. Furthermore, anything that interferes with the price system's ability to do so is, almost by definition, a detriment to social well-being.

To distinguish Smith's efficiency principle from other economic efficiency principles—such as the efficient market hypothesis said to apply to financial markets, or the sort of efficiency attributed to auction formats designed to transfer each property to the bidder who values it most—this one is ordinarily described as the principle of productive efficiency. To assert, as is sometimes done, that this or that nation's economy operated at 96 percent efficiency during a given five-year period is to say that its economy produced 96 percent of the "useful goods and services" it was capable of producing during those years.

The "market failures" mentioned in the foregoing formulation of Smith's efficiency principle require explanation. A butcher, baker, or candlestick maker must charge enough for his wares to support himself, his wife, and his children in relative comfort. To that end, the range of potential prices may be separated into those that are and those that are not acceptable to him. Orthodox economic theory declares the competition in his industry to be less than perfect if it allows suppliers to charge more than they absolutely must. Perfect competition is price competition so intense that it compels producers to sell what they do as cheaply as they collectively can. The profit earned by a producer charging the lowest sustainably acceptable price is called a normal profit, while any higher profit is called supernormal. In antitrust matters, the persistence of supernormal profits is considered evidence of market failure.

Smith himself once charged that "people of the same trade seldom meet together . . . but the conversation ends in a conspiracy against the public, or in some diversion to raise prices." Many of the more common diversions to raise prices, such as "predatory pricing," "discriminatory pricing," and outright price-fixing, along with "attempts to monopolize" and "unfair and deceptive" advertising claims, are prohibited by U.S. antitrust law. Others, such as idle productive resources, are not illegal per se, although a properly functioning market (price system) should in theory eliminate them.

Economists describe the market failures that seem at present to trouble them most as "sticky prices." The term refers to higher-than-necessary prices that, for whatever reason, refuse to descend to their theoretically sanctioned levels. Market fundamentalists argue that available resources would soon be fully utilized, and profits reduced to normal, if prices—especially wages—would but adjust as theory predicts. Were U.S. and European workers prepared to accept the same subsistence wages their global employers pay in other parts of the world, the story goes, the seemingly perpetual shortage of work would disappear as if by magic. The new work might have to be performed in sweatshops, twelve hours a day, six or even seven days a week, but there would be no lack of it. Or of slums, no doubt, for the exhausted workers to trudge home to late at night! Less worrisome, apparently, are the sticky levels of compensation enjoyed by CEOs, CFOs, and other high-ranking corporate officials.

A recent study selects, from among the scores of explanations that have been offered for the persistence of sticky prices, an even dozen that strike the investigators as particularly plausible.[2] It then analyzes the responses to questions posed in one-on-one interviews with ranking officials at midsize to large corporations concerning their pricing practices.[3] The plan was to test the proposed explanations against the responses received from living, breathing price makers—as opposed to the passive "price takers" who populate orthodox (neoclassical) price theory—to see which if any may be said to hold water. The strangest thing about the explanations selected for testing is that they failed to include the most obvious explanation of all. The simplest way to explain why certain expectations go unrealized is to point out that they are (and probably always were) unrealistic. Such could be true of neoclassical expectations only if orthodox price theory were seriously mistaken.

Idle resources are a well-known symptom of market inefficiency.

You can't achieve maximum output without full employment of available resources. Smith himself, along with most of the other early writers on political economy, was content to assume that resources left idle in one industry would soon find employment in another, much as a butcher or a cobbler might react to shifting demand by reinventing himself as a baker or glover. The current orthodoxy errs in assuming that a failing shoe factory can as easily be turned into a profitable glove factory, office building, or entertainment complex. By so doing, orthodox writers relegate excess capacity to the status of a mere transient phenomenon—an imbalance that can arise from time to time but can never become chronic.

In a recent study of chronic excess industrial capacity—which has in fact existed in most industrialized nations since the mid-1980s—James Crotty traces the alleged reluctance of free-market economies to harbor persistently idle resources to two "hidden assumptions" of orthodox (neoclassical) theory. The first, "which seems innocuous on its face, is that per-unit production costs rise rapidly as output increases."[4] The second asserts the above-mentioned possibility of low-cost transition from less- to more-profitable industries. Crotty explains how these (rarely acknowledged) hypotheses underlie much of conventional economic wisdom, and how they long enabled experts to dismiss the threat of chronically idle resources—including human resources—as a "logical impossibility."

Among the crown jewels of mathematical economics—along with the Black-Scholes equation and one or two other such tours de force—is the proof given by Kenneth Arrow in or about 1950 that a wide array of model economies are productively efficient. Arrow later joined forces with Gérard Debreu to extend this conclusion to a far larger class of economic models. The ones they treated collaboratively are far more complex than the single-product market depicted in figure 11.1. The number of goods and services involved may be arbitrarily large, and there may be many different ways to produce the same bundle of outputs. Yet for each such model economy there is at least one (and often only one) market-clearing price list £* to be found that coaxes forth—and transfers to end users—as many useful goods and services as possible from "available resources."

Arrow and Debreu had relatively little to say about the means by which the market-clearing price list £* might be found. They seemed to think that some variant of Walras's *tâtonnement* (or groping) process would suffice. It works as follows: An auctioneer (*tâtonneur*) prominently displays a price list £ specifying a price for each and every traded good or service. Buyers and sellers must then respond with lists of their own, specifying the quantities of each good or service each one is prepared to buy or sell at the listed prices. Positive quantities represent purchases, while negative ones represent sales. After tallying the responses, the *tâtonneur* compiles a new price list £′ on which the price of each good or service is higher or lower than on the preceding list, according to whether demand for it does or does not exceed supply. In time, the process should "settle" on a market-clearing price list £*. It works very well for single-product markets of the sort depicted in figure 11.1.

Participants in the *tâtonnement* process, other than the *tâtonneur* herself, have no direct influence on the prices at which they buy and sell. Their discretion extends only to the quantities they choose to acquire or offer for sale at the posted prices. They are, in short, passive price takers rather than active price makers. Such is the nature of the "perfect competition" alluded to in the foregoing formulation of Smith's efficiency principle. The term refers to a purely imaginary form of commercial competition between passive price takers, in which prices materialize in the minds of market participants as mysteriously as the biblical "handwriting on the wall," leaving them to react as they see fit.

The Arrow-Debreu results were warmly welcomed by the economics profession, both as a demonstration of technical virtuosity and as a confirmation of the esteemed principle of productive efficiency. In fact, Arrow and Debreu proved more. In addition to the plethora of goods and services they squeeze out of available resources, they showed that competitive free-market economies must allocate those goods and services in what has come to be called a Pareto optimal manner.[5] By that is meant that any subsequent reallocation assigning a more satisfactory share of physical GDP to one household would perforce assign a less satisfactory share to another. As a result, Arrow-Debreu economies are said to be "allocatively" as well as productively efficient.

Prodigious amounts of information are required to realize the efficiencies asserted by neoclassical orthodoxy. An unflinching appraisal of the information needed was offered by Professor Frank Knight, of the University of Chicago, in January 1946. During a debate between (youthful) attackers and (older) defenders of the neoclassical faith, he stated without hesitation or qualification that economic theory requires assumptions of "rational and errorless choice, presupposing perfect foresight," and of "foreknowledge free from uncertainty." With these and little else, he continued, the essence of economic theory—predicated on the concept of perfect competition—stands complete. "There is no possibility," he concluded, "that new laws will be discovered comparable in generality and importance with the basic principles long recognized."[6] To practitioners of any other science, "perfect foresight" and "foreknowledge free from uncertainty" sound like witchcraft. To economists they sound unremarkable.

It is often alleged, though never proven, that competitive free-market economies are fair as well as efficient. A fair economy would presumably reward each household with a pro rata share of physical GDP—one proportional to that household's contribution to community GDP, regarded as a measure of the common good. But no one has ever discovered a mathematical proof—comparable to the ones for which Arrow and Debreu were awarded Nobel Prizes in Economics—that competitive market economies do any such thing.[7] Nor has anyone assembled empirical evidence to that effect. It falls, accordingly, to the yellow economic press to establish that free markets are both fair *and* efficient.

It would be hard to imagine a clearer statement of the "fairness principle" than that contained in a brief by Paul L. Poirot titled "He Gains Most Who Serves Best." It originally appeared in the May 1975 issue of *The Freeman*, a monthly newsletter edited for nearly thirty years by Poirot and published by the Foundation for Economic Education.

Poirot's brief was reprinted in a collection titled *The Morality of Capitalism*.[8] The body of it does little more than repeat the title sentence—which it does no fewer than seven times in the space of four pages—and paraphrase it as follows: "A businessman's profits are a measure of his efficiency in the use of scarce and valuable resources to satisfy the most urgent wants of consumers." Later it explains that "if, for some reason, any present owner of scarce resources loses his touch, fails to serve efficiently, the *open competition* of the ongoing market process soon will bid the property into the hands of some new owner who serves better."[9] As always, competition is relied upon to produce the sanctioned result.

As an affirmation of faith before an audience already committed to individual liberty, private property, free-market competition, and constitutionally limited government, the article serves admirably. As an exercise in persuasion, however, it leaves much to be desired. Indeed, it fails even to explain why the new owner should be one who "serves better." Why, for instance, might she not be a corporate raider intent on running the company into the ground, disposing quietly of her shares, and skipping town before the market price can adjust to the truly desperate condition of the crushingly indebted shell of a once-profitable corporation she plans to leave behind?

Much popular economic writing is of this ilk, merely restating and rephrasing dubious beliefs as if all doubt concerning them had long since been laid to rest, with supporting documentation placed on file in every accredited library. Liberal and conservative think tanks alike are guilty of such misrepresentation, although the conservative ones tend—at present—to be better funded and more widely distributed.

The symptoms of market inefficiency are legion. Inflated prices, supernormal profits, built-in obsolescence, discriminatory pricing, predatory pricing, price signaling, price-fixing, product bundling, unfair and deceptive advertising, and chronically idle resources are but a few of the most common. Who can blame Gilles Raveaud for concluding, upon learning that orthodox economic theory deals almost exclusively with efficient markets free from all such abuses, yet home to "perfect foresight" and "foreknowledge free from uncertainty," that orthodox economic theory consists of little more than "complex mathematical models that only work in conditions that don't exist"?[10] In 2000, Raveaud was one of the student leaders of a highly publicized revolt against the teaching of economics in his native France.

Although the theory surrounding figure 11.1 seems clear and concise, one is entitled to wonder how accurate it is. Figure 11.2 displays three additional curves derived from the ones in the previous figure. The quantity $E(P) = P \times D(P)$ represents the amount buyers are prepared to *expend* on a day when P is the prevailing price. Likewise, $R(P) = P \times S(P)$ represents the *revenue* suppliers may expect to realize on such a day, and $V(P) = P \times Q(P)$ denotes the dollar value of the transactions that will actually take place at price P. The price P^* for which $E(P) = R(P)$ is evidently the same as the one for which $S(P) = D(P)$.

Because the point at which the revenue and expenditure curves intersect is also the highest one on the transaction value curve, the price P^* that maximizes the quantity of goods transferred from sellers to buyers serves also to maximize the revenues realized by the sellers. There is a harmony of interests in such a market, in that the best way for the sellers to make money is to sell the consumers as many goods as they can, which requires that such goods be priced as cheaply as they can. Even if allowed to conspire, the sellers in such a market would have nothing to gain by charging more than P^*. But such harmony is unusual. In most markets, conspiracy pays rather well, which is why price-fixing is considered a restraint of trade, in violation of U.S. antitrust law.

Figure 11.3 shows the expenditure curve of figure 11.2, derived as it was from the demand curve of figure 11.1, intersected by a different revenue curve—one derived from a more abundant supply curve—as would obtain if additional suppliers were to enter the market, making more goods available at every potential price. In it the point of intersection lies well to the left of the highest point on the expenditure curve, which is also the highest point on the transaction value curve, indicating that the sellers can earn more money by charging a significantly higher price P^{**} than the market-clearing one P^* for which $S(P) = D(P)$ and $E(P) = R(P)$. They can do so because the highest point on the transaction value curve now lies well to the right of the intersection of the revenue and expenditure curves. The suppliers in such a market have every reason to conspire to raise the price to the

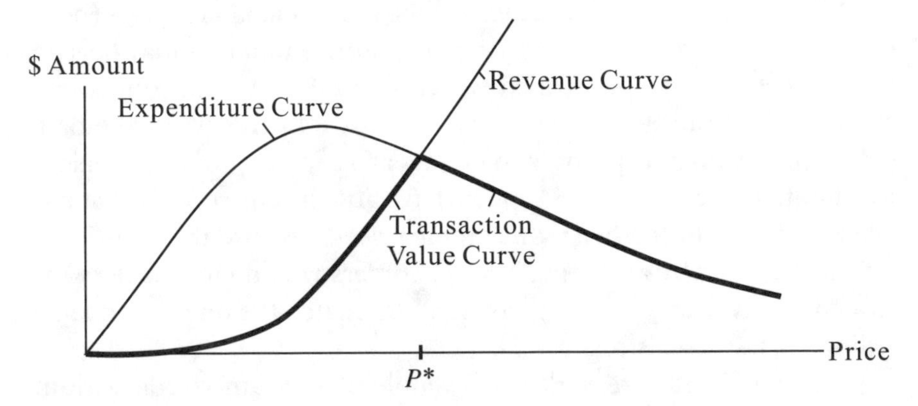

Figure 11.2. The associated revenue and expenditure curves cross above the same point P* *on the price axis, as do the supply and demand curves.*

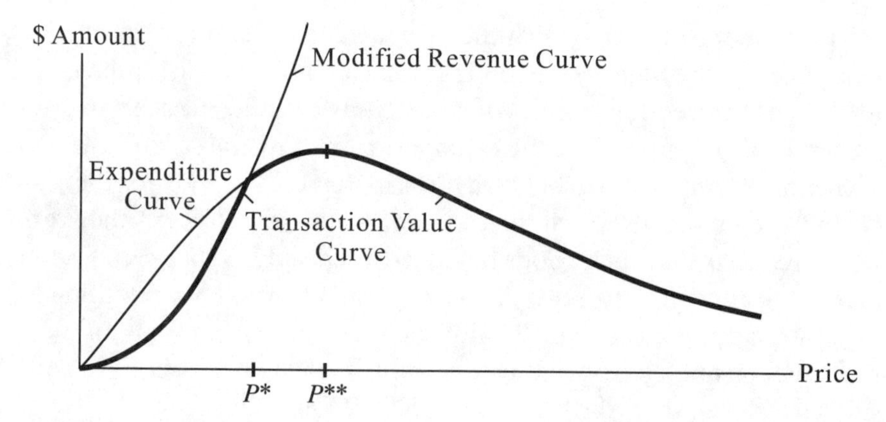

Figure 11.3. Under certain circumstances, sellers can make more money by selling at a higher price than P*. *Neoclassical theory regards it as axiomatic that, though aware of this possibility, they seldom if ever contrive to pull it off.*

revenue-maximizing level P^{**} while exploring ways to maintain it at that level without overt (criminal) conspiracy.

What distinguishes the conspiracy-friendly situation of figure 11.3 from the conspiracy-free geometry of figure 11.2 is the fact that the modified revenue curve intersects the expenditure curve to the left of the highest point on the latter. It does so because expanded production serves to elevate the entire supply curve, along with the revenue curve. The transition from conspiracy-free situations to conspiracy-friendly ones occurs naturally, as technological progress permits markets to be ever more deeply penetrated.

Market penetration is a seldom-discussed topic in orthodox (neoclassical) economic theory. As a rule of thumb, a market may be said to have been deeply penetrated when the unit cost of production is exceeded by a substantial majority (at least 80 percent) of the balk prices, permitting the product to be sold at a healthy profit to a majority of potential customers. Suppliers have little incentive, after that has been achieved, to further reduce their prices, because the additional revenues gained by attracting new customers would fail—as demonstrated in advance by market research—to match the ones forgone by selling for less to existing buyers.

If Adam Smith's remark that "people of the same trade seldom meet together . . . but the conversation ends in a conspiracy against the public, or in some diversion to raise prices," is to be taken seriously, the transition from collusion-proof to collusion-friendly markets must have taken place before 1776 in more than a few industries. Perhaps the most remarkable thing about all this is that it is possible to imagine a state of production so primitive that it *offers no incentive* to fix prices!

History's most dramatic upward shifts in supply occurred during the Industrial Revolution, with the onset of mass production. Smith saw the beginning of it, at the pin factory he described in his opening chapter. David Ricardo—about whom more later—saw factories springing up like mushrooms after a rain but doubted that the trend could long continue. Karl Marx saw mass production a century after Smith, deemed continued expansion an essential part of it, and wondered where it all might lead. Joseph Schumpeter (1883–1950), seeing it later still, was moved to coin the phrase "creative destruction."

Mass production permits a single producer to produce a substantial fraction—say 5–10 percent—of the amount an entire market will absorb. Smith's pin factory almost surely did that, along with Josiah Wedgwood's pottery factory and the dark satanic mills that housed row upon row of Richard Arkwright's spinning jennies. The effect of mass production on a supply curve like the one in figure 11.1 is indicated in figure 11.4. It inserts a small vertical segment into the otherwise smooth curve. The height of that segment is the mass producer's

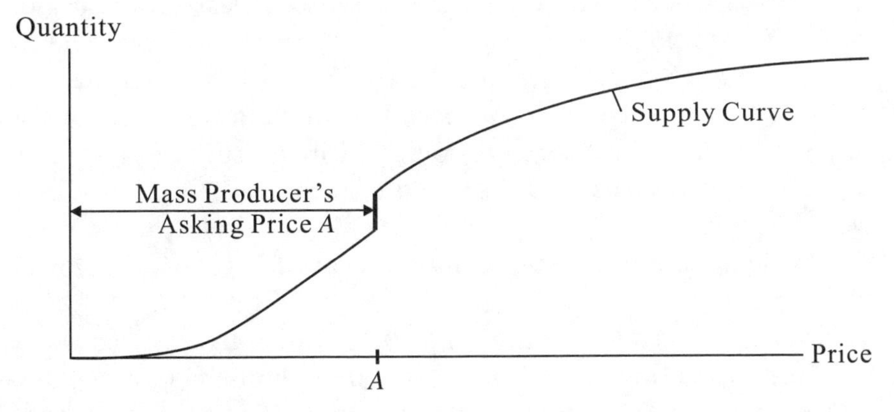

Figure 11.4. The usual analysis applies only to cottage industrialists, whose individual productive capacities are minuscule. Everything changes when mass producers enter the market, introducing vertical segments into the supply curve.

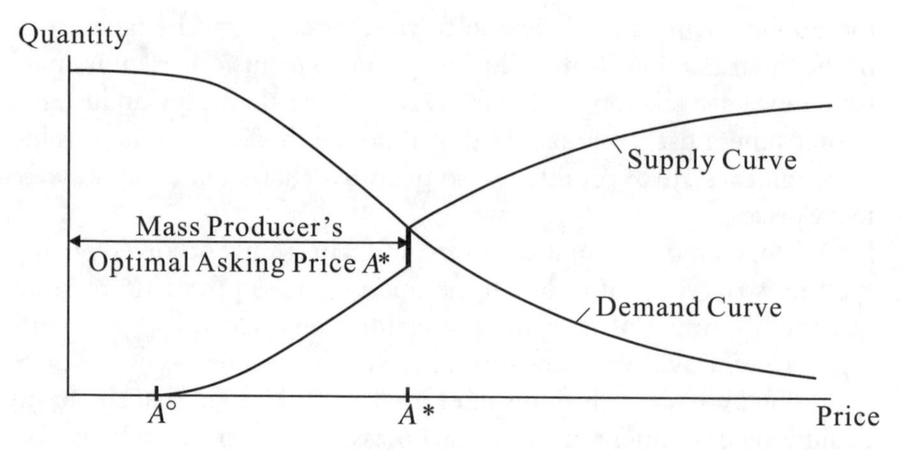

Figure 11.5. Because mass producers produce more cheaply than anyone else, they can sell more cheaply as well. But, as they soon discover, they have no need to.

productive capacity, while its distance to the right of the vertical axis is his asking price A. At any price P smaller than A, he flatly refuses to sell. But at any higher price, he stands ready to part with as much of his output as anyone wishes to buy.

The remarkable fact—the one that set the Industrial Revolution in motion—is that such producers, surrounded though they were by increasingly desperate competitors, possessed significant pricing power. Because mass production is cheaper than other kinds, they *could* sell for less than even their most efficient rival could match. Yet because their productive capacity was severely limited, at least in the beginning, they had no need to do so. They had only to depress the going price far enough to enable themselves to dispose of their entire output. For that to occur, the upper end of the little vertical window (a.k.a. riser) had to rest precisely on the demand curve, as it does in figure 11.5. Had they asked even a little bit more, the demand curve would have passed through the interior of the vertical window rather than its highest point, meaning that some of their goods would have gone unsold.

Had they asked less, the demand curve would have passed above the window, causing them to sell their entire output at a lower-than-necessary price. What fool would do either? For them, the extra money to be made by selling at the optimal asking price A^* instead of, say, the lowest price any competitor can afford to match (denoted A° in figure 11.5) is pure gravy.

The vast fortunes amassed at the dawn of the Industrial Revolution by the likes of Wedgwood and Arkwright suggest that the early mass producers did plenty of both. Their ability to produce a significant fraction of the quantity the markets they served could be expected to absorb meant that the early mass producers were anything but the "perfect competitors" (a.k.a. price takers) to whom orthodox (neoclassical) economic theory applies. Whereas that theory assumes all firms to be incapable of influencing the price at which their goods shall sell, even quite primitive mass producers have considerable latitude to choose their prices.

Such latitude is a mark of what the financial press describes as "pricing power," while economists speak of "monopoly power," perhaps in the belief that none save a monopolist could retain anything so potent. Sometimes, of course, an original mass producer will be able to preserve it by monopolizing the market served. The diagrams of Chapter 10 then identify the price his or her customers seem destined to pay. More often, however, their methods are successfully imitated, enabling other mass producers to enter markets originally served by just one. With such entry—because mass production permits suppliers to penetrate markets ever more deeply—is born the incentive observed in figure 11.3 to collude.

Without explaining precisely why, orthodox (neoclassical) price theory simply postulates that such incentives are without effect. No matter how anxious suppliers may be to realize the gains to be had by overt price-fixing—and perhaps in less flagrant ways as well—their raging competitive hormones allegedly prevent them from harvesting the fruits of collusion.[11] Some hold that the temptation to cheat on any price-fixing agreement by "secret discounting" will overcome any and all attempts at collusion. A few have even denied the need for antitrust laws that prohibit price-fixing, predatory pricing, and similar misdeeds on the grounds that the requisite "combinations in restraint of trade" are inherently unstable and may be relied upon to self-destruct in short order. Alan Greenspan went so far as to argue—in an article he contributed to a newsletter published by Ayn Rand during the 1950s and 1960s—that laws against fraud are superfluous, since free markets will in time punish (by damaging their reputations) all who even attempt to profit by such practices.[12]

Mathematically speaking, the little vertical segments in the supply curves of figures 11.4 and 11.5 correspond to "jump discontinuities" in the "functions" with whose graphs those curves coincide. Their effects on prices are seldom if ever addressed in works on price theory, in large part because Alfred Marshall saw fit to adopt (and to inscribe on the title page of his magnum opus) the Darwinian motto "*Natura non facit saltum*"—"Nature makes no leap." To Darwin it meant that evolution is an exceedingly slow, gradual process. To Marshall it signified that supply and demand curves may contain treads but no risers. Consequently, although Marshall himself (or any of his contemporaries) might have explored the significance of such jumps, they seem never to have done so.

In the ordinary course of events, mass production continues to expand until the capacity to produce exceeds foreseeable levels of demand. At that point, said Joseph Schumpeter, pricing power must vanish—allowing markets to return to the straight and narrow paths of perfect competition and productive-allocative efficiency. Only innovation enabled the early mass producers to reap supernormal profits. As soon as innovators cease to innovate, pricing power deserts them. Such was Schumpeter's theory of "creative destruction." To him, supernormal profits were but the temporary fruits of innovation. As one new firm after another rode successive waves of innovative technology into an established industry, old firms would either respond with comparable innovations or disappear.

Students of "imperfect competition" disagree. To them, it seems obvious that an oligopoly—by which they mean a small number of mass producers serving a well-defined market—may cling (without overt collusion) to pricing power more or less indefinitely. The new orthodoxy, which closely resembles its pre-Keynesian predecessor, strongly prefers Schumpeter's transient explanation of pricing power, supernormal profits, and idle industrial capacity in terms of "creative destruction" to the more iconoclastic claims of imperfect competition theory. We shall return to this question in Chapter 16.

An internal memo brought to light during the famous Microsoft antitrust trial of 1997–2000 suggested that while the company could make money by selling its flagship Windows operating system for $49 a copy, profit maximization would require that it charge $89.[13] Microsoft's ability to command the latter figure is a measure of its pricing power. In 1992, *Consumer Reports* estimated that the raw materials, processing, labor, freight, and packaging in a box of Kellogg's

Corn Flakes cost $0.66 and was sold at wholesale to supermarkets for $1.73.[14] The freethinking economist Thorstein Veblen—who died in 1929—used to point out that steel rails were quoted at $28 a ton for thirteen consecutive years, although it cost less than half as much to produce them. In short, there is an abundance of anecdotal evidence to the effect that free markets currently tolerate, and have long tolerated, all manner of supernormal profits.

No one seriously maintains that anything remotely resembling the *tâtonnement* process has ever taken place outside a classroom. Nor does anyone deny that the largest markets are deeply penetrated, that pricing power is ubiquitous, that corporate interests (say, in built-in obsolescence) often conflict with those of consumers, or that the government—which plays no explicit role in the neoclassical model—is almost every firm's best customer. Yet the champions of economic orthodoxy insist that free-market capitalism "works like" the neoclassical model thereof. This is what Paul Ormerod, in a penetrating critique of orthodox economic thought titled *The Death of Economics*, terms "the *as if* mantra" of economic orthodoxy.[15] It may be paraphrased as follows:

> Forget that tâtonnement never happens, that the most remunerative markets are deeply penetrated, that pricing power is commonplace, that excess capacity is ubiquitous, and that national governments are every firm's best customer. Free market economies still operate *as if* governed by the alleged laws of neoclassical economics.

Whenever doubts are expressed concerning the reliability of the neoclassical model, defenders of the faith fall back on this powerful catch-22. Their behavior is reminiscent of the tongue-in-cheek conditions of employment sometimes posted in workplaces, in which rule #1 asserts that the boss is always right, while rule #2 directs employees to refer to rule #1 whenever the boss is wrong. In a chapter titled "Professional Reservations," Ormerod summarizes a host of (rather technical) reasons—many of them identified by prominent theoretical economists—for doubting the neoclassical model. Chapter 12 will examine some of the more systematic evidence against which the neoclassical model has been tested.

CHAPTER 12

Evidence Pro and Con

Edward H. Chamberlin, who along with Joan Robinson very nearly engineered a revolution in economic thought during the 1930s, appears to have been the first to propose that the simpler predictions of the neoclassical model might be lab tested. In 1948, while teaching at Harvard, he designed a series of experiments using graduate students as "economic agents" to do just that. Dividing his "volunteers" into buyers and sellers (say, six of each), he issued a number of cards (typically two) to each. Every buyer's card was stamped with a retail price at which Chamberlin was obliged to redeem it, if and when paired with a seller's card of any denomination. Every seller's card was stamped with a wholesale price, which was debited to the seller's account, unsold seller's cards being redeemable at full face value upon completion of the experiment. Thus a pair consisting of a $3 buyer's card and a seller's card of any denomination was worth $3, less whatever the buyer had paid to acquire the seller's card, when the buyer presented the pair for redemption. If the price on the seller's card was $1, and the buyer had paid $2 to get it, then buyer and seller would each gain $1 from the transaction.

If the twelve sellers' prices begin at $0.25 and increase in increments of $0.25 to $3.00, while the buyers' begin at $0.75 and increase in similar increments to $3.50, the relevant supply and demand curves have the look of staircases, in which each step is one unit (seller's card) high and $0.25 wide, as indicated in figure 12.1. The supply graph

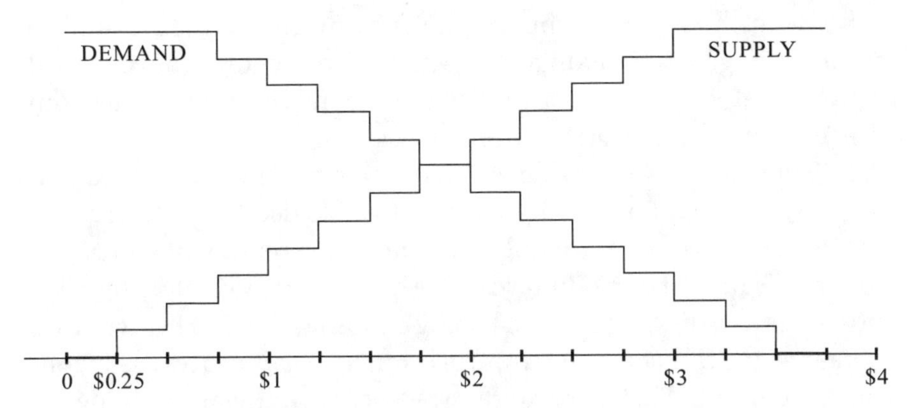

Figure 12.1. Chamberlin's 1948 experiments employed very simple staircase-like supply and demand curves.

leads up to the right, while the demand curve leads down from the left. Should the investigator retain the right to act as a *tâtonneur*, so that all sales take place at a common price of his choosing, the options are obvious. To find out how many sellers' cards will be available for purchase at a given price p, he need only count the number of wholesale prices smaller than p. To learn how many buyers can afford to buy at that price, he need but count the number of retail prices greater than p.

At any price up to \$1.75, the demand for sellers' cards will exceed the supply of them, while at any price greater than \$2.00, supply will exceed demand. But at any intermediate price, the holders of the seven highest-priced buyers' cards can afford to purchase sellers' cards, and the holders of the seven cheapest sellers' cards can afford to sell. Thus a *tâtonneur* would be content to halt his search at any price from \$1.76 to \$1.99 inclusive, since supply equals demand (equals seven units) at any such price. Elementary price theory predicts that the average transaction price in a free market for sellers' cards would fall in the foregoing range and that seven units would typically change hands.

By performing as predicted, the buyers and sellers together realize a gain from trade of \$12.25, far greater than the \$6.00 to be earned by exchanging all twelve sellers' cards. To verify that the larger figure is in fact the greatest possible, let N be the number of sales to be concluded by the several traders in the foregoing market experiment. If permitted to collude, the traders would naturally pair the N cheapest sellers' cards with the N dearest buyers' cards to maximize their combined profit. The group wouldn't care which of the chosen buyers' cards were paired with which of the chosen sellers' cards, since the combined profit would be the same in any case.

Having learned to decide which cards to trade when they're trading only N, the group should have no trouble deciding which N best serves their purpose. With only the values between $N = 0$ and $N = 12$ to test, trial and error soon reveals the twelve agents can together earn no more than the $12.25 to be had by choosing $N = 7$. The twelve can obtain as much via the decentralized *tâtonnement* process as by outright collusion. That is because, whether by accident or by design, figure 12.1 represents a collusion-free situation. Had Chamberlin imposed a more steeply rising supply curve—thereby creating a clear incentive to collaborate—he might have observed something markedly different. As it was, even without creating any such incentive, he did observe something worthy of mention.

Students permitted to circulate at will about the classroom in which the experiments were conducted, in what might be called a laissez-faire manner, had a tendency to do more deals than the neoclassical model would predict. Left to their own devices, they tended to break up into small clusters, within each of which a different average transaction price tended to emerge. And although the averages did not vary enough to "find a home" for every seller's card, they did produce an unexpectedly large number of sales, thereby casting doubt on the predictive power of the neoclassical model.

What might have caused a crisis of confidence in any other science had little impact on economics. Few practitioners seemed to know what to make of the new results, and none hastened to replicate them. No heated debate sprang up concerning the significance of the new results, and it was not until 1982 that James T. Hong and Charles R. Plott confirmed the occurrence, in a carefully controlled experimental setting, of supernormal trading volume in certain laissez-faire situations.[1]

Generations of economists had grown up believing that the experimental method was inapplicable to their science, and few welcomed Chamberlin's call to reconsider that conclusion. On the contrary, most appeared to sympathize with the Austrian expatriate Ludwig von Mises, who, in 1949, blithely asserted that "the ultimate yardstick of an economic theorem's correctness or incorrectness is solely reason, unaided by experience." Indeed, said he, the "particular theorems" of economics "are not open to any verification or falsification on the ground of experience."[2] In other words, economic principles are theorems and theorems are irrefutable. A child can complete the syllogism. Although it is hard to imagine a more unequivocal repudiation of everything that science purports to be, and although most practicing economists are aware of von Mises's assertion, surprisingly few have publicly disowned it.

When the Nobel Prize in Economics was awarded to Vernon Smith and Daniel Kahneman in 2002, for their work in experimental economics, it signaled the profession's long overdue—and probably still incomplete—embrace of the scientific method. Today there is an Economic Science Association, composed of economists who engage in experiments, and a journal called *Experimental Economics*. Other journals publish experimental results as well. Many now claim that the relevance of market experimentation for economic theory and for the policy process has been established beyond reasonable doubt.

As a beginning graduate student at Harvard in 1948, Vernon Smith took part in Chamberlin's experiments. Intrigued by the possibilities of the new method, he resolved upon joining the Purdue faculty during the early 1950s to revisit the results. He suspected that Chamberlin had at least partially misinterpreted his findings, and he sought to set the record straight. To that end, he imposed certain rules on the competition between buyers and sellers in his classroom laboratory. The rules he chose were those of what is now known as the double-auction model of a competitive marketplace.

Under double-auction rules, any buyer who wishes to make a bid must raise her hand and be recognized so that her bid can be known to all concerned. Sellers' offers are likewise made public. All bids and offers are recorded on a blackboard when made. Only the most attractive bid or offer has "standing" to be accepted. Any buyer is free at any time to accept the standing offer, and any seller may at any time accept the standing bid. Many experimenters add an "improvement rule" stipulating that any new bid or offer must improve on the standing one. This arrangement constitutes a double auction in the sense that while the bids are increasing, as at an English auction, the offers are decreasing, in the manner of a Dutch auction. Traders are ordinarily given no information whatever concerning the wholesale and retail prices offered other traders.

Smith's experiments generally produced average prices and trading volumes in close agreement with orthodox theory, although marginally profitable trades occasionally went unmade. Perhaps for that reason, the double-auction format is now the one most frequently employed in laboratory trading experiments. Smith's results have been replicated by dozens of researchers, in hundreds of sessions, under a variety of modified versions of the governing rules. While double auctions are by no means identical to *tâtonnement*, they certainly resemble it far more closely than did the laissez-faire ambience that prevailed in Chamberlin's classroom. So perhaps it is unsurprising that the results are all but identical to those to be expected of *tâtonnement*.

Nevertheless, such experiments produce only weak support for Ormerod's as-if mantra. Experimentation seems only to confirm that simple markets *can be rigged* to realize the predictions of orthodox economic theory by making such markets less free and more regimented, via the imposition of precise rules concerning the conditions under which trades may and may not occur.[3] Unrigged markets routinely fail to realize the predictions of the neoclassical model, and a cottage industry has sprung up to design efficient markets—many of them auction markets—in which to trade all manner of goods and services. Laboratory experiments seem ill suited to questions concerning the efficiency of entire economies, such as the ones considered by Arrow and Debreu, or the national markets in which are formed the prices we pay to feed, clothe, house, transport, medicate, educate, and entertain ourselves.

Among the more interesting theoretical developments of the post-Marshallian era was a clever technique for estimating the allocative efficiency of a national economy from readily available data. Had World War II not interfered, it seems likely that this novel technique would have been put to use before 1954, when Arnold Harberger of the University of Chicago finally applied it to data collected in the United States between 1924 and 1928.[4] Few expressed amazement when Harberger reported that the U.S. economy had operated with 99.92 percent efficiency during those years, apparently confirming for that time and place the oft-repeated claim that competitive free markets do indeed perform as the neoclassical model predicts.

Harberger's method involves the measurement of certain triangles, now known as Harberger triangles. The method has since been applied to many other economies and periods of time. William G. Shepherd summarized the findings of a dozen other such studies of the U.S. economy.[5] Six of his thirteen U.S. studies concluded that the nation's economy had operated with better than 99 percent efficiency during the study periods, while most of the others put the figure between 95 and 99 percent. His one study of the British economy concluded that it, too, had operated at better than 95 percent efficiency during the study period.

The sole exception in Shepherd's sample was a study by Keith Cowling and Dennis Mueller that, in the course of concluding that the U.S. economy may have operated between 1963 and 1966 at less than 87 percent efficiency, uncovered a significant flaw in Harberger's method.[6] The authors noted that Harberger's emulators, like Harberger himself, had treated advertising expenditures as involuntary costs of doing business rather than as voluntary investments in the rapidly and randomly depreciating asset known as market share. Both the U.S. and the British economies would have seemed significantly less efficient had the investigators adopted a different (and seemingly more accurate) interpretation of advertising data.

Imagine a corporate accountant instructed to separate her firm's costs of doing business from its investments. Because most of the people who buy advertising regard it as an investment, and most of the people who sell it present it that way, one might expect her to include advertising expenses in the investment category. But in practice, if only because the costs of doing business are payable from pretax dollars, while investment spending rarely is, she seems all but certain to enter advertising expenditures in the cost column, exactly as Harberger and others did. Should she do so, her firm's books will look a lot like those of a passive price taker. If she doesn't, they will look more like those of a firm with significant pricing power. Thus Harberger's conclusion that the U.S. economy operated at an astronomical level of allocative efficiency during the study period was due in large part to his willingness to classify advertising expenditures as (involuntary) costs of doing business rather than (voluntary) investments.

Advertising is an endlessly fascinating form of free-market competition, about which modern science has yet to learn very much. The brewing industry is as good a place as any to examine its effects. The process has been greatly facilitated, of late, by the publication of a comprehensive study titled *The U.S. Brewing Industry: Data and Economic Analysis*, by Victor J. Tremblay and Carol Horton Tremblay. The following account of advertising in the brewing industry follows theirs rather closely. These two specialists in industrial organization offer the most detailed account yet compiled of the industry and its history, along with the most comprehensive data set ever assembled concerning its many defunct (and few surviving) firms.

In 1938, five years after the repeal of prohibition—which forced no fewer than 1,568 once profitable brewers into bankruptcy—some 700 such establishments were again operating in the United States. By 1947, that number had fallen to 421, with more casualties to follow. By the early 1950s, with television advertising in its infancy, beer companies were among the most active users of the new medium. Many TV sets—including almost all of the ones in bars and taverns—were tuned to the Wednesday night fights sponsored by ("What'll ya have?") Pabst Blue Ribbon Beer. Between 1947 and 2001, the number of brewers of bulk lager beer—then as now the only kind consumed in quantity in the United States—declined from 421 to a mere 24. Meanwhile, the four largest firms saw their combined market share rise from 17 to 94 percent. Anheuser-Busch (AB) alone increased its share from less than 6 percent in 1950 to almost 55 percent today.

Among other things, AB was the first brewer to engage in market research. The project began in 1961, when the company's president, August A. Busch Jr., responded to an in-house proposal to increase advertising expenditures 8 percent by asking, "Is there any way I can find out at the end of the year whether I got what I paid for?" To address this question, the company hired Russell Ackoff, a professor of management and behavioral science at the University of Pennsylvania, to investigate the effect of advertising on Budweiser sales. This he did by analyzing the results of purchasing different kinds and amounts of advertising in different parts of the country.

Ackoff concluded that advertising is more effective when a variety of media are used and when the message is conveyed intermittently. He also found that television is the most effective medium for increasing sales, while outdoor advertising had almost no effect. Furthermore, different brands appeal to different personality segments of the beer-drinking public, and advertising is most effective when it targets a specific segment of the population. Now most brewers market several brands of beer, each one targeting a different demographic. By heeding Ackoff's advice, AB almost doubled its sales between 1963 and 1968, even as it decreased its advertising expenditures from $1.89 per thirty-one-gallon barrel—the standard unit of production within the industry since the nineteenth century—to a mere $0.80.

Little if any evidence suggests that advertising increases beer consumption. It does, however, seem to persuade the drinkers of "popular-price" beer—usually cheap and unadvertised—to switch to the premium brands, which are marginally more expensive and abundantly advertised. This well-documented effect is perhaps most clearly reflected in the combined market shares of what Tremblay and Tremblay describe as "third-tier" breweries—mostly local—which declined from more than 70 percent in 1950 to less than 5 percent by 1990. Few will be surprised to learn that television advertising has become a highly effective (if not indispensable) weapon in the "marketing warfare" in which U.S. brewers engage.

Appendix C of *The U.S. Brewing Industry* catalogs almost two hundred brewery mergers and acquisitions between 1950 and 2002. One of the most frequently repeated notations in that appendix states that the acquiring firm "purchased brands only," meaning that the plant and equipment belonging to the acquired firm were either not included in the sale or resold soon after. The lesson to be learned is that brands have value. In most cases, retail sales of the acquired brands—many of which had once been advertised brands—continued without interruption. The acquiring firms simply began to apply different labels to a fraction of the beer they shipped from their own breweries. The money previously spent to advertise the acquired brands had clearly functioned as an investment, helping to create assets of lasting (perhaps even permanent) value. In these cases at least, advertising expenditures cannot have constituted necessary costs of doing business, since the business continued for years—in some cases decades—after the ads ceased to appear.

Cowling and Mueller reveal that the data analyzed by Harberger and others are not what show the U.S. economy to have operated at upwards of 95 percent (by many estimates 99 percent) allocative efficiency during much of the twentieth century. It is the manner in which those authors interpret their data. Cowling and Mueller go on to remark that their own somewhat lower estimates are overestimates as well, since they result from the reclassification of advertising expenditures alone as investments rather than as costs of doing business. Plenty of other expenditures are equally deserving of reclassification, and every such reclassification reduces the assessed efficiency of the relevant free-market economy.

Cowling and Mueller do not explain how to obtain truly reliable estimates of the allocative efficiency of a typical postindustrial economy, nor do they venture a guess as to how far such an estimate is likely to fall below the astronomical figures currently accepted. They reveal only that it would require a great deal of hard work to find out. A further glance at the brewing industry will reveal the nature of the work to be done.

Tremblay and Tremblay reproduce a pie chart depicting the price-cost breakdown of mass-produced beer.[7] Figure 12.2 is derived from theirs. It suggests, among other things, that brewers spend far more for packaging, labor + production, and taxes + shipping than they do for advertising. In particular, the package in which consumers encounter a product is an important part of that product's appeal to the targeted demographic. It must distinguish the product not only from those of

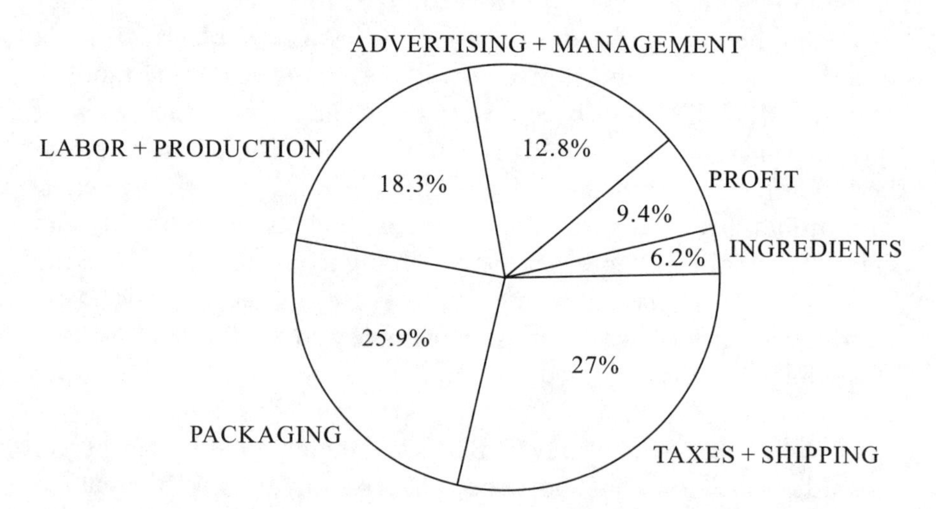

Figure 12.2. The breakdown of brewing industry revenues doesn't change very much from year to year, or even decade to decade to decade.

rival producers but also from its own producer's other brands. Brewers would spend significantly less than they do on packaging were it less important to the marketing warfare that is so ubiquitous in their industry. Much the same can be said for labor and production.

Both Anheuser-Busch and Coors have long attributed much of their success to their efforts at quality control. Those efforts are not inexpensive and would doubtless be curtailed if they were not so important for marketing warfare. The sums involved are not unavoidable costs of doing business but voluntary investments in brand building. Likewise for shipping. Brewers could surely reduce their shipping costs if the freshness of the product were less important for marketing warfare. If a mere 10 percent of the expenditures in each category except profit and advertising is assumed to be of the (voluntary) brand-building variety rather than the (involuntary) cost of doing business, the efficiency of the brewing industry would fall well short of 80 percent. And since the rest of the U.S. economy seems likely to be less efficient than the highly competitive brewing industry, it is hard to imagine that the efficiency of the economy as a whole exceeds 75 percent.

It would, on the face of it, require a great deal of hard work, in a great many industries, and virtually unlimited access to corporate books to determine the actual magnitude of the voluntary investments in market share that currently masquerade (in large part for tax purposes) as unavoidable costs of doing business. Until that work is done, the actual allocative efficiency of the U.S. economy will remain shrouded in mystery.

Fact-based science provides little support for the neoclassical model, including the indispensable as-if mantra. Market experiments fail to confirm model predictions unless carefully rigged to do so, and the data analyzed by Harberger and others enable competitive free-market economies to appear efficient only because they misclassify advertising and other voluntary brand-building investments as involuntary costs of doing business. Early in a scientific revolution, said the primatologist Shirley Strum, discrepancies between theory and fact "are either ignored or written off as the results of bad methods or 'bad science.' But these discrepancies eventually become overwhelming, and another version, model, or paradigm replaces the old one." Economics is still at the stage in which it is possible to ignore discrepancies between theory and fact, for the simple reason that such discrepancies are not yet overwhelming. Relatively few have been found, because relatively few attempts have been made—since Chamberlin's own—to find them.

Being economists first and experimentalists second, experimental economists seldom go looking for serious discrepancies. It would do their careers little good to find them. As soon as Vernon Smith investigated the double-auction format—which he candidly admits having done because he thought it more likely than any other to realize orthodox (neoclassical) expectations—it became the experimental market form of choice.[8] Few if any attempts have been made to reproduce or clarify Chamberlin's original (negative) results, using the best experimental practices developed by Smith, Plott, and the other pioneers of experimental economics. It was almost by accident that Hong and Plott became interested in the telephone markets that led them to confirm some of Chamberlin's original negative findings.[9]

Twenty-five years after Cowling and Mueller wrote, the avenue of research they pioneered remains largely unexplored. In pointing out that advertising constitutes a voluntary brand-building investment rather than an involuntary cost of doing business, they seriously undermined the indispensable as-if mantra. If their tentative verdict is ultimately upheld, it will be hard to go on pretending that the neoclassical model of economic activity is of more than historical interest. Anyone wishing to follow up on their groundbreaking investigation by finding out how many other ongoing investments—in addition to advertising—have been misclassified as involuntary costs of doing business for the purpose of measuring Harberger triangles has so far been discouraged by the magnitude of the task, the difficulty of obtaining adequate funding, and the uncertain reception that awaits the results.

Should they find that the misclassified expenditures are insignificant, and that free-market economies are only a little bit less efficient than previously believed, the investigators would gain publication in a prestigious journal—as Dennis Mueller notes, the highest efficiency estimates always seem to end up in the most prestigious journals—but face charges of unoriginality for having discovered nothing new.[10] Or, should they find that the misclassified expenditures are indeed significant, and that free-market economies are substantially less efficient than previously believed, they would face the indictments for "bad methods and 'bad science'" that are the inevitable lot of scientific revolutionaries. In neither case does the upside reward seem to outweigh the downside risk. Small wonder that younger investigators have consistently ignored the challenge posed by the findings of Cowling and Mueller.

The measurement of Harberger triangles makes it possible to assert the productive and allocative efficiency of modern industrial economies, and thereby the relevance of orthodox (neoclassical) economic theory for the policy process. But those who would do so have still to explain the fact that free-market economies appear to condone numerous market failures, including all manner of sticky prices, pricing power, and chronic excess industrial capacity.

Wikipedia recently reported that the "average economy-wide capacity utilization rate in the US since 1967 was about 81.6% according to the Federal Reserve measure. The figure in Europe is not much different, for Japan only slightly higher."[11] In India, according to the Hindu Business Line, the figure is only 70 percent. In China, it may be as low as 60 percent. According to *Wikipedia*, even the staid *Wall Street Journal* has conceded that "from cashmere to blue jeans, silver jewelry to aluminum cans, the world is in oversupply," while *The Economist* speculated that the gap between sales and capacity may be "at its widest since the 1930s." Furthermore, "excess capacity in steel hovers near 20 percent, and in autos it has been as high as 30 percent."

After pointing out that "there are no official data on global excess capacity," James Crotty goes on to assert that "reports from consulting firms and industry trade associations, as well as occasional studies by international organizations, agree that large excess capacity has plagued almost all globally contested industries for at least two decades." A few years ago, the CIA cautioned that "by the late 1990s a staggering one billion workers representing one-third of the world's labor force, most of them in the South, were either unemployed or underemployed."[12] How can the world's vaunted free-market economies be thought to be squeezing "as many useful goods and services as possible from the available resources" when so much capital and labor stand idle?

The standard explanation for both excess profits and idle industrial capacity involves sticky prices. While many economists consider such prices—along with the attendant inefficiencies—to be among the most important issues facing the emerging global economy, those who measure Harberger triangles seem to agree that all or most free-market economies operate at stratospheric levels of productive and allocative efficiency. How to resolve their apparent disagreement?

Unsupported though it is by fact-based science, orthodox (neoclassical) theory is highly consequential. In the guise of supply-side economics, and the more recent "Washington Consensus," it informs the quest for "free-market solutions" to socioeconomic problems that resumed—after Keynesian policies proved incapable of curing the stagflation of the 1970s—during the Reagan-Thatcher era. It directs that labor markets be made "flexible," that public property be privatized, and that taxes of every description be reduced or abolished. Financial markets are to be freed from irritating regulations, both nationally and internationally, until even the least onerous controls on international capital are lifted.

Orthodox (neoclassical) theory has often excused inaction in the face of tragedy, as during the early years of the Great Depression, when the Hoover administration stood idly by while the American economy proved itself to be anything but the self-adjusting, self-correcting paragon of efficiency portrayed by orthodox theory. Keynes—often described as Marshall's most brilliant student—once observed that his teacher's version of the neoclassical model amounted to nothing less than "a whole Copernican system, by which all the elements of the economic universe are kept in their places by mutual counterpoise and interaction."[13]

Coming as they did from acclaimed scholars, and conveying as they did a message welcome to the powers that be, such fanciful notions lent useful gravitas to the argument that the best policy for recovering from the Great Depression—during which unemployment gripped more than 20 percent of the British, German, and U.S. workforces—was no policy at all. Most members of the Anglo-American establishment, steeped as they were in laissez-faire teachings, genuinely expected free-market economies to cure themselves of whatever ailed them. Absent their support for do-nothing policies, it seems all but certain that New Deal–style programs would have been pursued sooner and with greater determination.

At present, one consequence of the neoclassical model—predicated as it is on the notions of perfect competition and allocative efficiency—

outweighs all others in its impact on the course of history. It is, of course, the doctrine of international free trade. That widely acclaimed instrument of public policy currently enjoys greater establishment support, and more determined grassroots opposition, than any other. Chapter 13 will examine the often bitter disputes surrounding globalization and free trade.

CHAPTER 13

Free Trade

"Free trade," as the term is used today, signifies an absence or bare minimum of import duties, export bounties, quotas, and nontariff barriers to the international exchange of goods and services. Although the founding fathers intended the U.S. government to generate the bulk of its revenues by collecting import duties—as it did, more or less, for perhaps a hundred years—free trade has come to seem as American as baseball, apple pie, and the right to bear assault weapons. It was one of Woodrow Wilson's Fourteen Points, a clause in the Atlantic Charter signed aboard ship by Churchill and Roosevelt in the summer of 1941, and a long-term goal of Anglo-American commercial diplomacy ever since. For a public figure to express doubt concerning the wisdom and ultimate necessity of free international trade in England or America today is to invite ridicule and (often bitter) denunciation.

Free trade is best understood as a management technique for nations, whereby they may allegedly increase their GDPs without acquiring additional resources. Like most such techniques, this one has been marketed aggressively by a large and powerfully motivated sales force. The economics profession strove for generations to "sell" free trade to the public at large, eventually succeeding with a small English-speaking elite. Acceptance in other circles, especially among non–college graduates, has always been lukewarm. Resentment of the perceived effects of free trade boiled over at the infamous "Battle in Seattle," permitting an unlikely coalition of grassroots protesters to forestall yet another incremental step toward free trade under the aegis of the World Trade Organization.

To understand how economists succeeded in selling free trade, we are better off examining the sales pitch employed rather than the merits of the product itself. No less an authority than Paul Samuelson has conceded that there are more arguments against free trade than for it, because "there is essentially only one [economic] argument for free trade." However, that lone argument "is an exceedingly powerful one: namely, *unhampered free trade promotes a mutually profitable international division of labor, greatly enhances the potential real national product of all countries, and makes possible higher standards of living all over the globe.*"[1] This, in essence, is the way economists have been pitching free trade since the time of Adam Smith to anyone willing to listen. Though polished assiduously, and occasionally reinforced, the pitch hasn't really changed since 1776.

That said, the historical strength of the (economic) case for free trade lies in a corroborating principle discovered by David Ricardo in 1817. Known as the principle of comparative advantage, it purports to guarantee "with mathematical certainty" that the benefits of restraints on international trade can never exceed the costs. Although Ricardo himself made relatively little of the principle, devoting less than a page to it in his 1817 book, later generations came to regard it as a towering scientific achievement, comparable in every way with Newton's laws of motion, Darwin's theory of evolution, Mendeleyev's periodic table, Maxwell's electromagnetic theory, Einstein's principle of relativity, and all the other jewels in the crown of modern science. Twentieth-century scholarship has converted "Ricardo's principle" into a mathematical theorem.

Born to a Jewish family active in the then-emerging European financial markets, David Ricardo (1772–1823) abandoned formal education at the age of fourteen to enter the family business. There he prospered, acquiring sufficient fortune to retire in less than thirty years to the life of a country squire. As such, with time out for a term in Parliament, he devoted much of his remaining time and energy to the study of political economy. His *Principles of Political Economy and Taxation* replaced Adam Smith's *Wealth of Nations* as the standard treatise on the subject until it was itself replaced after 1848 by John Stuart Mill's

Principles of Political Economy. No one in history has contributed more to modern economic thought, or is more admired by its current practitioners, than David Ricardo.

The essential content of Ricardo's principle can be expressed in terms of GDP. The strong form asserts that even if two or more nations have already expanded their GDPs as far as possible under a given set of restraints on international trade, they can almost surely expand them still further by relaxing those restraints. The weak (and most frequently proven) form of the principle refers only to nations not yet engaged in trade. Such nations, the principle asserts, have nothing to lose—and typically much to gain—by allowing trade to flourish. The costs of expanded trade, in the form of lost jobs and diminished wages, along with vanishing health-care and pension benefits, could conceivably equal but *can never exceed* the benefits of cheaper goods and services.

Perhaps the most remarkable feature of Ricardo's esteemed principle is that it applies to any and all groups of potential trading partners. By so doing, it allays the not unnatural suspicion that rich nations have nothing to gain from trade with poor ones, while the latter have reason to fear trade with the former. Ricardo originally illustrated the pristine logic by which he arrived at his conclusion with the aid of an example drawn from the wine and wool trade that, even in his day, had long flourished between England and Portugal. Because both goods could be produced more cheaply (with fewer man-hours) in Portugal than in England, thoughtful observers wondered why Portugal found it worthwhile to continue the trade.

Ricardo's eminently sensible answer was that because Portugal's advantage in wine production was far greater than its advantage in wool production, England could make it worth Portugal's while to produce more wine and less wool than Portuguese consumers really wanted, and allow England to restore the desired balance through trade. England, he argued, possessed a "comparative advantage" that was of value to both countries, since it enabled both to enjoy—through trade—more wine *and* more wool than either would otherwise have been able to do.

It would be difficult to exaggerate the importance of Ricardo's celebrated principle as a "showpiece"—a flagship brand—for all of orthodox economic theory. At the time of the infamous Battle in Seattle, the World Trade Organization's Web page prefaced its "Case for Open Markets" with the following anecdote:

> Nobel laureate Paul Samuelson was challenged, as a young man, by the mathematician Stanislaw Ulam to "name one proposition in all of the social sciences which is both true and non-trivial."
>
> It took Samuelson several years to find the answer—comparative advantage.
>
> "That it is logically true need not be argued before a mathematician; that it is not trivial is attested to by the thousands of important and intelligent men who have never been able to grasp the doctrine for themselves, or to believe it after it was explained to them."

Other accounts of the famous Samuelson-Ulam exchange omit mention of the years that elapsed between challenge and response. Ulam's memoirs mention only the challenge.[2] In his retelling of the incident, former Harvard economist Todd G. Buchholz describes Samuelson's response as "immediate," Ulam as "an insolent natural scientist," and the principle of comparative advantage as "the key to modern economic understanding." He then charges that because "few politicians . . . can follow the analysis . . . quotas, tariffs, and trade wars mar the world's economic history."[3]

Logic alone, according to Buchholz, justifies the opinion expressed by Milton Friedman and Rose Friedman that "international free trade is in the best interests of the trading countries and the world."[4] The 1928 edition of the *Encyclopaedia Britannica* agreed, deploring trends that at the time made it "all but impossible to secure a hearing for the *coldly rational* arguments for international free trade."[5] More recently, Nobel laureate James Buchanan complained that we "who do understand the simple logic of free trade have not been at all successful in disseminating our message."[6] In a stunning display of intellectual bigotry, orthodox economists seem to have convinced themselves that anyone who finds "the coldly rational arguments for international free trade" unconvincing has necessarily failed to understand them. Why, one wonders, has Professor Samuelson never written an article explaining the matter in words of a single syllable?

Paul Ormerod recounts an exchange that took place at a seminar held in March 1980 at the prestigious École des Hautes Études Commerciales in Paris.[7] An audience of five hundred MBA students was addressed on successive days by a series of distinguished French economists. The first to speak was Maurice Allais, a Nobel Prize winner in the subject, who horrified his audience by attacking the belief that free or freer trade is invariably beneficial. Contending that such trade is beneficial only when carried out between nations at comparable levels of economic development, he condemned efforts by the European Commission to force it on unwilling populations. Two days later Jacques Attali—then president of the European Bank for Reconstruction and Development, and Allais's former student—restored order to the proceedings by responding that "every obstacle to free trade is a factor which leads to recession" and dismissed dissenters as "*stupides*."

The foregoing exchanges typify the combination of flattery and intimidation by which free trade continues to be sold. Those who find the "logic of free trade" convincing are praised for their enlightenment and their ability to think abstractly, while skeptics are dismissed as "*stupides*" unable to "follow the analysis." What candidate for reelection can afford to be portrayed as an intellectual lightweight by the media? Even legislators who routinely vote against free-trade initiatives, such as Maryland's Senator Barbara Mikulski and Ohio's Senator Sherrod Brown, claim to "believe in free trade." Even labor leaders, whose followers are evaporating into thin air as jobs are shipped offshore, dare not speak out against free trade.

A recent Pew Research Center poll finds that, whereas the public at large remains ambivalent, a panel of the nation's most influential citizens (including military leaders, leading scientists and engineers, governors and mayors of large cities, experts in foreign affairs, and members of the news media, but excluding religious leaders) are persuaded that NAFTA has been good for the United States, that CAFTA will be, too, and—by implication—that all forms of free trade are worth whatever sacrifice they may entail.[8] Not coincidentally, the same panel of opinion leaders differed with the general public on the priority to be given job losses and factory closings, along with the ongoing influx of illegal alien workers, in the policy process. While the public at large considered job protection nearly as important as defending the country against terrorism, few opinion makers considered it a leading foreign policy concern.

The simple logic of free trade is at its simplest when reduced to pure mathematics. A complete proof of what might be called Ricardo's theorem need occupy no more than a paragraph or two of words and symbols, easily understood by any upper-class student of the mathematical or physical sciences. The proof involves equations of an unusual type, of which the one displayed in figure 13.1 is a particularly simple example. The left-hand sides of the equations in question may include any number of triangles, since there must be one for each trading nation. Those triangles may be "curvilinear" in the sense that their hypotenuses need not be straight but may bend outward, away from the right angle in the lower-left-hand corner of each one. And finally, the triangles need not always be triangles, but the multidimensional solids devoting a separate dimension to each traded good or service. Indeed, those of greatest practical interest reside in spaces of thousands if not millions of dimensions. Yet anyone who knows what Ricardo's theorem says, and can explain the proof of the two dimensional version, would seem to possess an uncommonly complete grasp of the subject.

That said, it seems unlikely that more than a handful of free trade's most vehement supporters could begin to explain what the cartoon presented in figure 13.1 might have to do with serious matters of public policy. Those who have studied economics will recall that the matter was explained to them on various occasions, and may even imagine that they understood it at the time. After all, they aced the exam, didn't they? The rest are simply parroting phrases they only pretend to understand.

It is easily forgotten, by people who don't use mathematics in their daily lives, that all mathematical statements are conditional in nature. If a body is rigid, then its equations of motion assume a particularly simple form. If a fluid is incompressible, a solid elastic, or a gas electrically neutral, the same can be said. And so on. The first thing to ask,

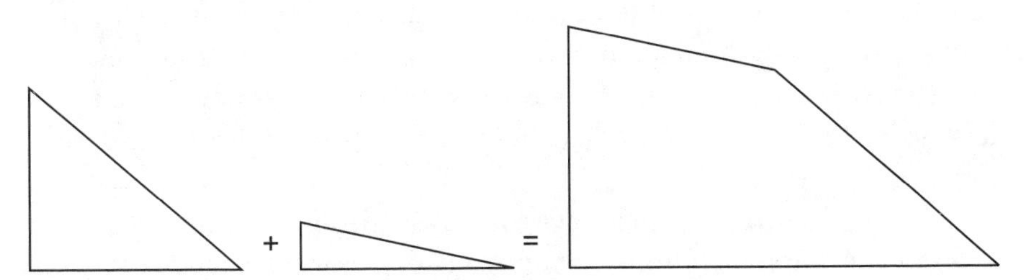

Figure 13.1. The mathematical proof of Ricardo's principle of comparative advantage makes use of a class of equations of which the above is among the simplest.

in connection with anything alleged to be "proven mathematically," is "under what conditions is the conclusion valid?" Checking the assumptions from which a theorem is deduced is like reading the fine print in a contract. If you haven't done so, the thing probably doesn't mean what you think it does.

Even among professional economists, few can enumerate the assumptions upon which Ricardo's conclusion rests. The most obvious asserts that the economies of all trading nations are in compliance with Adam Smith's efficiency principle. That means that the competition in each must be essentially perfect, to keep them uncorrupted by the market failures identified in U.S. antitrust law, including sticky prices, exorbitant executive compensation, and chronically idle resources. Moreover, Frank Knight's "perfect foresight" and "foreknowledge free from uncertainty" must be available to every individual in each trading nation. Should even one of the national economies in question fail to meet even one of the foregoing conditions, Ricardo's logic fails. It is no longer "mathematically certain" that no nation need fear free trade, meaning that the merits and demerits of free trade must be decided on a case-by-case and nation-by-nation basis, without recourse to any controlling abstraction.

Nations afflicted with inefficient markets may very well have more to gain by domestic reform than by expanding foreign trade. The ones plagued with abundant idle resources—especially human resources—may have more to gain by putting those resources back to work. Those hampered by undue pricing power may have more to gain by regulating or eliminating that power. Those in which CEOs and CFOs earn hundreds if not thousands of times as much as their employees may have more to gain by taxing or capping executive compensation. Those in which information is scarce or distorted may have more to gain by overseeing its collection and delivery. And so on. Small wonder that critical thinkers are wont to complain, as the spokesman for a student group protesting the teaching of economics in France quite recently did, that orthodox economic theory consists of little more than a collection of "complex mathematical models that only work in conditions that don't exist."

In addition to the efficiency-related conditions that the economy of each trading nation must satisfy, in order that the exalted "logic of free trade" shall remain valid, there is yet another prerequisite. International transportation must be cost-free. Had Portugal been located on the Australian continent, its wine-for-wool trade with the U.K. would not have been profitable in Ricardo's lifetime. Later perhaps, after iron-clad coal-burning steamships so greatly reduced the cost of interocean transport, but not during the age of sail. Likewise, if the moon were an oil sheikdom, trade between the earth and the moon would not pay, because gasoline brought back from the moon would have to sell for thousands of dollars a gallon. In the absence of transportation costs, Ricardo's theorem asserts that a nation might conceivably fail to benefit from trade but could suffer no actual harm by embracing it. In the presence of such costs, however, trade may very well prove burdensome. The celebrated logic of free trade is fatally flawed by the fact that it is predicated on an obvious falsehood. Transportation does cost money.

Defenders of free trade may reasonably argue that transportation is currently so cheap that its cost is rightly ignored. But practically speaking is not logically speaking, and it is the logic of free trade that is at issue here. Contrary to expert opinion, that logic does not establish beyond reasonable doubt that "international free trade is in the best interests of the trading countries." In a world of scarce resources, some of which must be diverted from production to transportation before trade can take place, such diversion may cause irreparable harm to some or all of the trading nations. Moreover, transportation is more expensive than it may at first seem.

So much of the cost of construction, maintenance, and even day-to-day operation of harbors, highways, railroads, airports, transport aircraft, and even oceangoing vessels is borne by the public (especially outside the United States) that the transportation charges shown on a shipper's invoice seldom represent more than a small fraction of the total delivery cost. Were shippers obliged to bear the missing costs, the volume of international trade would decrease markedly. A theory of international trade that overlooks transportation costs can only be described as deficient.

Free traders are also free to argue that sovereign nations need not engage in trade they deem harmful. But international trade doesn't work that way under the auspices of the World Trade Organization (WTO). Such trade is now conducted by—or in competition with—multinational corporations to which WTO rules grant a blanket entitlement to profit, regardless of any "ancillary damage" they may inflict on citizens of the nations in which they operate. Indeed, under WTO rules, such corporations may request that their own governments of convenience bring suit in their behalf in WTO courts whenever their corporate profits are threatened by actions taken in behalf of its own citizens by the government of a sovereign nation.

Economic theory refers to damages caused by commercial agreements to individuals external to those agreements—meaning that said individuals were powerless to block the agreements in question—as externalities. Most of the harm that befalls individuals as a result of international trade fits this description. Externalities are not unfairly described as the parts of orthodox (neoclassical) economic theory that make it safe to ignore all the other parts.

To make a long story short, attempts to salvage the "simple logic" of free trade seem doomed to failure. Ricardo's principle of comparative advantage applies only to a world in which information grows on trees, transportation costs nothing, resources are never chronically idle, pricing power is nonexistent, wages and prices are never sticky, and executive compensation is moderate. It is difficult to name a country in which even one of those conditions currently prevails, or will in the foreseeable future. Moreover, strictly speaking, logic alone can never establish the effectiveness, or even viability, of any economic policy. Hardy's truism applies to economics as well as astronomy. It is no more possible to prove mathematically that a given policy is viable, let alone advantageous, than to prove that there will be an eclipse tomorrow. Policies and their consequences belong to the tangible world of farms and factories, harbors and highways, money and media, while mathematical proof applies only to the imaginary world of numbers, equations, and precise geometric figures.

The pitch that sold international free trade to the Anglo-American elite, and has recently begun to appeal to other elites, is no less vacuous than the ones that sell fad diets, fad exercise programs, fad teaching methods, fad managerial techniques, fad investment strategies, fad energy sources, and so on. Ricardo's celebrated principle of comparative advantage is as irrelevant to a responsible policy process as a Ponzi scheme is to prudent personal finance. Yet the fact that the sales pitch is false and misleading does not mean that the product itself is worthless. To demonstrate that, one must consult the track record.

In fact, international free trade has a remarkably short and undistinguished track record. Only two important trading nations—Great Britain in the last half of the nineteenth century and the United States since World War II—have ever tried to implement the policy, and neither realized any clear-cut economic advantage from so doing. Sweden tried it briefly during the 1880s, but quickly abandoned the attempt. Although certain maritime city-states have prospered under free trade—Hong Kong and Singapore being obvious examples—full-service economies have been less well served.

The British experiment with free trade was already under way in June 1846, when Parliament repealed the notorious Corn Laws. Prime Minister Robert Peel had begun to chip away at the tariff almost immediately upon assuming office in 1842, when Britain imposed import duties on some four thousand commercial items. By imposing particularly onerous duties on barley, oats, and wheat, the Corn Laws had—since the end of the Napoleonic Wars in 1815—severely restricted the quantity of food entering the kingdom. The shortage became critical during the Irish potato famine of the 1840s. By gradually reducing the duties on those crops over three years to a nominal level, repeal dramatically increased the availability of bread in the kingdom.

To the surprise of many, the resulting fall in the price of bread was unmatched by a corresponding reduction in factory wages. As a result, free trade became wildly popular in industrial cities like Manchester, Birmingham, and Leeds and wildly unpopular in the grain-growing countryside. In between, opinions differed.

Heartened by this experience—and by the prosperity Britain enjoyed during the 1850s—advocates of free trade expanded their efforts in Parliament. The budget of 1860, prepared by then chancellor of the exchecquer William Gladstone, in collaboration with trade negotiator Richard Cobden, reduced the number of dutiable items from 419 to a mere 48, a figure later reduced to 12. Along with a liberal trade agreement with France, that budget placed the U.K. squarely on the path to free trade. It may not have looked like free trade, since the customs duties that remained in effect continued to provide an important fraction of the revenues upon which Her Majesty's government relied—a much larger fraction, for instance, than in either France or Germany.[9] But because the remaining duties applied to such commodities as coffee, tea, sugar, wine, cocoa, and other items not produced in Great Britain, they rendered the government immune to charges of protectionism. To this day, free traders rarely object to tariffs designed merely to generate needed government revenue. Their contempt is reserved for tariffs designed to protect noncompetitive domestic industries.

British students of political economy fully expected the unilateral adoption of free trade to trigger an unprecedented wave of prosperity. Richard Cobden—the leader in Parliament of the British free-trade movement—delivered a speech during the Corn Law debates in which he predicted that "if you abolish the Corn Law honestly, and adopt free trade in its simplicity, there will not be a tariff in Europe that will not be changed in less than five years to follow our example."[10] British leaders were genuinely convinced that, as Gladstone would later put it, Britain's unilateral conversion to free trade would "show the world that, while it is a good to substitute low duties for high duties, to change low duties into no duties at all is, in the view of national wealth, a still greater good."[11]

Continental nations—whose markets British manufacturers yearned to penetrate—had for some time been engaged in a sort of "competitive protectionism" wherein tariff reform was a more or less never-ending process conducted on a quid pro quo basis. Britain's unilateral rate reductions of 1860 merely led its natural trading partners to suspect that it had little left to concede. As those suspicions were confirmed, concessions from the Continent all but ceased. Foreign goods poured into British markets, while British exports to the Continent slowed to a crawl. The imbalance made it all but impossible for Great Britain to maintain its position as the world's mightiest industrial superpower. As early as 1890, by some accounts, both Germany and the United States forged ahead of the U.K. economically, with France and Italy hard on its heels.[12] During its commitment to free trade, which lasted until the outbreak of World War I, Great Britain managed to blow the very largest industrial lead any nation has ever possessed.

The lesson was not lost on delegates to the Bretton Woods conference of 1944, at which were established the institutions needed to guide economic recovery after the war. Among them was the General Agreement on Tariffs and Trade (GATT), designed to reduce international trade barriers through successive rounds of negotiation. Each individual round was to end in an agreement pledging each member nation to implement the negotiated reforms. Only after one round was complete, and the short-term effects known, would a new round begin. Every participating nation was expected to bargain aggressively in what it perceived to be its own national interest.

According to Ravi Batra, the key round of GATT negotiations was the so-called Kennedy Round, concluded in 1967.[13] Before that time, the U.S. economy had been relatively self-sufficient, with combined imports and exports seldom exceeding 12 percent of GDP. That fraction did not change appreciably in the years immediately following the Kennedy Round. Yet suddenly, in 1973, when the OPEC oil embargo fell like a wet blanket on the growth prospects of energy-poor industrial nations like Japan and South Korea, the fraction began to change abruptly. Since 1975, combined U.S. imports and exports have seldom fallen below 20 percent of GNP.

Needing large quantities of ready cash, and attracted by low tariff barriers, corporations domiciled in the most adversely affected nations invaded U.S. markets in force. Markets for low-tech products like textiles and footwear were the initial targets, soon to be followed by those for radios, televisions, automobiles, motorcycles, refrigerators, air conditioners, generators, turbines, cameras, and any number of others long dominated by U.S. firms.[14] Frequently in ways that would have violated U.S. antitrust laws had they been employed by domestic firms, foreign producers relentlessly targeted one U.S. market after another in their quest for cash to pay their energy bills.[15] According to Batra, virtually none of this "export-led growth" would have been possible without the draconian U.S. tariff reductions of the Kennedy Round.

The Kennedy Round was followed by the Tokyo Round of 1979, which brought further trade liberalization, and the Uruguay Round of 1992, which replaced GATT with the more potent World Trade Organization. NAFTA followed in 1994. These and other bouts of trade liberalization have enabled the leading practitioners of export-led growth to prosper as never before. Japan has built the world's second-largest economy—as measured by GDP—while China and India boast astronomical rates of economic growth. Several other "Asian Tigers" have fared almost as well, for almost as long, by selling more (typically far more) than they buy in U.S. markets.

The historical record seems remarkably clear on one point. Whether or not free trade is advantageous to the nations that practice it, it is a gold-plated windfall for the nations that don't. While foreigners are busy reaping the benefits of export-led growth, the citizens of the free-trading nations lose their jobs, their pensions, their health-care benefits, and—in many cases—their dreams. Before trying to decide whether the citizens of free-trading nations are fairly compensated for those losses, we should observe that there is a second question to which the historical record speaks with remarkable clarity.

Economic historian William Lazonick published a book in 1991 titled *Business Organization and the Myth of the Market Economy*. In it he examined the policies in force when such economic powerhouses as Great Britain during the nineteenth century and the United States during the twentieth were still building the economic dominance they later achieved. Perhaps surprisingly, he found that both had attained their dominant positions by violating the very guidelines they were later to advocate worldwide. Instead of practicing free trade, they nurtured their infant industries behind lofty trade barriers. Rather than allowing market forces to dictate the pace of progress, they hastened the building of canals, harbors, railroads, and the like with land grants and public funding.

At times, the British and U.S. governments chartered monopolies, erected trade barriers, and tolerated all manner of pricing power in an effort to encourage the public to save more of its income, and the banks to lend at more favorable terms, than undiluted market forces would have allowed. As a result, British and U.S. corporations enjoyed ready access to capital throughout the periods in which they became industrially dominant. In particular, the fact that the U.S. tariff remained high during the latter half of the nineteenth century, while the British tariff all but vanished, rendered U.S. corporations more profitable than their British counterparts, and thus more attractive to British capital. The American railroads were built with such capital. In no way did a combination of laissez-faire policy at home and free trade abroad enable Great Britain in the nineteenth century, and the United States in the twentieth, to gain economic dominance. Both did it with something closely resembling a deliberate industrial policy. So did Germany during the late nineteenth century, and Japan after 1870 under the Meiji emperors.

In 1989, development expert Alice Amsden published a book about the Korean economy titled *Asia's Next Giant*. In it she argued that (1) the first task of a nation bent on industrial development is to channel investment funds to its corporations, and (2) there is no better way to do so than by keeping the prices they pay artificially low, while the prices they receive remain artificially high. Suffice it to say, without a complete enumeration of the policies she advocated, that a substantial tariff was an indispensable part of her program.

Formal statistical analysis tends to confirm the foregoing observations. Paul Bairoch points out that when Germany—after toying for some years with the possibility of free trade—moved from mild to strong protectionism in 1885, its economic growth rate went from 1.3 percent in the preceding decade to 3.1 percent in the next.[16] When France did the same thing in 1892, its growth rate increased from 1.2 percent in the decade preceding the shift to 1.3 percent in the next. Sweden, too, saw its GNP growth accelerate in 1888, when it abandoned its flirtation with free trade. Only Italy experienced a discernible slowing of economic growth after joining (in 1887) the protectionist parade.

Elhanan Helpman quotes Bairoch to the effect that, except in Italy, "the introduction of protectionist measures resulted in a distinct acceleration of economic growth during the first ten years following the change in policy, and that this change took place regardless of when the measures were introduced."[17] Using data from ten European countries for the years between 1875 and 1914, Kevin O'Rourke estimated the effect of tariff levels on the growth of real per capita income.[18] His results confirm Bairoch's conclusion that increased levels of protection do indeed appear to have sped per capita income growth during the late nineteenth century. Michael Clemens and Jeffrey Williamson confirmed O'Rourke's finding for a sample of more than thirty countries between 1870 and 1913, but found the effect to be reversed during the post–World War II period.[19] Helpman cites a number of other studies that report similar findings regarding the postwar era, before concluding that "the evidence favors [albeit weakly] a negative effect of protection on rates of growth in the post–World War II period."

Dani Rodrik and Francisco Rodríguez would beg to differ.[20] By subjecting several oft-quoted statistical studies of international trade to close scrutiny, they were able to shed considerable doubt on the adequacy of the methods employed. They point out, among other things, that protectionism is a matter of degree rather than kind, so that methods that classify each nation as either a 0 (protectionist) or a 1 (practitioner of free trade) can hardly fail to overlook important differences. Moreover, nations committed to free international trade tend to pursue laissez-faire policies at home, making it hard to decide how much of their relatively rapid growth is due to their liberal domestic policies and how much to their similar international ones. They conclude their paper by quoting a famed nineteenth-century humorist: "It ain't the things we don't know that get us in trouble. It's the things we know that ain't so."[21] To contrarians like Rodrik and Rodríguez, the list of things professional economists purport to know, but which just ain't so, must include the blanket declaration by Milton Friedman and his wife, Rose, that "International free trade is in the best interests of the trading countries and the world."

Statistical studies of the sort cited above seldom address the matter of illegal cross-border cash flows. The fact that such flows go unreported, making their magnitudes difficult to estimate, does not make them unimportant. A book on the subject has recently appeared. Written by Raymond W. Baker, an experienced international businessman and student of global trade, the book explains the means by which most such flows are generated and estimates their various magnitudes.[22] He concludes, in particular, that illegal cross-border cash flows easily exceed $1 trillion annually, with about half the total originating in third-world nations.[23] He deems the resulting losses particularly unfortunate at a time when the seven largest economies generate fully two-thirds of the world's combined GDP. Such a drain on third-world income can only widen the gap between rich and poor nations, as well as that between rich and poor people around the world.

Baker distinguishes three distinct types of what he calls "dirty money": criminal, corrupt, and commercial. The criminal category is the most familiar, as it includes money from the sale of drugs, from human trafficking, from illegal arms sales, and from the counterfeiting of goods and currency, as well as the proceeds from smuggling and racketeering—activities featured every night of the week on prime-time TV. Corrupt money, which comes mainly from bribes and kick-backs—activities deemed too uninteresting to appear regularly on TV—constitutes the smallest of the three categories, accounting for only about 3 percent of the worldwide total. Commercial dirty money, which is also seldom seen on TV because it results from (boring) white-collar crime, is much the largest category. By Baker's estimate, it accounts for almost two-thirds of the dirty money total.

Mispricing is a particularly simple method of "liberating" money. Baker illustrates it with a story about a South American businessman who agrees to purchase an up-to-date piece of equipment from an American manufacturer for $1 million, then asks that his firm be billed for $1.2 million, the extra $200,000 to be deposited in his own personal New York bank account. The manufacturer, anxious to make the sale, agrees, perhaps after requiring the businessman to perform some nominal service (submit a report) in return for the extra payment. Baker suggests that some $200–$250 million are generated annually in this way, at least half of it being liberated from less developed countries.

When mispricing occurs between separate divisions of a single firm, located in different countries, Baker calls it "abusive transfer pricing." For example, millions of carats of diamonds were shipped out of South Africa during the twentieth century for next to nothing, to be cut, graded, and sold abroad. The receiving (typically European) division of the mining company paid the South African division very little for the gems, and virtually no taxes went to the South African government. Alternatively, if the company opened a division in a tax haven such as the Channel or Cayman islands, the gems could be sold to the new division for the same rock-bottom price, then resold to the European division for a great deal more, making the bulk of the profit appear to have been earned in the tax haven.[24] Since intracompany trade across borders accounts for 50 to 60 percent of all cross-border trade, there is plenty of scope for abuses of this sort. To facilitate such practices, which generate $300–$500 million annually, according to Baker, Ernst & Young circulates its *Transfer Pricing Global Reference Guide*, offering "creative and practical solutions for your transfer pricing needs."

Baker distinguishes abusive transfer pricing from "fake transactions," in which someone allows himself to be billed—usually for a kickback of some kind—for goods and/or services that are never in fact delivered. Such transactions have, he claims, sucked billions out of the former Soviet Union in the past fifteen years. He includes the billions pouring into the Bank of New York, and various European banks, in this category. It would be inappropriate to place them in the criminal category because, he says, the criminal origins of the funds are seldom if ever established. He estimates the volume of this flow at between $200 billion and $250 billion a year.

The dirty little secret about dirty money is that the global financial community, including the world's largest and most respected banks and financial institutions, is almost pathetically eager to handle it. As a result, member institutions make no discernible effort to determine the origin of incoming funds. Money derived from the sale of drugs, girls, or stolen goods is still money, and their job is to treat it with respect. The last thing they want is a diminution of the flow. On the contrary, they would doubtless welcome an increase. But, assuming Baker's estimates to be tolerably accurate, the flow of dirty money already represents 15 to 25 percent of the dollar value of international trade—as measured by the volume of total exports—and it's hard to imagine how that fraction could get much larger. A business owner can skim 10 percent of the proceeds from his operation, or maybe even 20 percent, but (even in Las Vegas) he can't get away with 100 percent, or (at a guess) even 75. Accordingly, the likeliest way to increase the flow of dirty money is to increase the flows of legitimate money with which it may mingle. That alone makes it hardly surprising that free trade has friends in high places.

A great deal of money is being made offshore, at present, due in part to the paucity of regulations in effect there and in part to lax enforcement. Moreover, the global financial community can barely conceal its eagerness to handle the tainted money, few if any questions asked. Accordingly, anything calculated to increase the volume of international trade finds strong support along the Washington–Wall Street axis, while anything expected to diminish that volume draws heated opposition. Yet the resulting free-trade lobby—liberally supported by the financial industry—escapes dismissal as just another greedy special interest group because of the pristine reputation free trade continues to enjoy.

It is remarkable how many otherwise skeptical opinion makers accept that reputation at face value. P. J. O'Rourke, author of the classic bestseller *Parliament of Whores*, devotes an entire chapter of a more recent book to his understanding of basic economic theory, including a rather fanciful endorsement of "Ricardo's Law of Comparative Advantage" involving the author John Grisham and the singer Courtney Love.[25] Scott Adams, creator of the comic strip *Dilbert*, coins the term "confusopoly" to describe a "group of companies with similar products who intentionally confuse customers instead of competing on price."[26] He observes that the telephone service industry, the insurance industry, the mortgage loan industry, the banking industry, and the financial services industry are natural confusopolies, "because they offer products that would be indistinguishable to the customer except for the great care taken to make them intentionally confusing." Nowhere does he betray the slightest doubt that the only legitimate function of a public corporation is to enhance the efficiency of the market(s) it serves by behaving as a perfect competitor.

How do two legendary skeptics, noted for their disrespect for all forms of authority, come to accept orthodox (neoclassical) economic theory on the authority of experts whose abstract technical arguments both claim to find incomprehensible? And how can the organized societies of skeptics, such as the Committee for the Scientific Investigation of Claims of the Paranormal and the Skeptics Society—publishers, respectively, of *Skeptical Inquirer: The Magazine for Science and Reason* and *Skeptic* magazine—resist exposing von Mises's claim in behalf of economic theory that "its particular theorems are not open to any verification or falsification on the ground of experience" to the ridicule that anything so preposterous would surely elicit from their readers? Why, in short, do so many challengers of other forms of conventional wisdom decline to challenge conventional *economic* wisdom? What have the purveyors of that wisdom ever done to earn such respect?

Why, a professional skeptic might ask, do Americans buy and sell water rights, mineral rights, hunting rights, fishing rights, grazing rights, patent rights, copyrights, translation rights, syndication rights, movie rights, book rights, sunshine rights, and rights-of-way, even as they trade more option rights than actual securities in financial markets, while insisting that the right to off-load steel, automobiles, garments, textiles, and consumer electronics onto U.S. soil must be given away free of charge? Why does the U.S. Constitution guarantee free speech, free worship, free assembly, and a free press, yet say nothing of free love, free lunch, and free trade? What do importation rights have in common with other items modern society deems it improper to buy and sell, such as sex, drugs, and babies, along with the outcomes of trials, elections, and athletic contests?

No one denies that protective tariffs and other barriers to trade elevate the consumer price index by forcing the public to pay higher prices for imported goods and services. In developed nations, imports now include most manufactured and many agricultural goods, along with growing portions of routine office work and other services. Such is the cost of protectionism. Nor does anyone seriously doubt that protectionism elevates per capita incomes by saving jobs, especially in manufacturing, where many of the better-paying jobs have traditionally been found. Such are the benefits of protectionism. Informed citizens tend—usually without knowing quite why—to suppose that the benefits in question far exceed the costs. Hopefully, the foregoing revelations will cause such citizens to reexamine their commitment to such suppositions.

Defenders of free trade are well aware that they have a lot of explaining to do. They must explain, first of all, why the large adverse balances of trade typically run by free-trading nations need not be a cause for concern. They usually begin, as does Buchholz, by pointing out that when Americans spend more abroad for cars, TVs, and other manufactured goods than their employers earn in foreign markets, they receive useful goods and services in return for useless paper.[27] That argument ignores the fact that exported cash, unlike imported cars, TVs, and articles of clothing, continues to circulate throughout the receiving economy, working its magic wherever it goes, and magnified many times over by the (Keynesian) multiplier effect. Figure 13.2 shows how cash flows between two trading nations.

If nations A and B did not engage in trade, the trade routes joining their economies would be absent, and the impact (Keynesian multiplier) of a dollar injected into either could again be computed as in Chapter 10. Each such dollar would be divided as before between savings and consumption, and each nation's multiplier would again be determined by its own "propensity to consume." The nation with the larger propensity would, of course, possess the larger multiplier.

If the nations did engage in trade, each dollar injected into either

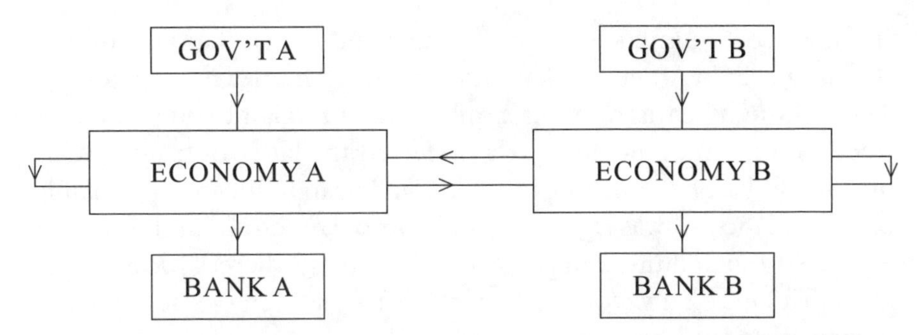

Figure 13.2. If the trade routes between economy A and economy B were cut off, the Keynesian multipliers for each individual economy could be computed in the usual way. Whereas opening the routes complicates the computations only a little, the results can differ significantly.

economy would be divided three ways instead of two. Some would again be saved, some spent for consumption of domestic goods and services, and the rest spent for imported ones. That is why the figure provides three exits from each economy. One of them leads to the associated bank, one leads back into the economy of origin, and one leads to the other economy. As a result of trade, each dollar injected by either government into its own economy will end up stimulating both national economies to some extent. Indeed, given each nation's propensities to save, consume domestically, and consume from abroad, a computation only slightly more complicated than the one described in Chapter 10 would enable each to determine how much self-stimulation and how much cross-stimulation will ultimately result from each dollar injected.

In particular, two nations with the same savings rate, but with different propensities to consume from abroad, are destined to achieve dramatically different results. Whereas the nation with the smaller appetite for imported goods and services will find it easy to stimulate and grow its own economy, the one with the larger appetite for imports will encounter difficulty. Money will exhibit a tendency to tarry in the economy of the reluctant importer, because there will typically be more cash in the pipeline flowing toward that economy than in the one flowing away.

The situation brings to mind a remark long attributed to Abraham Lincoln: "I don't know much about the tariff, but this I do know.

If I buy a coat in England, I have the coat and England has the money. If I buy a coat in America, I have the coat and America has the money." Aware though he was of Adam Smith, and of England's powerful free-trade movement, Lincoln clearly preferred the latter outcome. Buchholz scoffs, before launching into his own paean to Ricardo's principle, that "he [Lincoln] was right—he didn't know much about the tariff."[28] Seldom do such illuminati pause to reflect that, whereas Ricardo's exalted principle "only work[s] in conditions that don't exist," Lincoln's humbler insight is universally valid.

CHAPTER 14

Heterodox Economic Thought

The world continues, as it long has, to harbor diverse schools of economic thought. Yet only the one erected near the end of the nineteenth century under the leadership of Alfred Marshall, on the foundation laid by Adam Smith and David Ricardo, can be regarded as orthodox. The result goes by many names, including "mainstream economics," "neoclassical economics," and "laissez-faire economics," although "perfectly competitive economics" would be more descriptive. Of competing schools, the public knows little or nothing. Several have lately begun to band together in organizations like the Association for Heterodox Economics (AHE) and the International Confederation of Associations for Pluralism in Economics (ICAPE) to compete more effectively against the reigning orthodoxy, which is seen as hostile, misleading, and a fertile source of dysfunctional public policy.

Both AHE and ICAPE are umbrella organizations, offering membership only to other (recognizably heterodox) organizations. Each currently includes more than thirty member organizations, many of which belong to both conglomerates. The present chapter will attempt to describe the "family tree" of heterodox schools of economic thought, their current levels of activity, and the associations formed to promote them.

By the time of Adam Smith's death in 1790, his *Wealth of Nations* had been translated into Danish, French, German, Italian, and Spanish. Everyone in government on either side of the Atlantic, not least the framers of the U.S. Constitution, was conversant with laissez-faire ideology.[1] But conversant does not mean convinced, and many doubted the wisdom of Smith's policy prescriptions. Treasury Secretary Alexander Hamilton went beyond doubting to recommend, in his famous "Report on Manufactures," that the new nation directly contravene Smith by encouraging fledgling industries with tariffs and subsidies, at least until they could hope to survive in competition with the British.

Hamilton's chief opponent in the ensuing debates was Thomas Jefferson, who, like Smith, maintained that goods should be purchased wherever they could be obtained most cheaply. He was more than content to let factories and factory workers remain in Europe, while the United States developed along more agrarian lines. Students of economic history often remark on the irony that, although Jefferson's likeness appears on Mount Rushmore while Hamilton graces only the $10 bill, it was mainly Hamilton's plan that guided the nation through its first 150 years under the Constitution.[2] The Erie Canal was financed, at Hamilton's suggestion, with state-guaranteed loans; the western railroads received huge land grants to offset construction costs; and the military, beginning in 1798 with a purchase of ten thousand muskets from a still-struggling inventor named Eli Whitney, commissioned large quantities of arms and matériel from fledgling U.S. manufacturers.

The muskets themselves were the least of what Congress achieved by contracting with Whitney. The purchase also stimulated the development of mass-production methods and facilities in the United States. After working tirelessly to devise the equipment he would need to produce the required muskets within the allotted twenty-eight months, Whitney brought a quantity of parts to Washington and invited the startled congressmen to assemble their own musket locks. The age of standardized parts, they realized, had arrived in the New World. At about the same time, Congress founded West Point, ostensibly a military institution, and instructed it to train engineers skilled in the arts of cartography and road construction. When the time came to settle the West, it was done with the aid of maps drawn by Army officers, on roads built by Army engineers and protected by Army forts.

In 1825, a former member of the Württemberg Chamber of Deputies named Friedrich List (1789–1846) immigrated to the United States and, in the fullness of time, found work as a journalist. In that capacity he became familiar with Hamilton's writings. Soon thereafter, the discovery of coal on land he had acquired left him financially independent and enabled him to return to his homeland as U.S. consul at Leipzig. Once there he began to advocate—as Hamilton surely would have—extension of the German railway system and expansion of the customs union known as the Zollverein. Formed after the Napoleonic Wars, this union eliminated trade barriers between member states and imposed a common external tariff. Prussia soon emerged as the preeminent member and saw to it that the tariff remained low enough to expand the union's trade at Austria's expense.

Prussia's entire strategy in the aftermath of the Napoleonic Wars was designed to gain ascendancy over Austria in central Europe. The Austro-Prussian War of 1866 was just the last step in that campaign. Prussian leaders, who saw themselves as underdogs in a race to dominate the region, left no stone unturned in their zeal to prevail. They were, moreover, as attentive to industrial as to military policy. Indeed it was largely due to the fact that Austria's Prince Metternich, who dominated European politics between the end of the Napoleonic Wars and the revolutions of 1848, was notoriously indifferent to economic issues that Bismarck's Prussia was eventually able to bring southerly Württemberg and Bavaria—both of which initially leaned toward Austria—into the Prussian orbit that became the German nation.

In 1841, List published a book titled *Das nationale System der politischen Ökonomie*, which permanently cemented his reputation as an economist and expert on industrial development, first in Germany and later throughout the world. According to one biographer, "This book has been more frequently translated than the works of any other German economist, except Karl Marx."[3] In it List described a suite of economic policies predicated on national interest, and flatly contradicted the notion that a citizen's private interest is never in conflict with that of the nation. Although canals and railroads greatly stimulate national economies, he pointed out, they also bankrupt countless wagon drivers. And although slavery is ever a calamity to the nations that allow it, many of their citizens grow rich in the slave trade, or as slave-owning plantation owners producing sugar, tobacco, and cotton. Where Smith saw a world of common interests, and suggested that conflicts of interest are less real than apparent, List saw a world of genuine and abiding conflict between all but irreconcilable interests.

While approving of free trade among relative equals, such as the member states of the Zollverein, List denied that industrially backward nations could benefit from unregulated traffic with more advanced trading partners. At a time when Great Britain stood head and shoulders above all others in industrial power, List's policies were intended to enable the German union to play catch-up. James Fallows describes the determination with which the Meiji emperors in Japan, like Bismarck in Prussia/Germany, pursued Listian policies after 1870, with results deemed miraculous by many observers.[4]

List did not live to see his policy prescriptions enable the United States and Germany to surpass Great Britain in industrial might, as both did sometime between 1890 and 1910. List's health declined after 1840, to the extent that he felt obliged to decline the editorship of the *Rheinische Zeitung*, a post eventually filled by Marx. Government censorship and persecution soon caused Marx to flee the country. List lost much of his American property during a financial crisis and died by his own hand on November 30, 1846. Today, both he and Hamilton would probably be considered latter-day mercantilists, the name now given to a school of thought that emerged between the mid-sixteenth and the late seventeenth centuries to advise the crowned heads of Europe on the relationship between a realm's wealth and its balance of foreign trade.

Because he functioned as a public intellectual rather than as a teacher, List left behind no identifiable school of economic thought. But the German historical school (GHS) that came to dominate the teaching of economics in Germany from the middle of the nineteenth to the middle of the twentieth century borrowed far more from List than from the intellectual heirs of Adam Smith. Indeed the GHS tended to dismiss the English economists, with their emphasis on commercial freedom, as mere touts for an ideology that suited the needs of their nation at a particular moment in history. For latecomers to industrialization—such as Germany, Japan, and the United States—they recommended government planning and involvement in commercial affairs.

It is hard to identify any central core of shared belief among members of the GHS, beyond an emphasis on careful historical research, a willingness to deploy the power of the state to accomplish chosen ends, and an outspoken disbelief in the universality of economic laws. Toward the end of the nineteenth century, a bitter "method war" erupted between members of the GHS, led by Gustav von Schmoller, and members of the Austrian school of economic thought, led by Carl Menger. While the Schmoller camp dissected historical records, the Austrians pursued theoretical research of the sort practiced in England. The triumph of the GHS in Germany ensured that the English-speaking world would see little merit in any economic research conducted there until well after World War II. The German-speaking economists who flooded into the United States during the Nazi era, including Joseph Schumpeter, Gottfried Haberler, F. A. Hayek, Ludwig von Mises, and Oskar Morgenstern, came almost exclusively from Austria.

The GHS exerted significant influence on the development of American economic thought through the German-trained founders of the American Economic Association. Their early exposure to historical methods made them particularly receptive to mavericks like Thorstein Veblen, whose 1899 book, *The Theory of the Leisure Class*, called into question—indeed subjected to ridicule—many venerable orthodox teachings.

Born into a community of Norwegian immigrants on a remote Wisconsin farm, Veblen was educated at Carleton College in Northfield, Minnesota, the Johns Hopkins University, Yale—from which he earned a Ph.D. in philosophy—and later Cornell. He was thirty-five years old by the time he landed his first nonfarm job at the University of Chicago.

Veblen was a gifted wordsmith who enriched the English language with such phrases as "the higher learning," "captains of industry," and "conspicuous consumption." He is generally considered an institutional economist, in that he assigned less importance to individual wants, needs, and preferences than did his more mainstream contemporaries while emphasizing the importance of habits, customs, and conventions in every aspect of human behavior. Like other institutionalists, he distrusted mainstream laissez-faire economic theory, mathematics, and Marxism. It's harder to say what he and his compatriots did believe in, beyond "getting the facts." Indeed Frank Knight, who for many years taught a course on institutional economics at the University of Chicago, always began his first lecture by confessing that he didn't know what institutional economics was. Veblen never tried to construct a detailed model of an industrial economy and probably didn't think it could be done.

Institutional economics prospered in the United States until World

War II, probably reaching its peak of influence within the profession in 1917, with the election of John R. Commons to the presidency of the American Economic Association. The school's prestige declined after the Great Depression, which it proved powerless to explain, much less cure.

Though many of the early institutionalists were trained in Germany, they were less doctrinaire than GHS members and less given to personal vendettas. They were also blessed with leaders like Commons, Richard Ely, Francis Amasa Walker, John Bates Clark, Wesley Mitchell, and (more recently) John Kenneth Galbraith, who encouraged younger men—more often with words than with deeds—to rely on factual evidence in defending their theses and policy prescriptions. As a result, nearly all of the pioneering work in empirical economics done in the United States can be traced to the students of avowed institutionalists. During the Progressive Era, economists at the University of Wisconsin (a hotbed of institutionalism, where both Commons and Ely taught for many years) pioneered the then controversial notion that university-based social scientists could serve the public directly by designing government reforms. Under the progressive Republican governor Robert La Follette, they helped formulate the laws that created the country's first unemployment program and workers' compensation insurance and produced model regulatory bills for the railroads and utilities.[5]

The institutionalist movement marches today under the banner of the Association for Evolutionary Economics (AFEE), a name inspired by the title of one of Veblen's most famous articles, and also the International Thorstein Veblen Association.[6] AFEE publishes the *Journal of Economic Issues*, sponsors scholarly meetings (usually in conjunction with other learned societies), and maintains a Web page from which it is possible to download a representative sample of institutionalist literature. Members' interests have evolved in recent years to include the influence of diverse cultures on economic performance, domestic and international trends in income inequality, the impact of new technology on the biosphere, the influence of political power on economic outcomes, the increasing weight of multinational corporations on the global economy, and other issues upon which orthodox economics sheds more heat than light.

A distinction is now drawn between new and old institutionalists.

Whereas the old criticized orthodox economic theory for ignoring institutions, the new employ the methods of that theory to analyze institutional behavior. While some of the new are formally trained as economists, many are lawyers with a taste for economic theory. As early as 1916, Louis Brandeis wrote that a "lawyer who has not studied economics . . . is very apt to become a public enemy."[7] This has been true from the very beginning as it applies to antitrust law, but has become increasingly true in other branches of the law as well. Todd Buchholz asserts that negligence law, property law, criminal law, and the branch of the law that applies to corporate finance have been "dramatically transformed" by economists.[8]

Unsurprisingly, there is now an International Society for New Institutional Economics, from which AFEE is careful to disassociate itself by declaring in its mission statement that its interests extend only to the old institutional economics. There is also a European Association for Evolutionary Political Economy, a Japan Association for Evolutionary Economics, a Society for the Advancement of Socio-Economics, and an Association for Institutional Thought.

Similar in many ways to institutional economists are the self-styled post-Keynesians, who came together some years after the death of John Maynard Keynes to protect his legacy from the army of trivializers then seeking to erase what they could of the mark he made on economic policy and thought. The group originally included a number of Keynes's closest associates, including Richard Kahn, Joan Robinson, Nicholas Kaldor, Roy Harrod, and the Polish Marxist Michal Kalecki. Kalecki caused quite a stir when, upon his arrival in Cambridge, he turned out to have anticipated much of the argument contained in Keynes's *General Theory of Employment, Interest, and Money*.

Alfred Eichner, a leading chronicler of the post-Keynesian movement, traces its birth to the publication in 1956 of a book titled *The Accumulation of Capital*, by Robinson, and an article in *The Review of Economic Studies* titled "Alternative Theories of Distribution," by Kaldor.[9] Both, says Eichner, represented attempts to expand the gaping holes exposed by Keynes in the fabric of orthodox theory at a time when most in the profession—led by Samuelson—were salvaging as much as they could of the neoclassical creed by containing the "Keynesian heresy."

Eichner identifies five distinct points upon which post-Keynesian economics (PKE) differs with the reigning orthodoxy.[10] First, PKE recognizes and undertakes to explain the connection between economic growth and the distribution of income. It does so in part by focusing attention on the national rate of investment, both public and private, instead of fixating on the (consumer) price index. Second, it recognizes the dominant fact of recent economic history, the rapid if uneven rate of growth observed in all or most national economies, at least since the sixteenth century. Neoclassical models always suppose that even the most modern economy is forever in the process of coming to rest in a state of full employment and will inevitably do so in the absence of external interference, albeit less rapidly than politicians sweating reelection ordinarily wish.

Third, post-Keynesian theory presupposes the existence of sophisticated lending institutions with short-, medium-, and long-range goals and aspirations of their own. Fourth—and perhaps most important—PKE recognizes that wages and prices in the industrial core of a modern economy are mainly "administered prices," quoted on a take-it-or-leave-it basis, even if "special discounts" can occasionally be arranged. In other words, PKE recognizes both the existence and the potential importance of competition that is less than perfect. Fifth, PKE is designed to accommodate both noncompetitive market mechanisms (such as those involving single-source suppliers and no-bid contracts) and nonmarket methods of income distribution (such as public health care and entitlement programs) with relative ease.

Builders of the computable general equilibrium (CGE) models discussed in Chapter 10 might argue that they, too, have learned to incorporate such neoclassical incongruities into their models and might even question the adequacy of the treatment given them by PKE. But the post-Keynesians do at least deal with them openly, instead of acknowledging their existence with words while denying their importance with deeds, the way orthodox model builders too often do. So closely allied are post-Keynesian and institutional economics that many in both camps now describe themselves as "post-Keynesian institutionalists."

Another allied discipline is known as system dynamics. Yet another brainchild of Jay Forrester, a pioneer in the computer industry, system dynamics was developed during the 1950s as a multipurpose modeling tool amenable to use in the (indescribably primitive) computers of the day. It was adopted by the Club of Rome as the technology best suited to the long-term modeling required to produce the club's famous *Limits to Growth* report of 1972.[11] That report was based on output from a system-dynamics model known as World3. It enraged members of the economics profession by addressing economic issues without appeal to existing economic theory. Follow-up reports issued in 1992 and 2004 by three of the original four authors strengthen and clarify the main arguments.[12]

The more recent of these observes that "the highly aggregated scenarios of World3 still appear, after thirty years, to be surprisingly accurate." Specifically, "the world in 2004 had the same number of people (about 6 billion—up from 3.9 billion in 1972) that we projected in the 1972 standard run of World3. Furthermore, that scenario showed a growth in global food production (from 1.8 billion tons of grain equivalent per year in 1972 to 3 billion in 2000) that matches history quite well." Although no number of such confirmations can ever prove the model "correct," they do constitute tests that the model might have failed but didn't.

At the risk of oversimplification, it may be said that the report's take-home message concerns the possibility of overshooting the earth's carrying capacity. Like a remote island in the Pacific, the earth itself can sustain only so many cattle, horses, sheep, pigs, goats, and/or humans. Even though ecologists are as yet unable to pin down the earth's human carrying capacity as accurately as they might estimate the carrying capacity of a remote island for cattle, sheep, pigs, or goats—because we humans are better able even than rats to adapt to harsh conditions—it is self-evident that a limit exists. Sustained overpopulation leads in time to population collapse.

Perhaps the most dramatic example of such a calamity seems to have occurred on remote Easter Island, not long before the first Europeans landed there in 1722. Best known for its large and mysterious stone statues, the island was apparently populated by Polynesians about a thousand years ago. Archaeological evidence suggests that its now almost treeless slopes were then heavily forested and that the colonists soon developed a thriving and complex civilization. The population may have exceeded ten or even fifteen thousand people during the statue-carving era, which had all but ended by the time the first Europeans arrived. In addition to a number of unfinished statues, they found a population of only two or three thousand. Admittedly sketchy evidence suggests that gradual deforestation led in time to a sudden famine, followed by population collapse.

Jared Diamond has called the Easter Island collapse "the clearest example of a society that destroyed itself by overexploiting its own resources."[13] Both Diamond and the *Limits to Growth* authors maintain that humanity itself is now flirting with a similar disaster. The 1992 update of *Limits to Growth* claimed that the global population had already overshot the earth's carrying capacity and would eventually collapse unless appropriate steps are taken. In the 2004 publication, the authors voice their disappointment that humanity has never—as it seemed prepared to do at the 1992 global summit on the environment in Rio de Janeiro—decided to deal seriously with looming environmental threats. They deplore the fact that the Rio + 10 conference held at Johannesburg in 2002 produced even less action, and express a wish to "inspire the world's citizens to think about the long-term consequences of their actions and choices—and muster their political support for actions that would reduce the damage from overshoot."

System dynamics is a grab bag of computer-friendly techniques for modeling virtually any system that undergoes change. It has been used with varying degrees of success to model chicken farms, fish migrations, wild animal populations, urban growth, corporate sales, various microclimates, pressure-control devices on steam engines, and (by the *Limits to Growth* authors) the global economy itself. Unlike the mathematics employed in the construction of macroeconomic and CGE models, designed as they are to produce a single "snapshot in time," these techniques yield an entire "moving picture" of a system in evolution.

The output of system-dynamics models is frequently presented in the form of "evolution diagrams." The one in figure 14.1 shows how the hiring and departure rates in the sales force of a start-up corporation might evolve with the size of that force. Both the size of the force and the rate of departures from it rise rather steadily to their stable equilibrium sizes, while the hiring rate rises well above its eventual equilibrium value before falling back to it. The builders of CGE models are of course entitled to reply that they, too, can produce moving picture film by stringing together "stills" in the usual manner. But system-

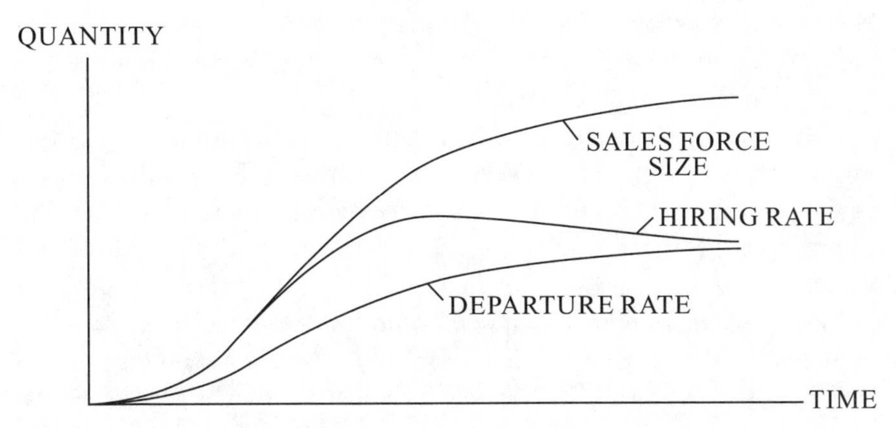

Figure 14.1. Sales force size at a start-up corporation.

dynamics modelers claim to achieve equally reliable predictions in less time, with less effort, and more affordably.

System dynamics is well known to environmental and ecological economists, who use it to explore the effects of proposed policy changes and to explain in words and pictures how the phased implementation of a particular option will play out over time. The usual goal of such investigations is sustainability—for the East Coast deer population, the Atlantic cod, the Pacific salmon, the African rhinoceros, the Asian tiger, the mountain gorilla, the Northwest timber harvest, the Oriental rice crop, and so on. Herman Daly finds it amusing that there is now a fight to control the meaning of the word "sustainability."[14] No sooner had activists like Rachel Carson and the *Limits to Growth* authors managed to stir up public awareness of impending threats to the ecosystem than business interests began to introduce oxymorons like "sustainable growth" into the dialogue. In Daly's opinion, the human race possesses but a single asset—the planet we live on—and our fate depends on the magnitude of the return we are able to coax from that (fixed) asset.

The concept of sustainability in a finite, nongrowing, and materially closed ecosystem turns out to be a difficult one to incorporate into orthodox economic thought, in which absolutely nothing is permanently finite, nongrowing, or materially closed.[15] Daly argues that if sustainable development is to mean anything at all, it must mean that the economy is but a part of the earth's materially closed ecosystem.

Hence mankind must abandon the ideal of eternal economic growth. We must learn instead to become healthier and happier in ways that do not entail indefinite expansion.

There is an Association of Environmental and Resource Economists (AERE), which currently boasts some eight hundred dues-paying members from more than thirty nations, as well as a smaller International Society for Ecological Economics and a United States Society for Ecological Economics (USSEE). AERE publishes the *Journal of Environmental Economics and Management*, as well as a semiannual newsletter, while USSEE publishes *Ecological Economics*. AERE also sponsors sessions at the winter meeting of the Allied Social Science Association (ASSA), the annual meeting of the American Agricultural Economics Association, and the World Congress of Environmental and Resource Economists. Members find employment in government, industry, and the university system, where they inhabit departments of agricultural economics, forestry, and natural resources, as well as economics per se.

The International Network for Economic Method aims to serve all who—like the participants in the Austro-German method war of the late nineteenth century—wish to debate the appropriateness of various methods to the study of economics. To that end, it publishes *The Journal of Economic Methodology*, whose editorial board includes specialists in game theory, econometrics, economic history, and experimental economics, as well as historians, philosophers, and sociologists of science, in an effort to revive the once vital debates between critics and defenders of perfectly competitive, laissez-faire, neoclassical orthodoxy.

Agent-based computational economics (ACE) is among the newest economic methods. Alan Kirman, then based at the European University Institute in Florence, used ACE to explain a curious experiment involving ants. Two identical food sources were placed at identical distances from an ant colony and were constantly replenished, so that they remained identical at all times throughout the experiment. Since there was nothing whatever to choose between the two sources, one might think the ants equally likely to visit either one. And in the long run, that was indeed the result: roughly equal numbers of visits were ultimately made to each site. In the short run, however, something quite different was observed.

At any given instant during the course of the experiment, each individual ant was en route to or from a food source, either A or B. But the number en route to or from A was seldom equal—even approximately—to the number en route to or from B. For a while A would be the more popular destination, but then B would become the more fashionable, then A again, and so on.

The fraction of the colony en route to or from A continued to oscillate, sometimes including two-thirds of the colony, sometimes less than one-third, for as long as the experiment continued. These puzzling observations presumably had to do with the well-established fact that ants are able to follow one another to known food sources. Once an ant has found such a source, it is likely—though not certain—to return. Moreover, during its food-laden return to the colony, it secretes a substance that alerts other ants to follow. Some breeds secrete more copiously than others, causing the details of ant behavior to vary from one breed to another. Yet the oscillatory behavior seems present in all breeds. Kirman demonstrated that the degree of randomness present in a breed's behavior explains this puzzling fact.

The model he constructed required each member of an imaginary ant colony to exercise one of only three departure options: a colony member could (1) return to the most recent site at which it found food; (2) follow another ant to its fruitful destination; or (3) elect to visit, at its own discretion, a different site. Kirman learned how to control his virtual ants' oscillatory behavior by varying the probabilities with which they exercised the three options. The graph in figure 14.2 shows the fraction en route to or from food source A at any time during a particular run of his experiment.

When a high probability was assigned to option 3, whereby an ant switches destinations at its own discretion, Kirman found that the fraction of the colony en route to or from food source A tended to oscillate between rather narrow limits (say, 45–55 percent), seldom escaping from that range. But when he assigned a low probability to option 3, the amplitude of the oscillations increased. In the latter event, the fraction en route to or from A was either larger than 80 percent or smaller than 20 percent during much of each run. The fact that actual (as opposed to virtual) ant colonies exhibit relatively large oscillations indicates that the instincts to follow a successful forager and to return to proven food sources are strong. But the persistence of sizable fluctuations across all breeds indicates that random behavior also increases biological "fitness."

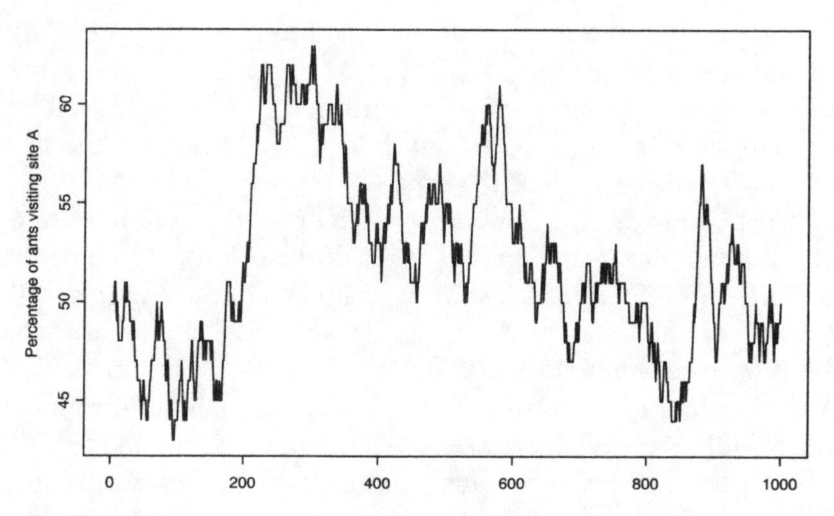

Figure 14.2. A typical run of Kirman's ant colony simulation.

The ants in Kirman's experiment are examples of the independent decision-making agents ACE is designed to investigate. The probabilities with which they exercise the available options amount to inbred strategies, likely to have evolved over countless generations for exploiting the kinds of information ants are genetically equipped to process. ACE is a technique for analyzing the interactions of such agents in environments containing many of them and in which a variety of distinct opposing strategies may well be present.

Although there is as yet no organized association of ACE practitioners, there is an active ACE website accessible through both ICAPE and AHE. There is, moreover, a Society for the Advancement of Behavioral Economics, many members of which make regular use of ACE methodology. Behavioral economics has to do with the rules by which every organism—including humans—makes everyday decisions. The (randomized) rules employed by Kirman's virtual ants seem fairly typical.

Also deemed heterodox are the Marxist and socialist economists once common on both sides of the Atlantic. Many in the United States now belong to the Union for Radical Political Economics, a venerable organization currently headquartered in Amherst, Massachusetts, which publishes the *Review of Radical Political Economics* as well as a quarterly newsletter, holds an annual summer camp meeting, and contributes several sessions to the ASSA convention. The members—many of whom are confirmed activists—are busily absorbing the harsh lessons of the twentieth century and preparing themselves to assume a leading role in the next popular uprising. There is also a Conference of Socialist Economists and a journal titled *Rethinking Marxism*, both accessible through ICAPE and AHE.

The Economic History Association publishes *The Journal of Economic History*, while the Economic History Society puts out *The Economic History Review*. Though neither belongs to any heterodox umbrella organization, few if any members of either group seem prepared to concede the orthodox claim that mathematics offers a higher road to economic understanding than does the study of historical records.

On the contrary, most economic historians would probably agree with the conclusions reached by William Lazonick in his 1991 book, *Business Organization and the Myth of the Market Economy*. That book challenges the long-accepted wisdom that orthodox (laissez-faire) principles have been responsible for capitalism's short list of unambiguous success stories. The catalog of such stories must surely include the industrial growth of Great Britain during the eighteenth and nineteenth centuries, that of the United States during the nineteenth and twentieth, and those of Germany and Japan from the final decades of the nineteenth century until perhaps 1990. Beyond them, it is less than clear what else to include.

More to the point, capitalism's success stories are not necessarily laissez-faire's success stories. As Lazonick points out, capitalism's greatest successes were faits accomplis by the time the governments involved even began to embrace the (very possibly false) laissez-faire principles presented as gospel to students of Economics 101. Great Britain was already an industrial colossus by 1846, when it repealed its protective Corn Laws, and had only increased its dominance by 1860, when it finally took the plunge into (almost) complete international free trade. The United States was in an even more dominant position vis-à-vis war-torn Europe and Japan in 1945, when it began at last to work toward—rather than bloviate about—global free trade. And so on. It would be impossible in a single chapter of a single book to do justice to the elaborate argument presented by Lazonick, which is itself a summation of more specialized arguments developed by a small army of earlier economic historians. Suffice it to say that the historical record contains little if any support for what, if memory serves, Paul Samuelson has been known to call the "eternal verities" of economic theory.

The most striking fact about the heterodox economics community is the sheer anonymity in which it operates. Neither the public nor the press—which in a democracy is expected to inform (and even advise through its editorial function) the public on matters affecting the common good—seems to know that such a community exists. The governments, universities, and newspapers of Great Britain and the United States seem anxious to keep secret the fact that contrarian views are not only held but convincingly defended by large numbers of learned men and women. Robert Rubin, secretary of the Treasury under President Clinton, explained how it's done in a 2006 interview with William Greider of *The Nation*. "This stuff," he said, in reference to evidence that free trade and globalization have damaged standards of living worldwide, "is really interesting. Unfortunately, almost nobody in the press asks this stuff . . . or tries to do it in a thoughtful way. So there's almost no public understanding of any of this stuff."[16]

A golden opportunity to promote the requisite understanding was lost in 1994, when James Fallows's book *Looking at the Sun* rolled off the press just after the Japanese economy fell into a lengthy recession. Fallows, now a national correspondent for *The Atlantic Monthly*, whose first book, *National Defense*, won the National Book Award in 1982, lived with his family in Asia for four years while he learned at first-hand about the workings of the Japanese economy. On the strength of his journalistic connections, he was granted interviews with members of the Japanese government, academic, and corporate elites. And, calling on his experience as an interviewer, he was able to form as clear an understanding of the thinking behind their policies and actions as any Westerner is likely to obtain.

His conclusion was that by reverting to the practices that had spawned the first "Japanese miracle"—predicated as those practices were on the mercantilist advice of List and Hamilton—Japanese leaders had fashioned a society that worked about as well for the Japanese people as any on record. He therefore found it offensive that "Britons and Americans often act as if theirs were the only possible principles [of politics and economics] and that no one else, except in error, could choose any others." When that happens, he says, "political economics becomes an essentially religious question—leading to the standard drawback of any religion, the failure to understand why people outside the faith might act as they do."[17] Had the Japanese kept the United States running scared (and in those days the United States really was running scared) just a little bit longer, the far-reaching debate Fallows hoped to ignite might have taken place. In the event, the establishment was spared that embarrassment by the decline of the Japanese threat.

Other than students, few in the West seem aware that a potentially important debate is being suppressed. On June 21, 2000, the influential Paris daily *Le Monde* featured a long article under the headline "Economics Students Denounce Lack of Pluralism in Teaching Offered," in which it explained that economics students at the École Normale Supérieure (ENS), France's premier institution of higher learning, were circulating a petition calling for a reduction in what they (the students) considered "uncontrolled use" by their economics instructors of "mathematical formalism disconnected from reality" and deploring the omission from their course work of historical facts, the functioning of institutions, the behaviors and strategies of economic agents, and other heterodox topics.

The original petition, along with commentary on it, is downloadable from the Post-autistic Economics Network (www.paecon.net). The name of that network was suggested by the dictionary definition of the word "autism," which *Webster's New Collegiate Dictionary* (1961) takes to mean "absorption in fantasy, to the exclusion of reality." How better to phrase a complaint against theories dominated by "complex mathematical models that only work in conditions that don't exist"? Gilles Raveaud—a leader of the ENS revolt and source of the latter quotation—compares neoclassical economics to Marxism and the Bible in that it, too, "purports to explain everything, rather than admitting that there are many things it hasn't figured out." To pose as a science today, a discipline must at least be up-front about the questions it can't yet answer.

The AHE Web page contains, among other things, a list of twenty-one universities worldwide (including twelve in the United States) that offer graduate (and, in most cases, some undergraduate) training in economics and political economy of other than the mainstream, orthodox, neoclassical sort. None of the U.S. institutions listed belongs to the Ivy League or otherwise ranks among the nation's most prominent research universities.

After Raveaud and a friend visited the Workshop on Realism and

Economics in the U.K., frustrated Oxbridge students issued similar manifestos—one of which attracted 750 signatures—and started websites of their own. Students at Harvard then issued a manifesto reminiscent of the ENS declaration and formed Students for a Humane and Responsible Economics to agitate for an alternative to Ec10, the only introductory economics course offered in the university. Such a course, they hoped, would offer a balanced perspective instead of the "conservative propaganda" put forth by Martin Feldstein, Ronald Reagan's former chief adviser on economic matters, who has taught Ec10 since 1984 and who (according to one student) habitually assigns readings "in support of tax cuts for the rich, the privatization of social security, and . . . blaming the poor for their poverty." Another "saw it as hypocritical—given that Harvard values critical thinking and the free marketplace of ideas"—that all social science majors were required to enroll in so doctrinaire a course.

Students at Columbia, in New York, have yet to revolt against the narrowness of the offerings of their economics department. Until quite recently, however, they did have Jacob McKean to plead their case. As a senior majoring in history and African-American studies, McKean contributed a column titled Juice Is Stranger Than Friction to the student newspaper, where it ran on alternate Thursdays. On February 16, 2006, he chided the administration for the lip service it gave to "academic freedom" and "intellectual diversity," even as it sought to cleanse one department of professors critical of Zionism, sling right-wing mud at the institution's few outspoken leftist professors, and silence as many campus war critics as conveniently possible. But his most disdainful remarks were reserved for the economics department, whose students learn little or nothing about heterodox schools of economic thought.

After speaking to a variety of students, he had heard many versions of the same complaint:

> "We don't talk about different schools of economic thought. I don't even know what they're called, or what other economic ideas would be. The professor mentioned . . . different schools of thought . . . once in Principles . . . but we weren't tested on it" . . .
>
> "No one would ever ask a question like 'why is unemployment natural.' There's just no discussion of anything like that" . . .

> "There's no discussion of whether the models we're studying are right or wrong, or if there are different models, or if they [the ones we study] even apply to the economy."

Many of the students interviewed said they would prefer a more diverse curriculum. Those who found the lack of one at all troubling reported an intimidating and isolating environment for discussing divergent viewpoints of any kind. Did the administration, McKean wondered, tolerate the ideological uniformity of the economics department because it happened to mirror its own right-wing convictions? He closed with a call for the university to overhaul the economics department.

McKean is quite correct. Juice is strange indeed. And the strangest thing about it may be that orthodox economists actually have quite a lot. They have enough to keep reporters from asking Robert Rubin about the way globalization degrades living conditions in developing nations; they have enough to keep those same reporters contemptuous of the demonstrators who turn up by the thousands to protest any and all attempts to impose additional free-trade agreements on yet more defenseless nations; and, above all, they have enough juice to prevent alternate schools of economic thought from establishing beachheads on the campuses of the more prestigious colleges and universities. In the chapters to follow, we shall examine some additional debates quashed by mainstream juice.

CHAPTER 15

Spontaneous Cooperation

Many-player game theory, as explained in Chapter 6, comes in two flavors. The cooperative flavor developed by von Neumann and Morgenstern allows the players to communicate freely and form binding agreements, while the noncooperative flavor proposed by Nash absolutely forbids the players to communicate or collude. But human endeavor is seldom so black and white. Much of it is guided by customs, fashions, mores, unwritten rules, and gentleman's agreements, which, though nonbinding and unenforceable, are often more scrupulously observed than most statutes or decrees. The following examples are more fully described in chapters 4 and 5 of Robert Axelrod's oft-quoted book on the genesis of cooperation.[1]

During World War I, units of both armies observed certain unwritten rules. Almost as soon as the highly mobile campaigns of August and September 1914 gave way to static trench warfare, high-ranking officers on either side began to observe what they considered unacceptable amounts of "fraternization with the enemy."[2] It was the result of a passive live-and-let-live attitude that arose spontaneously among British, French, and German soldiers alike. It continued almost until the end of the war, despite the best efforts of senior officers to stamp it out, despite the passions aroused by combat, despite the compelling logic of kill or be killed, and despite the ease with which the various high commands were able to repress scattered attempts to arrange formal truces. One can hardly help wondering:

1. How did such behavior ever get started?
2. How was it sustained?
3. Why did it break down toward the end of the war?
4. Why have similar patterns emerged in so few other wars?

Fortunately, many of the diaries, letters, and reminiscences of the actual participants have been preserved. In fact, a book-length study of them is available, exploring field and home-front attitudes toward fraternization.[3] It describes numerous manifestations of this seemingly inexplicable phenomenon.

As early as November 1914, an English noncom observed that little shooting was done by either side in the hours immediately after dark. His own quartermaster usually brought up rations at about that time, and it seemed likely that the enemy did likewise. So accustomed had the men become to this routine that they laughed and joked on their way to and from the chow line. Elsewhere, the hours between 8:00 and 9:00 a.m. were set aside for latrine visits. Indicated by flags, those areas were not targeted by snipers on either side.

A British officer recalls his astonishment when, while facing a Saxon unit of the German army across a few hundred yards of no-man's-land, his command was struck by sudden and unexpected artillery fire. Afterward, a brave German got onto his parapet and shouted out, "We are sorry about that; we hope no one was hurt. It's not our fault, it's that damned Prussian artillery."

Such behavior enraged senior officers, who made numerous attempts to instill a more aggressive attitude in both armies. On the British side, courts-martial were convened, and entire units—as well as individual soldiers—were punished. All to little avail. It was not until the very end of the war, when both sides adopted the practice of raiding and infiltrating opposing trenches, that the live-and-let-live attitude began to wane.

Fraternization between opposing armies during World War I was facilitated by the static nature of the fighting. The same units might face each other for months on end across the same piece of real estate, affording each side ample opportunity to learn the other's habits and to reflect on the essential sameness of the positions in which soldiers on both sides of the line found themselves. They were quick to discover "nice" strategies—whereby each side resisted the temptation to open fire—as well as the importance of prompt retaliation when fired upon. Response was often in some easily recognized ratio, so that two or three shots came back for each one fired. But the exchanges rarely escalated. Enlisted men understood that opposing officers sometimes had to be placated with shows of aggression and regularly returned harmless fire in kind. Even the replacement process didn't necessarily upset a pacific equilibrium, because new men could be—and regularly were—clued in to the status quo. "Mr. Boche ain't such a bad sort," one English soldier recalls being told, "you leave 'im alone and 'e'll leave you alone."

More recently, in their bestselling *Freakonomics: A Rogue Economist Explores the Hidden Side of Everything*, Steven D. Levitt and Stephen J. Dubner reveal that the members of rival drug gangs in Chicago have taken to shooting one another in the buttocks—instead of to kill—in their territorial disputes.[4] It seems to act as an adequate deterrent, while reducing the considerable level of danger a gangland "foot soldier" confronts on a daily basis. It should not be supposed that *limited* spontaneous cooperation arises only in times of global conflict, or ceased in 1918.

The intelligence and foresight of World War I soldiers had much to do with the tacit live-and-let-live arrangements that emerged at various times and places along the western front. But studies of the animal kingdom reveal that neither characteristic is required to recognize and respect a community of interest. Robert Axelrod devotes an entire chapter of his 1984 book, *Evolution of Cooperation*, to cooperative arrangements among nonhumans.[5] Animal strategies consist not of consciously formulated plans but of genetically programmed patterns of behavior. The leonine practice of hunting cooperatively seems an obvious example. Leopards, cheetahs, tigers, catamounts, and lynx are solitary cats that live and hunt alone. It is likely that ancestral lions did so as well, and no one seems to know where the instinct to hunt cooperatively first came from. But once in existence, it is easy to see how individuals in whom such an instinct resides might prove fitter than those in whom it does not.

The same can be said of the roaring, chest beating, and tree shaking all great apes seem to indulge in before initiating physical combat, in hopes of securing a favorable result without risking injury. Likewise the parallel walk ritual performed by male deer before locking horns.

Limited—rather than all-out—competition seems to be an important part of nature's way.

The practice rests on reciprocity. A lioness unwilling to share her kill would face expulsion from her pride, probably to perish during the dry season. Another practice that depends on reciprocity is observed among sea bass. Endowed with both male and female sex organs, these fish form pairs that more or less take turns at being the high-investment partner (laying eggs) and the low-investment partner (providing sperm to fertilize those eggs). Up to ten spawnings occur in a day, and only a few eggs are produced each time. Pairs tend to separate if the roles are unevenly shared. Other examples of reciprocity involve small fish or crustaceans that remove and eat parasites from the body—or even from the inside of the mouth—of larger and potentially predatory fish. These aquatic partnerships occur mainly in coastal waters, or on reefs, where the inhabitants occupy established territories. They are all but unknown in the open ocean, where recognition would prove difficult. Other examples include a hermit crab and its sea anemone partner, a cicada and the various colonies of microorganisms housed in its body, or a tree and its mycorrhizal fungi.

Much of what Axelrod has managed to learn about cooperation has emerged from studies of the game known as iterated prisoner's dilemma (IPD). It consists of a succession of games of ordinary prisoner's dilemma (PD), much as the World Series consists of a succession of baseball games. The reader is reminded that PD is a symmetric two-player game in which each player must choose one of only two possible actions: Each may either cooperate C or decline to cooperate D. When both cooperate, each receives R points. When both decline to cooperate (a.k.a. defect), each receives $P < R$ points. On all other occasions, the decliner (or exploiter) receives $T > R$ points, while the cooperator (or victim) receives $V < P$ points. The foregoing information is frequently presented in tabular form, as shown in figure 15.1

The table entries exhibit the manner in which the rewards to WE, the row chooser, correspond to the four possible collective decisions. The corresponding rewards to THEY, the column chooser, are obtained by interchanging the letters T and V. T stands for temptation, R for reward, P for punishment, and V for victim. Naturally, V is smaller than P, which is smaller than R, which is smaller than T. We also require that $2R$ shall exceed $T + V$, for reasons that will become evident

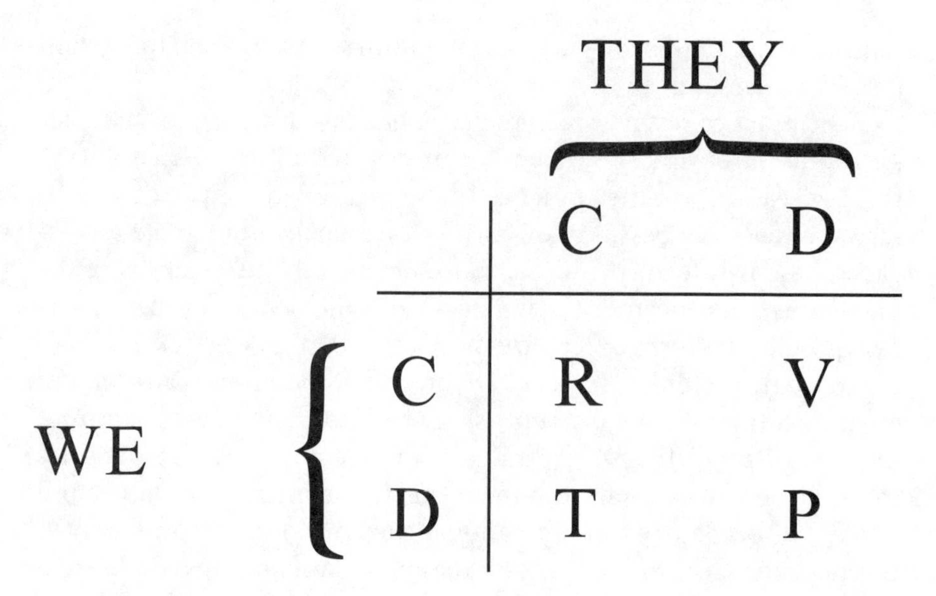

Figure 15.1. The reward matrix for prisoner's dilemma, with littoral instead of numerical payoffs.

shortly. IPD differs from PD in that the players of IPD may consult the record of previous games in an effort to divine the likelihood of future cooperation. Few strategic situations are simpler to describe, have been scrutinized in greater detail, or can be examined from more points of view.

Social psychologists were among the first to recognize the value of knowing how to exploit situations that resemble IPD to the extent that the key decision—which must be made repeatedly—is whether or not to trust, cooperate, or do business with a given individual. In experiments with human subjects, the amount of cooperation achieved, and the manner of achieving it, have been found to depend on a wide variety of only peripherally relevant factors, such as the circumstances in which the game takes place, the attributes of the individual players, and the prior relationships (if any) that exist between them. Over the years, countless studies of IPD have been described in the social science literature. As Axelrod puts it, IPD has become the *E. coli* of social psychology.

Unfortunately, few such studies furnish reliable guidance to those who wish to improve their performance in situations resembling IPD, in part because the human subjects involved are typically inexperienced players. How much could you improve your chess game by watching novice players? On the other hand, determined efforts to discover demonstrably superior chess strategies via mathematical analysis have proven all but fruitless. Accordingly, Axelrod resolved to organize a tournament between expert players of IPD, large numbers of whom were invited to submit their most trusted strategies. As tournament organizer, he undertook to match every entry against every other, to determine the scores of the resulting matches, and to publish the results. The contestants were ultimately ranked according to their average scores over the fifteen matches in which they participated.

The invited participants were social scientists and mathematicians who had published on PD, IPD, or some closely related subject. Each was invited to submit a strategy in the form of a computer program capable of communicating its individual decisions—cooperate *C* or defect *D*—to a master program acting as referee. Each IPD match was to consist of two hundred games of PD, and the master program was to maintain an up-to-date record of the games already played. There are several ways to construct such a record. A commendably concise option consists of a single two-hundred-letter word constructed from the alphabet {*T, R, P, V*} in which the first letter represents White's winnings in the first game, and so on, up to game 200. For simplicity, Axelrod substituted the numerical values {*5, 3, 1, 0*} for the more abstract {*T, R, P, V*}. The requisite records may be kept in either literal or numerical form. In either case, the referee must update the record after each game of the match by adding a single letter or digit to the right-hand end of the ever-growing word. Alternatively, at the expense of some duplication of effort, the players can maintain their own records.

A record of Black's winnings may be constructed from a record of White's by interchanging the letters *T* and *V* or their numerical equivalents, 5 and 0. Moreover, a complete record of each team's decisions may be inferred from a record of either player's winnings, since there is only one pair of decisions that can cause a given player to receive a specific reward. If, for instance, White should win *T* in the seventeenth game, we know that Black cooperated while White defected in that game. And so on. Each team must be free to consult the entire record, or any part of it, before making its next choice between *C* and *D*. Fourteen completed entries (playable strategies) were received, and each one of them was matched against every other, as well as itself and a strategy called Random, which merely flipped a coin (electronically, of course) whenever its turn came to decide between *C* and *D*.

A substantial fraction of the submissions were variants of tit for tat, a remarkably simple strategy well known in advance to all contestants. Tit for tat (a.k.a. T4T) is the strategy that cooperates in the first game of a match, then duplicates the opponent's most recent decision in all subsequent games. By so doing, T4T both encourages cooperation and discourages defection. Moreover, it is easily interpreted. This is a virtue against human opponents, who often misread or overlook signals encoded in an opponent's decision history. Finally, T4T is "nice" in the sense that it never defects without provocation. Nice strategies performed significantly better than non-nice ones in Axelrod's tournament.

Because the contestants were fully aware of T4T, and many considered it the "team to beat," it is hardly surprising that many of the entries were specifically designed to perform well against T4T. Yet T4T was the surprise tournament winner, in large part because the strategies designed to excel against it fared poorly against each other, while T4T performed creditably against all comers. So pleased were all involved in Axelrod's tournament (call it Axelrod I) that he offered to organize an Axelrod II. The results of both tournaments are reported in appendix A of his book.

In view of the fact that the scores in an individual game of PD range from $V = 0$ to $T = 5$, the individual scores in a match consisting of two hundred games must lie between 0 and 1,000 points inclusive. Individual match scores in Axelrod I ranged from a low of 276.3 to the high of 504.5 recorded by T4T, which is little better than half the highest possible score of 1,000 points. On the other hand, because the highest combined score a pair of opponents can achieve in a single game is the $R + R = 6$ they obtain by mutual cooperation, while the lowest is the $P + P = 2$ they obtain by mutual defection—the only remaining alternative being the $T + V = 5$ the two of them earn when one cooperates and the other defects—their combined scores over a two-hundred-game match must lie between 400 and 1,200 points inclusive. Hence the total of 15,187 points scored in the fifteen matches in which T4T took part—a complete record of which will be found in figure 15.2—is better than 84 percent of the possible 18,000. Finally, T4T earned an almost equal share of the total recorded by itself and its fifteen opponents, since its own total of 7,579 points was better than 49.9 percent of their combined winnings. It did so, moreover, without

#	1	2	3	4	5	6	7	8	9	10	11	12	13	14	15
T4T	600	595	600	600	600	595	600	600	597	597	280	236	279	359	441
OPP	600	600	600	600	600	600	600	600	597	597	285	241	284	362	442

Figure 15.2. The scores of tit for tat against each opponent in the first of Axelrod's two tournaments.

defeating a single opponent, having tied eight of its fifteen matches and finished within five points of the winner in its other seven.

Whoever designed T4T obviously realized that there are more points to be scored—and sometimes more money to be made—by finishing a close second to an opponent who scores a great many points than by finishing a few points ahead of one who scores only a few. Forty-nine percent of a nine-inch pie is more nourishing than fifty-one percent of an eight-inch one. That's why it's so important to choose a nice strategy in IPD. Such strategies give opponents no reason to fear that their willingness to cooperate will be used against them. In fact, the top eight strategies in the rankings that emerged from Axelrod I were nice. Each of them earned at least 470 points—better than 93 percent of the winning score—while none of the not-so-nice entrants earned more than 401 points, or 80 percent of the winning score. To exceed 600, a player must average better than $R = 3$ points per game over the course of the match, which none of the participants in Axelrod I and II even approached.

The only way to average better than $R = 3$ points per game in IPD is to earn $T = 5$ points on numerous occasions while holding victimizations $V = 0$ to a minimum. All participants in the Axelrod tournaments seem to have dismissed the possibility of scoring more than 600 points and chosen accordingly to lift their per-game average as close as possible to the $R = 3$ point level. The strategies submitted were therefore designed, as is T4T, to encourage cooperation and punish defection.

To improve substantially on T4T's performance in Axelrod I, one would need to improve its score in one of the five matches in which it failed to accumulate even 450 points. One such match took place against Random, the strategy that flips a coin (electronically, of course) before each turn and cooperates only if a head is obtained. T4T doesn't do nearly as well against Random as does the strategy that always defects, in part because Random is wholly unresponsive to signals. T4T's match against Joss was more revealing. Joss is the strategy that always cooperates in the first game, and always defects after its opponent defects, but cooperates with only 90 percent probability after its opponent cooperates. The match began with an initial run of five consecutive *R*s, followed by nineteen alternate *V*s and *T*s, followed in turn by 176 *P*s. T4T earned 60 points during the first 24 games, while Joss earned 65, for averages of 2.5 and 2.7 points per game, respectively. Thereafter, each earned a single point per game, as play degenerated into 176 consecutive mutual defections. Neither strategy contained any provision for trying to reestablish cooperation once the pattern of mutual defection set in.

How much would it have hurt either player to throw in a few short gratuitous bursts of cooperation in an effort to realize another burst of mutual prosperity, such as the one that began the game? Had the opponent failed to respond to such an overture, the player making it would have reduced his match score by one point per gratuitous cooperation while his opponent was gaining four points per. But had the gesture succeeded, it would have resulted in another run of *R*s, *T*s, and *V*s, like the one with which the game began. The fact that the original run of *R*s, *T*s, and *V*s—a result of the fact that both strategies cooperated initially—increased each player's match score by more than thirty points rather strongly suggests that subsequent bursts of gratuitous cooperation might well be worth the attendant risk, even though they can increase an unresponsive opponent's eventual margin of victory. If one such gesture results in a run of *R*s, *T*s, and *V*s, it should probably be repeated in due course. If it doesn't, it possibly should not. Variations of T4T that take a few such risks should perhaps be described as "entrepreneurial" forms of T4T.

The results of Axelrod II tend to confirm conclusions based on Axelrod I, although the tournament rules were changed in one important respect: instead of ending each match after two hundred games, the stopping time was turned into a random variable by rolling (again electronically) a pair of heavily loaded dice after each game to determine if the match would continue. By designing the dice to show "snake eyes" on 0.346 percent of all throws, and ending each match the first time snake eyes appeared, Axelrod contrived to make the length of each match equally likely to exceed or be exceeded by two hundred games. By preventing the players (strategies) from knowing when a particular match would end, he discouraged wholesale defections near the end of the match. It was felt that such eleventh-hour defections had, to some extent, clouded the results of Axelrod I. In any case, T4T won again.

Gratified though he was by the turnout for his second tournament—in which there were sixty-two playable entries—Axelrod realized that the number of future IPD tournaments to which he could attract thoughtfully chosen strategies was limited. So, in an effort to subject his conclusions to more stringent testing, he resolved to simulate an entire sequence of tournaments on a computer. The least successful strategies in the early tournaments would not be admitted into the later ones. Instead, they would be replaced by additional copies of those that enjoyed early success. In this way, he could simulate survival of the fittest on a computer.

Axelrod's virtual tournaments were meant to mimic biological evolution in a region where many animals of a single species regularly interact with one another. When two members of the species meet, both may cooperate, both may decline to cooperate, or one may cooperate while the other declines. If individual members of the species can recognize those with whom they have previously interacted, and remember the outcomes of those interactions, they may utilize T4T—or any other such history-dependent decision rule—in their struggles to survive and (hopefully) prosper. It is both convenient and arguably legitimate, when performing such experiments, to regard each successive tournament as a simulation of one generation of the species—one in which each decision rule is employed by numerous individuals.

To simulate survival of the fittest, Axelrod decided that the number of players using a given strategy in a given tournament should be proportional to the number of points scored by that strategy in the previous tournament, expressed as a fraction of the points scored by all the strategies in simultaneous use. When this was done with the strategies entered in Axelrod II, the lowest-ranking eleven lost half their initial popularity after a mere five generations, while the middle-ranking ones more or less held their own, and the high-scoring elite gradually gained adherents. By the fiftieth generation, the rules that appeared in the bottom third of the Axelrod II rankings had fallen into virtual disuse, while most of those in the middle third had begun to disappear, and those in the top third continued to gain adherents. Again, T4T won going away. The performances of the other high-ranking strategies are indicated in figure 15.3. Six of the strategies prospered throughout, and a few others managed to hold on to their initial adherents, while the overwhelming majority were driven to the brink of extinction or beyond.

Strategy 8, Harrington, furnishes an instructive example of ecological extinction. The only non-nice strategy among the top fifteen finishers in Axelrod II, the predatory Harrington actually managed to gain adherents during the first 150 generations, almost as quickly as the eventual winners were doing. But then its fortunes began to change. With ineffective strategies becoming extinct, potential victims became ever harder to identify. By the thousandth generation, Harrington had become as extinct as the strategies it had initially victimized. Although non-nice behavior can succeed in the short run, it tends over time to destroy the very resources upon which its success depends. Still, Axelrod was not satisfied. To test T4T against more (and possibly stronger) opponents, he designed yet another simulation, this one involving "genetic algorithms."

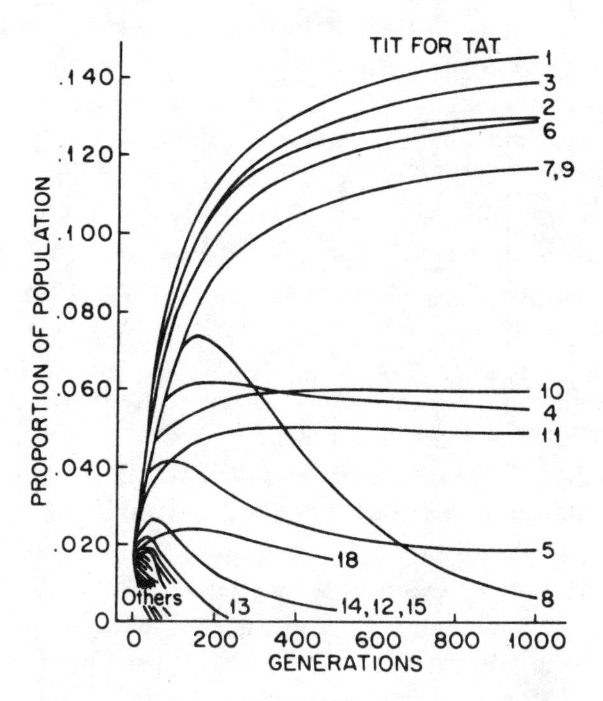

Figure 15.3. A few of the strategies in Axelrod's virtual tournament prospered throughout, while others prospered briefly before falling on hard times.

Genetic algorithms are the brainchild of John Holland, a pioneer in the field of artificial intelligence.[6] They are among the simplest of the so-called complex adaptive systems (CAS). As Holland conceives them, CAS exhibit multiple levels of organization, with "agents" at one level serving as building blocks from which agents at a higher level are constructed: a group of proteins, lipids, and nucleic acids forms a cell, a group of cells forms a tissue, a group of tissues forms an organ, a group of organs forms an organism, and a group of organisms forms an ecology. In much the same way, a group of workers forms a department, a group of departments forms a division, a group of divisions forms a corporation, and so on, through industries, sectors, national economies, and the global economy itself.

The emerging science of complex adaptive systems offers a growing bag of (mostly computerized) tricks, often motivated by biological and economic considerations, with which to study all manner of natural phenomena. It applies most readily to systems containing niches, each of which can be occupied by appropriately adapted agents. As the world of commerce harbors doctors, lawyers, computer programmers, plumbers, and schoolteachers, so tropical rain forests accommodate trees, flowers, and clinging vines, as well as monkeys, snakes, tree sloths, and butterflies. The very act of filling one niche tends to open up others, so that complexity has a pronounced tendency to breed additional complexity.

CAS agents come adorned with "tags," not unlike the messages found on T-shirts, which encode capabilities, patterns of behavior, and strategies for survival. When one agent encounters another whose tag reveals it to be of the same species and opposite sex, the two may mate. When one band of agents meets another, each must decide whether to flee or fight. And so on.

The decisions made by individual or composite CAS agents are governed by strategies encoded on the (potentially elaborate) tags identifying those agents. Strategies tend to be simple in the beginning and to grow more complex with the passage of time. The initial strategy for a frog on a lily pad might direct that it remain as motionless as possible until either a small flying object comes within striking distance or a large animal approaches. But that strategy might be refined over time as the frog learns to recognize and ignore certain unpalatable types of flying objects and as it learns that passing trucks are not threatening. Holland explains how all this, and more, can be modeled by tag interactions.

Games between competing species of CAS agents suffer from a drawback, shared with games like chess, checkers, and Go, in that their strategy spaces are too large to be searched exhaustively. Genetic algorithms search them in a cursory but purposeful fashion. Because a completely specified strategy stipulates which of many allowable responses should be made in each and every situation that could conceivably arise, the strategies discovered by practical algorithms tend to make liberal use of "default options" specifying an action to be taken in "all other" situations—something like "hunker down and hope for the best" or the handyman's "spray and pray." The frog's option to "remain as motionless as possible" in all but a few well-defined circumstances is a typical default option. On some occasions, the strategies discovered by genetic algorithms can include entire default hierarchies, as encountered in psychology and other branches of behavioral science. The name "genetic algorithm" comes from the manner in which interacting tags may "cross over" to form new tags. Thus the tag ABCDEFG may combine with the tag *abcdefg* to form ABC*defg* and *abc*DEFG. Holland got the idea from real genes and chromosomes, which behave in a somewhat similar fashion.

In his search for sound strategies to use in IPD, Axelrod elected to include only those that "remember" just the three most recent games of PD.[7] The results of all the remembered games in a given match could then be encoded in a word consisting of three letters from the alphabet $\{T, R, P, V\}$. Because $4^3 = 64$ three-letter words can be made from a four-letter alphabet, a strategy from the class in question consists of at least sixty-four instructions of the form "In situation ___ cooperate with probability ___." The first blank should contain one of the sixty-four possible three-letter words, serving as situation identifiers, while the second contains the probability (expressed as percent) with which the user is to cooperate in the situation identified.

Actually, the strategy in question is not yet completely specified. To finish the description, there must be twenty-one additional statements relating to the opening moves, when the word containing the record of past encounters consists of fewer than three letters. Because twenty-one words of length two or less can be formed from a four-letter alphabet, each completely specified strategy consists of a list of $21 + 64 = 85$ instructions, only sixty-four of which remain relevant after the third game of the match. If only nonrandom strategies are allowed, so that each probability is either a 0 or a 1, there are still $2^{85} = 3.87 \times 10^{25}$ such strategies.

In any case, Axelrod began his simulation with a more or less random selection of such strategies emblazoned on the tags worn by the participating agents, and allowed every one of them to play against every other, for several rounds, before beginning to reassign players with low-scoring strategies to more successful ones. He also allowed strategies to mutate randomly from time to time, by altering a single one of their eighty-five conditional instructions, and to cross over genetically, in the manner explained above.

Often the populations evolved away from cooperation and toward defection, at least in the beginning. Usually, however, the trend toward defection later reversed itself, in favor of reciprocity. The frequency of cooperation rose steadily in such cases, until far exceeding the initial levels. Moreover, the new population seemed immune to victimization, as the evolved strategies were quick to punish defectors.

All of the foregoing computer simulations, and most experiments involving human subjects, convey the same fundamental message: that reciprocity and cooperation are among the expected results of evolution. They often emerge, out of the ashes of greed and envy, even in mindlessly selfish populations. They can survive and even flourish in apparently hostile environments—environments they themselves tend to modify in ways favorable to their own survival. Axelrod draws four conclusions from the results of his own and other tournaments and simulations:

1. Don't be envious.
2. Don't be the first to defect.
3. Reciprocate both cooperation and defection.
4. Don't be too clever.

The fourth point is emphasized by the results of debriefing human opponents of an IPD-playing computer programmed to execute T4T.[8] Few of those questioned even noticed that after making a single independent choice, the machine slavishly copied their own selections. Excessive cleverness can easily disguise an essential willingness—even eagerness—to cooperate.

Many of the benefits sought by living organisms are disproportionately available to cooperating groups. Individuals in a position to benefit from cooperation too often yield to the ever-present temptation to victimize potential allies. The lure of defection in PD echoes this siren song. Would-be cooperators must be nice, but not too nice. Lest they fall victim to freeloaders, they must be quick to punish violations of even tacit agreements to cooperate. Axelrod cites three additional attributes of a strategy or organism that seem to encourage cooperation:

1. *Robustness.* A sound strategy must thrive not only against others like itself but in an environment in which all manner of other strategies are in use.
2. *Stability.* A sound strategy must be able, once it has achieved local dominance, to fend off invasions of its niche by rivals.
3. *Initial viability.* A sound strategy must be able to establish itself in a population dominated by other strategies.

He adds that cooperation is unlikely to materialize unless time horizons are long enough to make initial investments in cooperation seem promising. Long matches reward risk-tolerant strategies like T4T, along with its more entrepreneurial relatives.

Introductions to economic thought, such as the textbook by Paul Samuelson and William Nordhaus, frequently compare the prisoner's dilemma with situations in which two competing entrepreneurs—say two gas stations located across the street from each other—must decide how much to charge for their respective wares.[9] If both charge the same low price, each will earn a moderate profit. If both charge the same high price, each will earn a substantially larger profit. But if one charges a low price while the other charges a high one, the former will earn the largest possible profit while the latter incurs a loss.

In the language of prisoner's dilemma, charging a high price constitutes cooperation (*C*), while charging a low one constitutes defection (*D*). Nash equilibrium behavior therefore requires both players to defect. The expectation that both competitors will do so conforms with traditional economic theory, in which the threat of being undersold is presumed to exert a moderating influence on prices.

If, on the other hand, one were to assume that both station operators have the opportunity to post new prices every morning, the situation more closely resembles the iterated prisoner's dilemma, in which each match consists of a lengthy sequence of games of prisoner's dilemma. As a result, both competitors seem likely to adopt strategies like T4T designed to reward cooperation (*C*) and discourage defection (*D*). That being the case, both station operators seem likely to charge the same high price they would charge if one of them owned both stations. Such a conclusion, which would contradict traditional economic theory, is rarely mentioned in introductory texts.

It would be a mistake to jump to conclusions about the behavior of the competitive pricing mechanism based on nothing more than the foregoing analysis. After all, gas station operators (like other entrepreneurs) have a lot more than two alternative prices to choose among. But the mere suggestion that the emerging science of competition, committed as it is to empirical testing, may contradict long-held beliefs concerning the merits of free-market competition is intriguing to say the least. Chapter 16 will follow up on that suggestion.

CHAPTER 16

Imperfect Competition

Nineteen thirty-three was a year to remember. Hitler became chancellor of Germany, Franklin D. Roosevelt entered the White House, the Washington Senators played in the World Series,[1] and two young economists published books about less-than-perfect competition. Aided and abetted by the Great Depression—then raging on both sides of the Atlantic—Edward H. Chamberlin in Cambridge, Massachusetts, and Joan Robinson in Cambridge, England, obliged the profession to concede not only that some competitive markets are more competitive than others but that the least competitive can behave as if cartelized or monopolized. Even after 1936, when John Maynard Keynes launched his own more policy-specific revolution, professional economists continued to speak of an "imperfect competition revolution."

Whereas Robinson took it more or less for granted that firms in an imperfectly competitive industry would charge cartel (a.k.a. monopoly) prices and sought to discover how grievously such behavior would distort the income distribution, Chamberlin asked how such firms might reasonably decide which prices to choose, and where those prices might lie in the interval between monopolistic and perfectly competitive prices.[2] To find out, he was obliged to analyze what is in effect a dynamic many-player game, and to do so at a time when even two-player zero-sum game theory was little more than a mathematical curiosity. Not until the end of the decade did von Neumann—

and later John Nash—even begin to think about many-player games. It is quite remarkable that with no formal theory to guide him, Chamberlin was able to make the progress he did.

Among other things, Chamberlin invented the scheme economists have used ever since to classify markets. In addition to "perfectly competitive markets," the terms "monopoly" and "monopsony" had long been in use to describe markets controlled by a single seller or buyer, respectively. To these Chamberlin added the terms "duopoly" and "duopsony" to describe markets dominated by just two sellers or buyers, as well as "oligopoly" and "oligopsony" for markets dominated by a mere handful of one or the other. It is natural to arrange these in the form of a pyramid, with the allegedly more (allocatively) efficient structures positioned above those deemed less so:

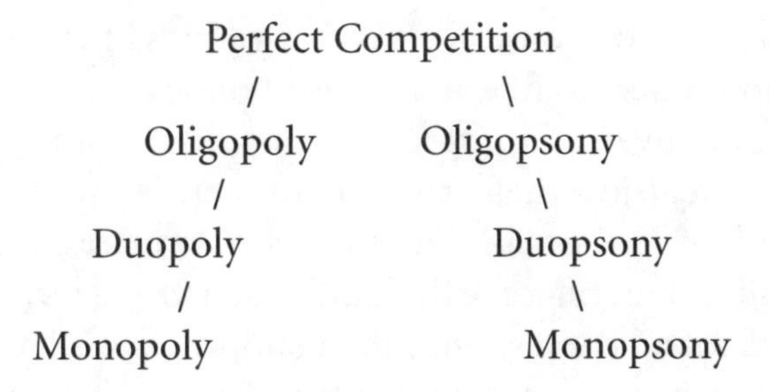

It has long been acknowledged that the vast majority of U.S. industries are oligopolies.[3] Some writers further distinguish between "tight" oligopolies and oligopsonies and "loose" ones, while some describe the latter types as "monopolistically competitive" and "monopsonistically competitive," respectively. For Chamberlin, monopolistic competition was a catchall term for what went on in any market in which the fewness of either buyers or sellers appeared to place potential trading partners at a disadvantage. In ordinary speech, monopsonies, duopsonies, and oligopsonies are described as "buyers' markets," while monopolies, duopolies, and oligopolies are called "sellers' markets."

It is not unusual for markets that originally seemed almost perfectly competitive to degenerate, with the passage of time, into buyers' or sellers' markets, or for markets of any description to slip—however slowly—from higher to lower levels of allocative efficiency. It is, on the other hand, quite rare for a market of one sort to be transformed into one of a higher (more allocatively efficient) sort, even under duress.

When antitrust agencies break monopolies up into oligopolies, for instance, the resulting competitive firms (think Baby Bells) often struggle to preserve their independence.

Champions of the perfectly competitive model (PCM), such as Milton Friedman, have been known to scoff at the notion of an imperfectly competitive revolution, on the grounds that the study of such competition has added little more to the world's understanding of matters economic than a glossary of rather uninformative technical terms.[4] Unless duopolies/duopsonies and oligopolies/oligopsonies degrade market efficiency by commanding significantly higher prices than perfect competitors, he asks, why bother talking about them? And how, in a world in which free-market economies operate at astronomical levels of allocative efficiency, could any important industry remain inefficient? Friedman, in short, saw nothing even remotely suspicious in the astronomical efficiency ratings attributed by Arnold Harberger and others to free-market economies.

Chamberlin invented the term "monopolistic competition" to emphasize his conclusion that firms in competitive markets can—and may on occasion—command near-monopoly prices. Markets served by cartels and monopolies are difficult to analyze in terms of ordinary supply and demand curves, since their supply curves have no slope. A monopolist, for instance, typically offers to supply all anyone wants at a price of his own choosing and nothing whatever at any lesser price.

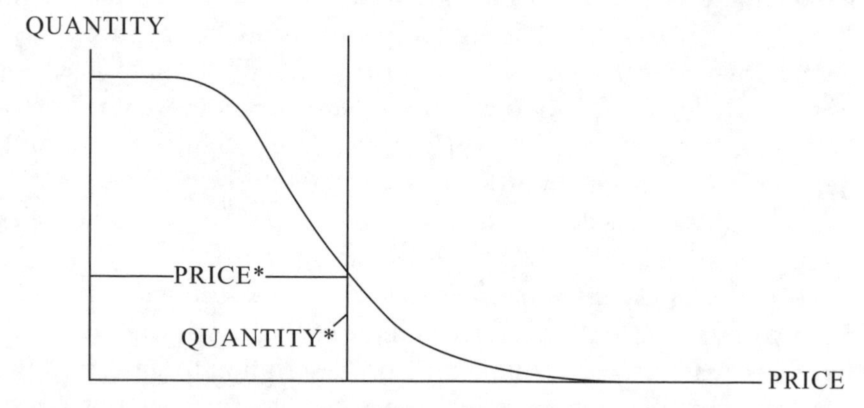

Figure 16.1. When a market is served by a small number of mass producers, a customer can obtain all he or she wants at their (necessarily) common price, but nothing at all at any lower price. Hence, the aggregate supply curve is a vertical segment extending upward indefinitely.

Accordingly, his supply curve is a vertical line extending endlessly upward, as indicated in figure 16.1. It reveals the quantity demanded at the monopoly price but fails to explain how that price might be arrived at.

Antoine-Augustin Cournot, the apparent originator of supply and demand curves, was the first to explain how monopoly prices are chosen. The process is best illustrated with the aid of an auxiliary diagram, such as the one in figure 16.2. The curve depicted plots a monopolist's profit against her price. The shape of the curve suggests—as does common sense—that she will lose money if she charges too much or too little but will prosper mightily if she prices her wares appropriately.

The coordinates of the highest point on the curve in question—call them Price* and Profit*—represent the profit-maximizing price and maximum profit obtainable by the monopolist in question. Before Cournot, economic writers had been so vague about the evils of monopoly as to leave the impression that laws of supply and demand don't apply to monopolists. If nothing else, Cournot made clear that such laws do indeed apply to even the most secure monopolist, who can price herself out of the market as easily as any competitor.

Chamberlin noticed, as others apparently had not, that figure 16.2 also depicts the conditions under which two, three, and possibly more identical firms—selling products as indistinguishable as the eggs, fruits, and vegetables one buys at a farmers' market—necessarily operate. Because they all sell identical products, they must all charge a

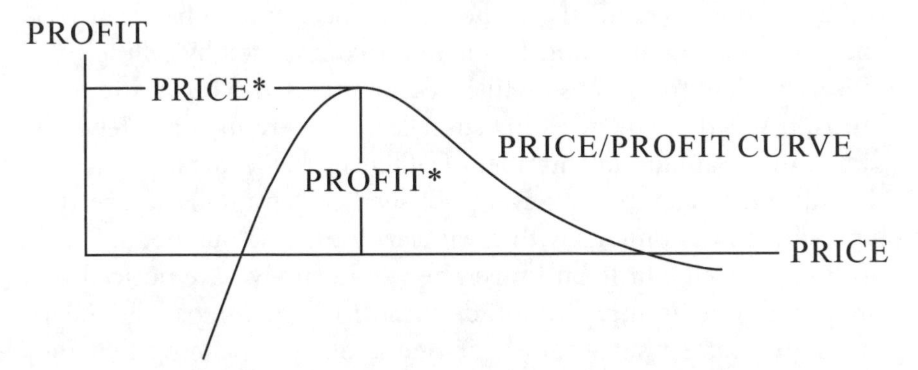

Figure 16.2. Several identical firms can share a common price-profit curve, making it counterproductive for any of them to charge less than their common profit-maximizing price.

common price. If they also command equal market shares, the profit curve in figure 16.2 continues to depict the relationship between their common price and their common profit.

Those common profits are, moreover, still maximized by charging Price*. Once the firms have discovered as much by bidding an excessive starting price down to the point at which no firm stands to gain from further price reductions, the bidding will cease. Should one of them blunder by driving the common price still lower, the rest will be obliged to match the unwelcome cut, harming the initiator no less than the rest. This was the simple argument with which Chamberlin shook the economic establishment to its very roots in 1933. By it he demonstrated for the very first time that—at least in certain circumstances—enlightened competition need not diminish what differs in name only from a monopoly price. Independent though they are, the firms involved wield as much pricing power as would a cartel or monopoly. With or without competition, Price* remains the natural market price.

The question before the managers of a firm currently unable to exercise price leadership is not unlike the one facing the players of iterated prisoner's dilemma when rewards are denominated in U.S. dollars: Is it better to earn $200 while finishing first or $600 while finishing second? For-profit competitors ordinarily select the latter alternative. That doesn't mean price followers are doomed to finish in second place forever. It merely obliges them to accept that result temporarily, until they're able to reverse the order of finish through advertising, product development, or other non–price initiatives. Anheuser-Busch's gradual conquest of the beer market between 1950 and 2000 is a case in point. It was not accomplished by selling Budweiser for less than Pabst, Miller, Schlitz, and the other premium brews. It was done by investing so-called excess returns astutely. The same can be said for any number of other market leaders.

Although Chamberlin never published a diagram like figure 16.2, he described its contents with great clarity. Had he pursued his train of thought even a little bit longer, he would surely have noticed that similar arguments apply to nonidentical firms. As long as the sellers' wares are "perfect substitutes" for one another—meaning that they cannot sell simultaneously at different prices—the natural market price can be identified from a diagram not unlike figure 16.2. Figure

16.3, for instance, contains three different price-profit curves. Each one peaks above a different point on the (horizontal) price axis. Firm 1 is the natural price leader because, while it has every incentive to undersell any price higher than *P′*, no firm has any incentive to undersell that. Given a choice between accepting *P′* or accepting less, they seem all but certain to accept *P′*.

The prices *P″* and *P‴* preferred by firms 2 and 3 are of negligible consequence here, since firm 1—merely by pricing its wares at *P′*—can prevent firms 2 and 3 from selling for more. Neither has anything to gain either by selling for less or by asking more. Firm 1 is the natural price leader.

One can never completely rule out the possibility of a price war. Firm 1 could attempt to purge the market of opposition by charging a price *P* between *B′* and *B″*, the break-even prices for firms 1 and 2, in the hope that firm 2 or 3 or both will abandon the market. That, if it came to pass, would promote firm 1 to the rank of monopolist and render it free to restore the market price to roughly its pre-price-war level in an industry free from competitors.

It should be borne in mind, however, that any attempt to eliminate a rival might backfire. It could do so if firm 3 had a much larger war chest than firm 1, enabling it to lose money longer than firm 1 can tolerate reduced earnings; or if a powerful firm 4 stood poised to enter the industry as soon as firm 3 withdrew; or if the industry were subject to antitrust laws that—like U.S. antitrust law—plainly prohibit predatory pricing.

The magnitudes of the preferred prices *P′*, *P″*, and *P‴* relative to

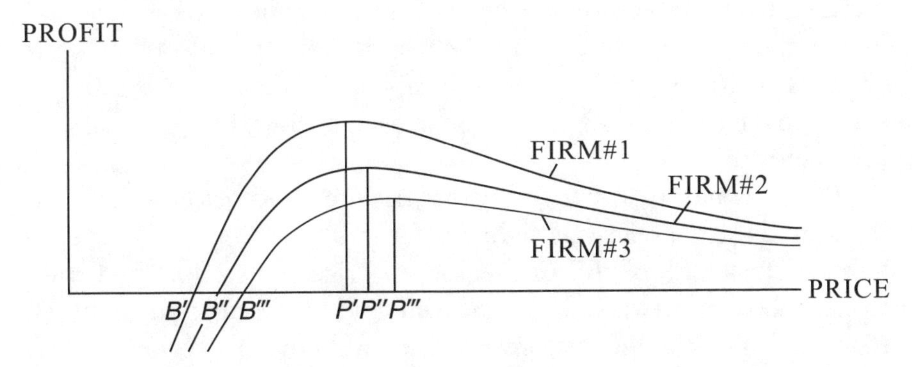

Figure 16.3. Several nonidentical firms can obtain almost identical results.

the break-even prices B', B'', and B''' are of critical interest here. Other things being equal, firm 3 is the most vulnerable to bankruptcy, because its break-even price B''' is the highest in the industry. If P' were only a little larger than B''', firm 1 could put firm 3 out of business simply by charging less than B''' until firm 3 shuts down. It seems natural, therefore, to describe the gap between the largest break-even price (in this case B''') and the smallest preferred price (here P') as the "vulnerability gap" for the industry. It measures the difficulty (cost) of putting the most vulnerable rival out of business. Marketing warriors seldom seek to eliminate rivals until the vulnerability gap narrows. Only when the expected benefits promise to overwhelm the attendant cost do the predators become interested.

If the accounting department is able to persuade management that ABC Corporation can cut its prices by 15 percent across the board and still make payroll—even if rival firms all match the cut—the news will be most welcome. The lead accountant will be commended, and her year-end bonus increased. But the firm will be in no hurry to make any such cut. The ability to do so is money in the bank to management. Why squander it on a price reduction almost certain to be matched by rival firms? As adult bears seldom do battle over food, as bucks in the wild seldom fight to the finish over an estrous doe, and as generals in the field are reluctant to engage an alert well-supplied enemy, so giant corporations are loath to attack rivals capable of fighting back. Only when the odds favor a quick and relatively bloodless victory do the mighty initiate decisive combat. Yet orthodox (neoclassical) theory is predicated on the belief that any entrepreneur blessed with the power to cut prices significantly will exercise that power without delay. As a result, billions of people and jobs remain hostage to a remotely abstract theory that portrays the captains of industry as buffoons incapable of deploying the power vested in them as judiciously as bellicose generals, hungry bears, or bucks in rut!

It is only natural, by the way, that the preferred prices P', P'', and P''' should form a relatively tight cluster well to the right of the (even tighter) cluster B', B'', B''' of break-even prices. Any effort by firm 3 to close the gap between B' and B''' will tend to close the one between P' and P''' as well, without in any way narrowing that between B' and P'. Like the court-confirmed gap between the $49 Microsoft believed it could afford to charge for its Windows software and the $89 it eventually did charge, the gap between a firm's break-even price and its preferred price will ordinarily be large.[5]

A moment's reflection confirms that a monopolist (trust) owning all three firms would prefer to charge a price P^* intermediate between P' and P''', since the price that maximizes total industry profit can lie nowhere else. The monopoly price P^* would then exceed the competitive price P', as predicted by traditional economic theory. Yet the difference would be insignificant unless P''' were far larger than P', which seems unlikely in a world full of industries composed of similar firms, producing similar products, using similar technologies, and following similar managerial practices.

Here is one reason why free markets tend to evolve into tight oligopolies over time. Whereas the American markets for automobiles, farm tractors, and mass-produced beer were once served by scores (indeed hundreds) of independent suppliers each, none now supports more than a handful. The related fact that competitive market prices like P' tend to resemble monopoly prices like P^*, while far exceeding break-even prices like B', also helps explain the oft-noted but seldom explained fact that antitrust remedies rarely if ever have the intended effect.

After AT&T was broken up in 1982, long-distance calls didn't change much in price, while local ones became more—rather than less—expensive.[6] "The breakup of the aluminum industry from a near monopoly under the domination of Alcoa to an industry in which four firms have big shares of the market," wrote Robert Heilbroner and Lester Thurow in 1987, "does not appear to have changed aluminum prices or aluminum pricing policies."[7] After the Standard Oil Trust was broken up in 1911, the prices of kerosene and other petroleum products changed very little.[8] And so on.

In each instance, the prices charged by the newly created competitive industries were about the same as the ones charged by the near monopolies they replaced, just as the foregoing analysis of figure 16.3—according to which P^* barely exceeds P'—would predict. In no case did the resulting price reduction even begin to close the vulnerability gap between the competitive price P' and the largest break-even price. As Donald Dewey, who studied the subject for the better part of fifty years and eventually wrote an entire book on the subject, repeatedly expresses doubt that antitrust activity has ever contributed anything to "economic welfare."[9]

The repeated failure of antitrust remedies to perform as advertised—meaning as predicted by orthodox economic theory—is an undeniable "discrepancy" between accepted theory and established fact. As such, it fits neatly into the sequence of events described by Shirley Strum in Chapter 4. When found, she writes, such discrepancies "are either ignored or written off as the results of bad methods or 'bad science.'" The failures of various free-trade initiatives to perform as expected—an obvious example being the failure of NAFTA to generate the promised 300,000 new jobs in the United States—constitute more such discrepancies, as do the well-documented historical successes of protectionism. The yawning gap between the prices people pay to obtain manufactured items and the often far smaller cost of producing them is yet another.[10] There are plenty of other such discrepancies.

It should be noted in passing that there is no obvious limit on the number of curves that could appear in figure 16.3. In principle, the industry in question could include any number of firms, each with its own individual price-profit curve exhibiting a preferred and a break-even price. Alternatively, the curve identified with firm 1 in the figure could represent an entire class of firms—call it class 1—each with similar if not identical price-profit relationships, while the curves identified with firms 2 and 3 could represent other such classes, each containing an arbitrarily large number of firms.

Although price leadership in the industry would then be shared among the several members of class 1, little else would change. In particular, no firm in any class would have the slightest incentive to charge less than P' for its wares, while every firm in class 1 would have a powerful incentive to undersell any higher price. Logic alone does not—as economic orthodoxy maintains—require that firms in an industry be large and few in number to command near-monopoly prices. Small wonder that such prices have become the rule rather than the exception in modern markets.

The distinction made by Heilbroner and Thurow between aluminum prices and aluminum pricing policies deserves amplification.What they call an "aluminum pricing policy" is what in game theory would be called an "aluminum pricing strategy." It constitutes a rule—computer programmable no doubt—for maintaining the prices of various grades of aluminum at levels appropriate to prevailing market conditions. If and when conditions change, prices must be free to change with them. A well-chosen pricing strategy will specify advantageous changes. Typically there is a basic grade of aluminum (or steel, or sugar, or beef, or crude oil) from which the prices of other grades may be obtained by applying conventional markups. So, once the price of the basic grade is determined, the prices of all related grades typically follow by convention.

In markets governed by the law of one price, there are two questions a firm's pricing committee must ask with regard to the currently prevailing market price *P*: Can we afford to match it, and if so, should we try to beat it? If the answer to the first is negative, the second need never be asked. If the first answer is affirmative and the second negative, the committee should be content to match the prevailing price. Only if both answers are affirmative should the firm attempt to seize price leadership by underselling *P*. In this fashion, the responses to the foregoing questions partition the entire range of possible market prices into distinct intervals, as shown in figure 16.4.

The endpoints of the middle (matching) interval can be read directly from figure 16.3. For firm 1, the endpoints are B' and P', as shown in figure 16.4. And so on. The interval labeled *N/?* contains all the values of *P* that a particular firm cannot afford even to match, let alone beat, while the one labeled *Y/Y* contains all the prices that the firm should hasten to beat by charging the price it prefers. The one in the middle, labeled *Y/N,* contains all the prices the firm should be content merely to match. Such strategies may be called matching-interval strategies, since each one identifies an interval of prices that the firm in question should be content to match.

A computer programmed to execute an appropriate matching-interval strategy will play the market games depicted in figures 16.1 and 16.2 with considerable skill, giving no opponent cause to lower the common price while allowing few if any to profit by so doing.[11] Should such a machine's opponents insist on charging a rock-bottom

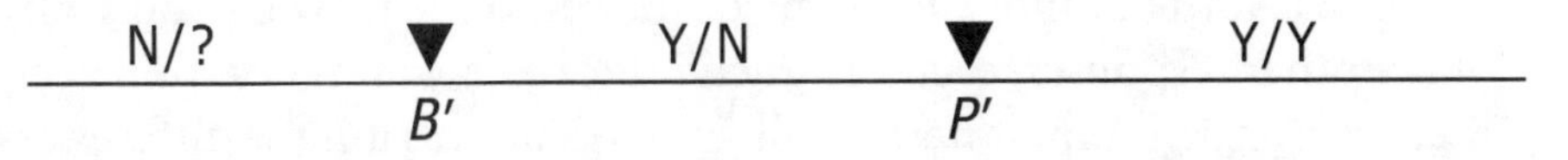

Figure 16.4. Interval matching strategies undersell very high prices, match intermediate ones, and abandon the market when faced with impossibly low ones.

price—in order, presumably, to sell as much as possible—the computer will do likewise as long as it can without losing money. But should the opposition exhibit a willingness to earn more by selling less, such a program will "go along to get along." Chamberlin's analysis of undifferentiated product markets was unique in that a computer programmed to play as he proposed could do so effectively against almost any array of opponents. In contrast—as will be seen—computers programmed to play in the manner endorsed by economic orthodoxy are woefully inept.

The curves appearing in figures 16.1 and 16.2 are computer generated. The market shares controlled by firms 1, 2, and 3 were important inputs to the program that generated them. When one firm lures customers away from another, the winning firm's entire price-profit curve rises, while the loser's sinks. Marketing warfare, defined by Ries and Trout to be the practice of "taking business away from somebody else," is an unending—and not inexpensive—battle for most U.S. firms.[12] The American Marketing Association boasts 38,000 members and estimates that some 750,000 marketing professionals are currently active in the United States and Canada.

Such professionals rely heavily on questionnaires asking—in one way or another—that respondents check just one of the following boxes:

☐ I would pay a large premium for brand X in preference to Y or Z;
☐ I would pay a small premium for brand X in preference to Y or Z;
☐ I would pay a large premium for brand Y in preference to X or Z;
☐ I would pay a small premium for brand Y in preference to X or Z;
☐ I would pay a large premium for brand Z in preference to X or Y;
☐ I would pay a small premium for brand Z in preference to X or Y;
☐ I would pay no premium for any brand.

Ordinarily, a large majority of the respondents will choose either the second, fourth, or sixth option, thereby indicating a willingness to pay what they deem a small price premium—though not a large one—to obtain a favorite brand. Few if any of the respondents will choose any of the other responses, each of which indicates a willingness to pay either a large premium or none at all. In particular, as long as differences in prices between one brand and another remain small, only the fickle few customers who choose the final response will switch brands in response to price changes.

A small price cut is one that isn't worth responding to, because it will cause only the fickle few to switch brands. A substantial price cut is one that does require a response, because it threatens a firm's home turf—its loyal customer base. Undifferentiated product markets, ruled as they are by the law of one price, are simply markets in which all price cuts are substantial. In reality, there are no such markets. They are nothing more than a convenient fiction, introduced for the sake of clarity and simplicity.

Marketing warfare does not concern itself with the fickle few customers who either have no favorite brand or refuse to pay any premium at all to obtain it. There aren't enough like that to bother with. Firms engage in such warfare in hopes of persuading customers that their own brands are worth the small premium they sometimes command. To that end, modern firms employ radio, network and cable television, direct mail, and magazine, newspaper, and billboard advertising, along with large sales forces, cash rebates, mail-in contests, and other promotional activities. Package design and point-of-sale material are important as well. Some firms have lately begun to experiment with word-of-mouth marketing campaigns, in which consumers either volunteer or accept compensation to plug a product to their friends and acquaintances.[13] Surprisingly, such "marketing agents" (shills?) seem to experience no loss of credibility when they confess their involvement in an organized campaign. Indeed their "conversational partners" (marks?) actually seem more likely to pass the message on when informed that it is part of an organized marketing campaign.

Marketing warfare is expensive. Yet by following the natural price leader, all three firms (or groups of firms) in figure 16.3 will earn healthy profits, leaving them with cash to invest in such combat. Since the more profitable firms in an industry are better able than the rest to afford such investment, they tend over time to take business (market share) away from their less profitable rivals. This route is the second—and more usual—by which loose oligopolies become tight ones.

The rules of marketing warfare are fairly simple. The first and most important asserts that the fog of war hangs as heavily over the marketing battlefield as any other, while the second affirms that the more one invests in market share, the more of it one may expect to win. The key word here is "expect," since chance is an important determinant of marketing success and apparently reasonable expectations frequently go unrealized. Marketing budgets have to be divided among various kinds of advertising, promotion, and other activities believed to stimulate sales, and some divisions invariably prove more effective than others.

In a market served by four firms with equal (and equally well spent) marketing budgets, each firm should eventually control 25 percent of the market. But should one of those firms double its marketing budget while the other firms don't change theirs, the more aggressive firm should gradually increase its market share to 40 percent, at the expense of the other three, each of which seems likely to retain only 20 percent.[14] And so on. The relationship between a firm's long-term market share and its steady-state marketing budget should, by the foregoing logic, satisfy the equation

$$\text{MARKET SHARE} = \text{WE} \div (\text{WE} + \text{THEY}),$$

where WE stands for a firm's own marketing budget and THEY represents the sum of all competing budgets. The competition for market share takes place on a completely different time scale from price competition. Whereas firms must be prepared to react to rival price cuts within weeks, days, or even hours, their overall marketing budgets require only quarterly or even annual review.

Marketing warfare games operating on the WE ÷ (WE + THEY) principle have been studied in some detail.[15] Given such a game, it is no trick to write an optimal-response program that accepts as input a particular value of THEY and returns as output the exact value of WE needed by team WE to earn the largest possible profit against the given value of THEY. But the possession of such a program does not, by itself, guarantee the profitability of team WE. For one thing, team WE has no way of knowing in advance what THEY is going to be. For another, THEY may be so large that team WE has no chance of turning a profit. In that event, the program will return the nothing-ventured-nothing-gained budget WE = 0.

Optimal-response programs share certain features. They all contain a smallest value THEY* of THEY against which WE = 0 is the best possible response, and they all contain a largest value WE* of WE that the program will ever return (recommend) under any circumstances. As a result, the potential values of THEY can be partitioned into three nonoverlapping intervals, as is done in figure 16.5. The rightmost in-

FORGIVING ▼	DEMANDING ▼	HOPELESS
?	THEY*	

Figure 16.5. Ongoing investment games operating on the WE ÷ (WE + THEY) principle are all but hopeless against heavy opposition but are forgiving of budgeting errors against light opposition. There is an intermediate level against which profit is possible but tricky to obtain.

terval, in which THEY exceeds THEY*, is labeled "hopeless" because team WE has no chance of turning a profit against such large values of THEY. Shares of such a market are overpriced, and team WE should avoid them by choosing WE = 0. The leftmost interval, in which THEY is significantly smaller than THEY*, is labeled "forgiving" because shares of the market are then so cheap that even a carelessly chosen (far from optimal) marketing budget WE will earn a handsome profit. Unless THEY is very close to 0, the value of WE associated by the optimal-response program with a forgiving value of THEY will be only slightly smaller than WE*, say on the order of 85 to 95 percent as large as WE*. Such values of WE are sometimes called robust, because they produce near-optimal profits against a wide variety of THEY values.

The middle interval in figure 16.5, where THEY is almost as large as THEY*, is labeled "demanding" because team WE's potential profits are small against such values of THEY, and its marketing budget WE must be very carefully chosen (near optimal) to earn even half that much. As a practical matter, team WE may decline to compete in a market in which it is likely to face demanding values of THEY. The question mark in figure 16.5 is intended to indicate that the dividing line between forgiving and demanding values of THEY, though critically important, is imprecisely defined. The difference between the two is a matter of degree, not of kind.

To play the game in question, team WE must decide what value of THEY the opposition is likely (collectively) to choose, and supply its computer with the best guess it can of that unknown quantity. The optimal-response program will then return the value of WE that would be optimal if "guess" were 100 percent accurate. As long as THEY falls in the "forgiving" range, the value of WE returned by the optimal-response program using "guess" as input in place of THEY will prove entirely adequate. In fact, it will earn team WE a near-optimal profit on all such occasions. But when THEY lies in the "demanding" range, the difference between "guess" and THEY is not to be ignored. Although underestimates are marginally less damaging than overestimates, either can turn potential profits into outright losses.

Like prisoner's dilemma, games that function on the principle WE ÷ (WE + THEY) can be played more than once. The Axelrod experiments clearly demonstrate that the repeated and one-time-only versions of a particular game can differ as night from day, in part because "the power of nice" applies only in repeated-play situations. A computer programmed to play an oft-repeated version of the marketing warfare game described above by choosing some robust value of WE, such as WE = 0.85 × WE*, on all occasions would play moderately well against as many as three opponents.[16]

To further improve its performance in repeated play, the machine would need a program for computing sensible guesses as to THEY from its record of past encounters. So equipped, it could input its best guess to the optimal-response program and play the resulting value of WE in the next repetition. The results would remain satisfactory as long as "guess" remained comfortably within the "forgiving" range, but might plunge team WE into an ocean of red ink in apparently demanding situations. Accordingly, team WE could be well-advised to modify its response program to return WE = 0 when "guess" exceeds, say, 90 or 95 percent of THEY* before entrusting its corporate fate to a suite of computer programs.

Here is a second mechanism by which an initially competitive industry may devolve into an oligopoly. Even firms whose break-even prices remain well below the market price may be vanquished in marketing warfare. It famously happened to Schlitz, and has doubtless happened to countless others. Meanwhile, Anheuser-Busch was increasing its market share from about 6 percent in 1950 to better than 50 percent by the year 2000. Anheuser-Busch didn't do that by selling Budweiser for less than Schlitz, Pabst, Miller, Coors, or any of the other premium-quality mass-produced beers sold in the United States during the last half century. The firm became dominant by marketing its products more astutely than did its competitors. The same can be said of the automobile industry, the farm implement industry, and a host of other industries that grew into tight oligopolies during the twentieth century. The brewing industry is of particular interest, due to the admirable record of its (d)evolution furnished by Tremblay & Tremblay.

An influential school of economic thought would deny the importance of the distinction between "guess" and THEY. The principle of "rational expectations" is most frequently invoked in the macroeconomic debates that spawned it, yet surfaces occasionally in microeconomic disputes as well. Wherever it goes, the underlying idea is disarmingly simple. It is that "rational" economic agents—meaning flesh-and-blood as well as corporate persons—possess incommunicable knowledge enabling them to predict future economic events, without being able to explain to the rest of us how they do it. The oft-demonstrated fact that the perpetuators of this belief possess no comparable prescience is somehow deemed irrelevant.

Deemed equally irrelevant is the fact that the principle of rational expectations can be tested—at least as it applies to marketing warfare—by organizing tournaments along the lines of those organized by Axelrod. Each participating team would be required to submit a Java program, capable of running on virtually any computer, to play some version of the foregoing market-share game repeatedly. The tournament organizer would then cause every possible group of, say, six competing teams to play a lengthy series of games to determine the leading money winner. Because a market served by six suppliers is marginally overcrowded, it is all but certain that some of the teams would exhaust whatever credit they were given at the outset and be forced to withdraw from the competition.

No team submitting a program capable of predicting THEY will incur any actual losses, at any stage in the series, whether or not its THEY approaches or even exceeds the limit THEY*. Indeed, any team whose program incurs one or more losses can be accused of harboring irrational expectations. Furthermore, because prisoner's dilemma can be interpreted as a marketing game, it may be argued that the Axelrod tournaments have already tested and found wanting the principle of rational expectations. Perhaps biologist Paul Ehrlich had that principle in mind when he scoffed, "Economists think that the world works by magic."[17]

A nineteenth-century analysis of price competition suggested that sellers take turns raising or lowering—as need be to increase their own profit—their prices unilaterally until the law of diminishing returns ceases to permit further improvement. The array of prices so arrived at is said to be in Bertrand equilibrium, in memory of the French mathematician Joseph Bertrand (1822–1900), who first proposed such behavior. Although his proposal leads only to confusion in markets governed by the law of one price, it does lead to an unambiguous conclusion in markets where different brands of the same thing can sell at (slightly) different prices.

Were such a market served by two firms only, one could record on paper the process whereby Bertrand expected their prices to gravitate toward equilibrium. Such a record appears in figure 16.6. In it x and y represent the prices of brands X and Y, respectively. Curve HH' separates points (x, y) at which x is too large relative to y from those at which x is too small, while VV' has like significance for y.

Only if the point (x, y) actually lies on HH' is team X unable to in-

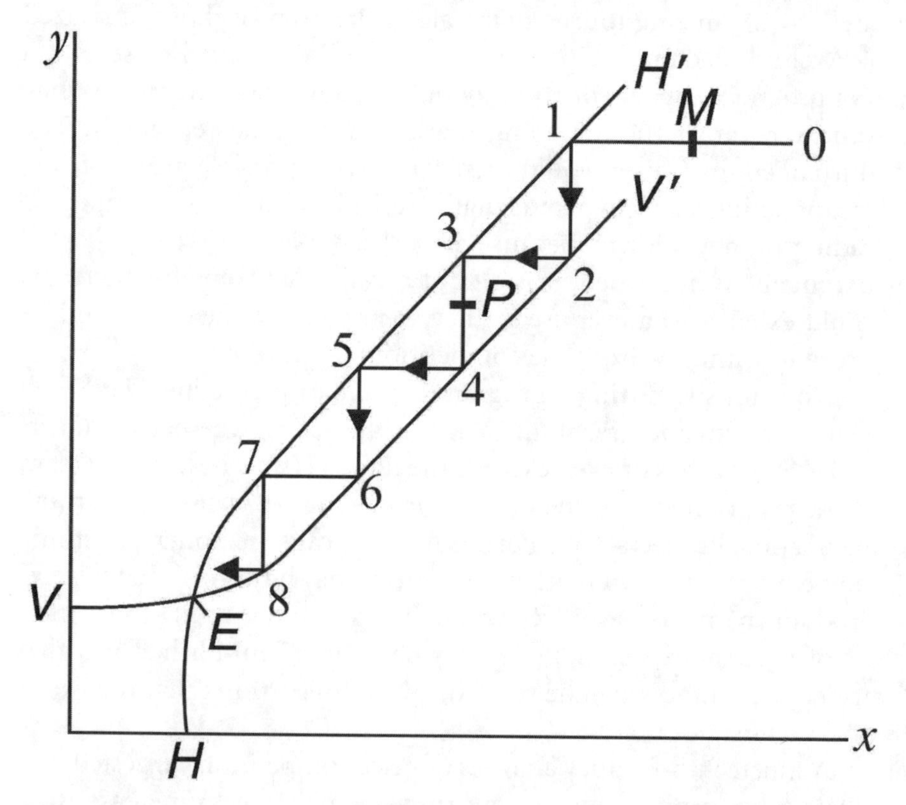

Figure 16.6. The process by which Bertrand expected Nash equilibrium prices to emerge in markets served by oligopolists closely resembles that by which Nash equilibrium fishing fleet sizes were supposed to emerge in Chapter 6.

crease its own profit by varying x. Likewise, only if (x, y) lies on VV' can team Y not increase its own profit by varying y. Hence, only if (x, y) lies at E, the intersection of HH' and VV', can neither team increase its own profit by varying its own price. The "staircase" 012345678 . . . leading down from the starting point 0 to the immediate vicinity of E illustrates the series of modifications Bertrand expected an initial pair (x, y) of prices to undergo. It begins when team X transfers the point (x, y) from 0, which lies on neither curve, to 1, which lies on HH'. It continues when team Y transfers it from 1 to 2, which lies on VV'. And so on. Each horizontal segment represents a reduction in x, while each vertical corresponds to a reduction in y.

Had the process begun from a starting point S close to the origin O of coordinates, the staircase of adjustment leading from S to the immediate neighborhood of E would have consisted of a series of horizontal and vertical line segments representing price increases instead of price reductions. The reader can construct the Bertrand staircase leading from an arbitrary starting point S to the immediate vicinity of E as follows: Place the tip of a pencil at S and move it vertically to VV'. Then move it horizontally to HH' and vertically again to VV'. After the first move, each subsequent pair corresponds to a single step of the desired staircase leading up, down, or perhaps sideways to E.

In terms of pricing power, the Bertrand analysis suggests that duopolists serving a market for branded products possess very little. Cournot's rather similar 1837 analysis suggests much the same thing, as does Chamberlin's 1933 analysis of what he called "differentiated product" markets. A host of more recent analyses, allegedly inspired by game theory, do likewise. Could that be the reason why mainstream economists rely on such analyses—when circumstances prevent them from avoiding the subjects of duopoly and oligopoly altogether—while ignoring Chamberlin's 1933 treatment of undifferentiated product markets in terms of price/product curves like the ones in figure 11.3? Do they reject Chamberlin's analysis because evidence contradicts it, or because the others are more readily compatible with the existing orthodoxy? Despite abundant evidence to the effect that pricing power abounds in the world's most important markets, leaders in the field seem determined to ignore the possibility that Chamberlin was onto something.

The Bertrand adjustment process requires team X (a.k.a. firm 1) and team Y (a.k.a. firm 2) to know the locations of HH' and VV'. If team X knew only how to decide whether HH' lay to the right or left of a given point (x, y), while team Y knew only whether VV' lay above or below such a point, they could reach the same equilibrium by alternating short horizontal steps directed toward HH' with vertical ones directed toward VV'. As a result, there is little doubt that even poorly informed firms could reach Bertrand equilibrium without collusion if content to do so. There is, on the other hand, every reason to doubt that profit-oriented competitors behave this way. We shall return to the issue presently.

Bertrand originally proposed the process that now bears his name in 1883, while reviewing a book published forty-five years earlier by Cournot. In it Cournot had analyzed a model not unlike the one depicted in figure 16.6, save for the fact that Cournot's x and y represented the *quantities produced* of undifferentiated products X and Y rather than their prices.[18] Cournot's model presumes that the rival firms will decide independently how many units of X and Y to produce and will permit the (necessarily common) price to rise or fall to whatever level it must in order to sell $x + y$ units. Bertrand doubted that profit-oriented firms would ever surrender that much control over their prices and suggested a revised model in which the prices of X and Y, rather than the quantities produced, would function as decision variables.

As mentioned above, Bertrand's suggestion leads only to confusion as it applies to markets for undifferentiated products. Accordingly, it was not until the twentieth century—when manufacturers began to affix brand labels to everything from bananas to gasoline pumps—that Bertrand's suggestion received the attention it deserved. Today, however, Cournot and Bertrand equilibrium points—along with the adjustment processes that lead to them—dominate conventional economic thinking about imperfect competition.

While inapplicable to games played only once, the adjustment rules governing the construction of the staircase 012345678 . . . in figure 16.6 constitute workable—if inept—strategies for repeated versions of those games. Every action depicted in figure 16.6 can be carried out in that context. But why, one wonders, would profit-oriented competitors be content to descend the Bertrand staircase, to the land of low, low prices, when they can so easily ascend it instead?

Starting from E, for instance, team X cannot increase its own profit by unilaterally increasing x. It can hope, however, to do so by announcing a price increase—to take effect in, say, thirty days—and then waiting for team Y to respond by announcing a similar increase. Team X's announcement constitutes an invitation to earn more by selling less. Should team Y decline so princely an invitation, team X can rescind the announced increase at little cost to itself. But why, pray tell, would team Y decline? Would not shareholders revolt against a management team that behaved in a manner so contrary to their interests? Would team X (a.k.a. firm 1) not get word to the investment banking community that team Y (a.k.a. firm 2) seems ripe for a hostile takeover?

Team Y could risk aggravating team X by announcing a slightly smaller increase than did X, so that the designated destination (x, y) lies on or near VV' instead of midway between HH' and VV'. But team X is then at liberty to revise its own announced increase downward, making the result less beneficial to both firms, or to cancel the announced increase altogether. So a strategically minded team Y seems more likely to accept team X's invitation with gusto—no doubt in the hope that more like it will be forthcoming—than to discourage team X in any way.

Such "virtual price adjustments" are useful in two ways: they minimize the frequency with which prices must actually be raised—as is unavoidable from time to time in an age of perpetual inflation—and they encourage the concerted action required to raise prices. They are useful even to natural monopolists, such as regional suppliers of natural gas and electric power. When such firms feel the need to increase their service rates, especially after an extended period of price stability, they often begin by announcing unnecessarily large increases before bowing to pressure from local politicians to moderate their demands. That way everybody wins. Local newspapers enjoy increased sales, local people are made aware that energy price increases are imminent, local politicians are hailed as champions of the common man, and the local utility amasses favors it can call in from the politicians at a later date.

The Bertrand adjustment process cannot, of course, run forever in reverse. At some point, matching price increases necessarily cease to benefit both firms. When that finally happens, either participant can halt the process by refusing to match an announced increase. This can be done either by announcing a more modest increase or by announcing none at all. In the former event, the initiator seems likely to respond by reducing its announced increase, to match the more modest counterproposal. In the latter, it has no choice but to cancel the announced increase entirely. But neither of those events seems likely to occur at or near *E*. On the contrary, prices seem likely to ascend in lockstep until the indicator (x, y) of current prices reaches a position such as *P*, well above and to the right of *E*, before coming to a halt. In such a location, both firms earn far more than they would at *E* while selling significantly less.

Having arrived at such a location, the firms of course run the risk of a price war breaking out, causing the point (x, y) to descend the Bertrand staircase once more toward *E*. But strategically minded firms can—and easily do—avoid that pitfall. Instead of decreasing x until (x, y) rests on or near HH', team X can announce a sale of specified duration, during which (x, y) will rest on HH', before returning to *P*. And instead of announcing even deeper cuts to take effect immediately, team Y can wait until X's sale is over to announce a sale of its own. In this way, the firms can maintain the illusion of vigorous price competition while sharing all or most of the profit a cartel or monopoly might extract from the same pool of customers. And of course, much the same thing can happen when a third brand—call it brand Z—enters the market.

Just such a three-phase cycle has long prevailed in the beer industry, where Anheuser-Busch products go on sale one month, followed by Miller and Coors products the next, and all other brands the third. This practice affords the mass producers of beer a welcome opportunity to control their inventories without destroying the golden goose they clearly recognize their shared monopoly to be. Every advanced economy contains a host of shared monopolies, whose industry profits are scarcely less than those a formal cartel might command. Brand loyalty certainly facilitates the achievement of such profitability but is by no means necessary for the purpose. Firms whose products (such as gasoline) command no such loyalty have learned to exploit other features of the markets in which they operate to similar ends.

The effective clairvoyance economic orthodoxy attributes to marketing warriors tasked with choosing a budget (WE) appropriate to a seemingly unknowable THEY contrasts vividly with the lack of vision said orthodoxy attributes to price competitors unable to escape low, low prices by ascending Bertrand's staircase. It suggests a pseudoscientific willingness to accept almost any hypothesis that supports the desired conclusion—namely that the world has to work more or less the way the perfectly competitive model says it does—while ignoring hard evidence to the contrary.

That contrast is by no means the only ground upon which the economic establishment has been accused of practicing pseudoscience. The same accusation has been made many times before, on a variety of other grounds. Perhaps the most credentialed accuser was Frederick Soddy, a Nobel Prize–winning chemist who devoted the second half of his working life to the study of economics, which he considered a pseudoscience in need of a totally new beginning. In addition to exploring the oft-neglected implications of the second law of thermodynamics for economics, he argued that modern banking systems foster a dangerous confusion between debt—a symbolic form of wealth—and true (tangible) wealth. He compared the notion that mankind can continue indefinitely to live off the interest generated by mutual indebtedness to a perpetual motion machine, and warned that whereas debt can grow at an unlimited rate for an unlimited length of time, tangible wealth can only keep up for so long. The inevitable lag, when it comes, can only lead to class warfare by debt repudiation.

Soddy was one of the few career scientists who, when economic ideas were explained to him, did not shake his head politely and resolve to have nothing more to do with the subject. Most trained scientists react to economics as Benjamin Franklin—much of whose old-world reputation rested on his contributions to the science of electricity—reacted to metaphysics: "The great uncertainty I found in metaphysical reasonings disgusted me, and I quitted that kind of reading and study for others more satisfactory."[19] One need only substitute "economics" for "metaphysics" to obtain a sentiment almost any scientist could subscribe to. Biologist Paul Ehrlich, quoted above, is another critic of the subject, as, reportedly, was Richard Feynman.

The critical difference between oligopolies and secure monopolies lies not in the prices they are able to charge their customers but in the freedom the latter enjoy to dispose of their earnings as they will. Whereas an oligopolist or insecure monopolist must reinvest all or most profit in the business—merely to maintain market share—a secure monopolist is by definition relieved of any such requirement. His or her profits can be used to acquire other businesses, to fund philanthropies, to ape the lifestyles of the rich and famous, or to indulge in expensive hobbies.

CHAPTER 17

Policy Implications

Suppose, for the sake of argument, that the central theses of the previous chapter are essentially correct: oligopolies are able to command virtually the same prices that monopolies and/or cartels would extract from the same customer pools, and no conceivable form of antitrust legislation will significantly diminish the pricing power that enables them to do so. What then? How should the policy process respond? Which if any old laws should be repealed, and what new ones enacted in order to make the best of the newly assessed situation?

The late John Kenneth Galbraith was among the first to ask such questions. In *American Capitalism: The Concept of Countervailing Power*, he was led—presumably by his reading of Chamberlin—to equate oligopoly with "crypto-monopoly," and to jump from the acknowledged fact that "oligopoly was general in the economy" to the conclusion that pricing power "akin to that of a monopolist" was at work in many if not most mid-twentieth-century markets.[1] He then asked who was likely to be harmed by such power and what might be done to assist the victims.

His answer, as the title suggests, was that the victims should be allowed to amass "countervailing power" with which to protect themselves against the possessors of pricing power. Laborers should be allowed to form strong unions, while farmers should either be allowed to form cooperatives authorized to enforce minimum prices on staples like cotton, wheat, soy, and tobacco—along with feed for cattle, hogs, and chickens—or be guaranteed that the government will do it for them via surplus-perpetuating price supports.

First published in 1952, *American Capitalism* became an instant bestseller. It brought immediate celebrity to its author and remains in print to this day, more than fifty years and 400,000 copies later. The initial reviews in professional journals were almost uniformly favorable. *The American Economic Review* commended Galbraith's new theory to the "careful consideration of his professional colleagues," while *The Review of Economics and Statistics* considered that he had "produced a theory of capitalism which cannot be disregarded by anyone, although it will disturb many." Britain's prestigious *Economic Journal*—normally dismissive of colonials—pronounced it "shrewd, witty, and forceful."[2] Others compared its growing influence to that of Keynes's *General Theory of Employment, Interest, and Money*, and the American Economic Association convened a blue-ribbon panel to discuss the content at its 1953 national convention. There, it was less well received.

After a brief opening statement from Galbraith, future Nobel laureate George Stigler rose to speak. Though a member of the Columbia University faculty, he was a rising star in the ultraconservative Chicago school of economic thought. He accused Galbraith of "playing with blocs"—an unpardonable sin in the ruggedly individualistic tradition of orthodox economic thought—and denounced Galbraith's proposed theory as mere prejudice, bearing little resemblance to a coherent body of "scientifically defended" propositions. After that, it didn't much matter what the other panelists had to say, because Stigler had already spoken the magic words: there would be no "Galbraithian revolution." Others in the profession could safely return to their accustomed pursuits.

Having entered the field as an agricultural economist, Galbraith was well acquainted with the history of agricultural cooperatives. While many of them succeeded as bulk-purchasing agents, none has ever accomplished its primary mission to stabilize the prices farmers receive for their crops and livestock at levels sufficient to shield members from bankruptcy. The reason, as he explained in a chapter on agriculture, is clear. Although the cooperatives were free to set target prices, and could exhort their members to sell for no less, they could neither prevent them from doing so nor punish the ones who did. As a result, renegade members could and routinely did exploit their compliant brethren—the ones who did "hold the line on price"—by selling in secret at prices inflated by those in compliance. In the event, none of the cooperatives managed to retain price control for more than a year or two.

As a result, the Agricultural Marketing Act of 1929 created the Federal Farm Board with resources to capitalize and sponsor a system of national cooperatives backed by government stabilization corporations. Had these worked as intended, wheat, cotton, dairy, and tobacco farmers would have been represented in their markets by at most a handful of high-volume sellers. When that plan also failed, the Agricultural Adjustment Act of 1933 levied a "processing tax" on all growers of designated "basic crops." The proceeds were then distributed among those who complied with the act's voluntary production limits, so that noncompliant growers were obliged to subsidize compliant ones. The annual farm surplus was thereby eliminated until 1937, when the Supreme Court declared the processing tax to be unconstitutional. Since then, with a brief hiatus during World War II, farm and nonfarm taxpayers alike have been obliged to subsidize agricultural overproduction.

To solve the current farm problem, one might try to reinstate the New Deal processing tax by overturning the Supreme Court decision of 1937. Or, even without revisiting that decision, one might institute a schedule of fines for overproduction and require farmers to post bonds to be forfeited in case of infraction. The mechanism need not be complex. Each farmer would be required to negotiate a "legal capacity" with his or her county agent, based on past performance, while the U.S. Department of Agriculture would be required—by the end of each year, well before the next planting season—to authorize the utilization of a stated fraction of each farmer's negotiated capacity.[3] The farm sector could once again, as in 1933–36, earn more money by producing less, as other oligopolies routinely do. A similar plan worked for many years in the oil industry, with the Texas Railroad Commission doing the authorizing.[4] An only slightly more elaborate version could work as well in agriculture, under the USDA.

Then again, one might actually turn agriculture into a tight oligopoly. It could be done without title to a single acre changing hands. Congress would begin by issuing half a dozen licenses to raise and sell either animals destined for human consumption or feed (hay, corn, and such) destined for consumption by such animals. Said licenses would be available only to government-sponsored cooperatives owned and operated by the member farmers. The cooperatives would be territorial and overlapping—call them North, South, East, Mideast, Midwest, and West cooperatives—so that each farmer might choose which of at least two to join. Some might even join both, by dedicating some acreage to one and the rest to another. Then, as intended by the act of 1929, farmers would be represented in their markets by a mere handful of high-volume sellers.

The cooperatives could decide, as all oligopolists do, how much of each commodity to produce in a given year and issue corresponding futures contracts—obligating the cooperative to purchase a specific quantity of a given commodity from the holder on a specific date at a specific price—to each member. The issuance of such contracts must of course be done on a pro rata basis, so that each farmer receives a bundle of contracts proportional in value to the amount of land—adjusted for productivity—he or she has chosen to dedicate to the issuing cooperative. Such contracts could be made exchangeable, at least within cooperatives, so that members wishing to plant more corn and less wheat, or to raise more hogs and fewer chickens, could exchange one sort of contract for another. To that end, the cooperatives might operate resale markets (electronic, of course, since every self-respecting farmer has long since become Internet connected) for their own futures contracts.

Either approach would encourage farmers to take land out of agricultural production. Remarkably low-cost incentive programs would then induce them to choose their most environmentally sensitive land, such as that along riverbanks or in potential dust-bowl areas, for removal. Those with much environmentally sensitive land could sell their futures contracts—or USDA allotments—to others with less such land. Still other farmers would choose instead to scale back production by using fewer expensive fertilizers and herbicides, thereby reducing their yields per acre instead of their cultivated acreage. In time, this could ease the plight of the many Midwestern farmers who no longer dare to drink water from their own wells, due to seepage into the water table of chemical residues from decades of high-intensity farming.

The proposal to oligopolize agriculture offers several additional advantages. Since the cooperatives would soon become highly profitable member-owned and member-operated organizations, their presence in rural areas would revitalize small villages and towns. They would also develop deep pockets, inviting suit for punitive damages if and when the Frankenfoods they will surely experiment with are shown to cause genuine harm. To insure themselves against such risk, the proposed cooperatives will presumably test with increasing rigor the products they bring to market. Indeed, it seems likely that much of the testing—and many of the other tasks—now performed by the USDA could safely be delegated to such cooperatives. If, over time, they were seen to become unduly powerful, they could be subjected to government regulation the way railroads, airlines, and utilities once were.

Laborers, like farmers, have historically lacked pricing power. Galbraith saw their salvation in unions. After all, vast numbers of unions were already in existence, and—with critical assistance from New Deal legislation—they seemed to be growing in influence and prosperity. Moreover, they seemed to raise the wages and improve the working conditions of union and nonunion laborers alike, because factory owners are more disposed to grant wage and other concessions when they can by so doing preserve the nonunion status of their shops. Galbraith had little to say about the massive internal corruption that has historically invaded all or most unions, and though keenly aware that "nothing so weakens a union as an employer who cannot afford to pay and is closing plants or going entirely out of business," he never betrayed any suspicion that free-trade policy might one day erode U.S. corporations' ability to pay a living wage.[5]

Whatever their accomplishments, unions were never an ideal cure for the ills they were designed to combat. They took the form they did because nothing better could be arranged without government sponsorship, at a time when even democratic governments were generally hostile to their aims. Only during the New Deal did organized labor receive even a fraction of the support it needed to survive and prosper in the United States. The infamous Taft-Hartley Act of 1947 signaled the end of such support.

A more effective cure for the same set of ills would replace unions with a mere handful of licensed employment agencies. Corporations with more than half a dozen employees would be forbidden to hire any worker not represented by a federally licensed agency, unless willing to guarantee said worker steady employment leading to a secure retirement. Agency employees would be obliged to contribute a substantial fraction of their wages either to Social Security (as agency temps now must) or to an agency-funded pension plan via withholding. In time—and with appropriate legislative support—employees could perhaps be weaned away from Social Security and into more remunerative agency-administered programs overseen by the government.

Funds held in agency-administered retirement accounts should be transportable, so that model employees could afford to leave their agencies when offered employment by firms impressed with their work. Most entry-level workers would presumably enter the workforce through the employment agencies, hoping in time to earn promotion into the exalted corporate world. As a group, workers consider on-the-job performance a more valid claim to advancement than classroom achievement.

Union contracts would be replaced with agency contracts. Every employer would be free, as one such contract expires, to negotiate a new one with any licensed agency. No longer would firms be forced to decide every three years whether 'tis better to close their doors than to deal with a monopoly (a.k.a. a union) inclined to forget—as monopolists so often do—that the customer is always right. The competition between agencies would probably involve little or no price competition, but would surely involve "the power of nice," dedicated as it is to forging mutually profitable long-term business relationships. Experienced labor negotiators on both sides of the table might welcome such a revolution in labor-management relations.

Economics at its best can function as a science of scarcity. But such a science has little to offer a policy process incapable of distinguishing between that which is truly scarce and that which is only rumored to be. Essentials like food, clothing, and shelter are by no means scarce in today's developed nations. Equally plentiful are luxury items like skis, diamonds, Hummers, and trips to Las Vegas. Cuban cigars and estate-bottled Burgundies are scarce, to be sure. But few other things are.

Jobs, in contrast, are scarce wherever you go. In Mexico—and in the lands farther south—people are willing to pay thousands of dollars, and then endure weeks of hardship leavened by mortal risk, to be smuggled across the Sonoran Desert into the United States, knowing as they go that only the most menial tasks await them. Desperate natives of other continents risk long sea voyages—often concealed in the sweltering holds of cargo ships—with equally unenviable prospects. China watchers warn that hundreds of millions of jobs must be created there to feed the peasants about to be evicted from their ancestral lands—typically held in packages of five acres or less—to make way for modern (tractor-, herbicide-, and chemical-fertilizer-based) agriculture.

Worse still, the job shortage is destined to become more acute. One study showed that in European firms, computer-aided design increased the output of draftsmen between 200 and 6,000 percent, with an average gain of around 500 percent, when introduced during the 1970s and 1980s.[6] Computer-integrated manufacturing will have even more dramatic effects when it comes to pass but will take longer to implement, due to the expense and technical difficulty of automating entire factories. But let there be no mistake. The "lights-out" unmanned factories of science fiction, churning out endless streams of flawless goods and services "untouched by human hands," will eventually be built.[7] No known law of nature stands in the way of their construction. Such factories may even prove energy efficient, thanks to the millions who will no longer need to drive to work each day. Only one real technical breakthrough is needed to realize this vision.

One fine day, in the not too distant future, a team of bright young persons will teach a robot to catch and throw a baseball. The contraption needn't possess Nolan Ryan's fastball or Bert Blyleven's curve, or be able to catch Hoyt Wilhelm's knuckler. A simple game of backyard catch, at the level of Little League all-stars, will be more than adequate. A team of robots endowed with the hand-eye coordination required for such recreation could operate any factory in the world better and more cheaply than the most skilled and dedicated human workforce. Until it crosses that divide, automation will remain a work in progress. Robots endowed with ordinary human hand-eye coordination will not be built tomorrow, or even next year, because technical difficulties intervene. But built they will be.

Some tasks will always remain beyond the reach of robotics. Only a curious few will ever pay to see a robot perform Hamlet, or to hear one sing the blues. Robots won't pitch in the World Series, box for a heavyweight title, or tee it up at the Masters. Nor will they run giant corporations, be elected to public office, or prosper as prostitutes. At the other end of the spectrum, they will probably never bag groceries, muck stalls, pick fruits and vegetables, or sweep floors. While machines could surely be designed to perform those and other undemanding tasks, they seem unlikely to do so as well or as cheaply as needy men and women. Automation tends to attack and "hollow out" the middle of the income distribution, where the greatest potential savings lie. Automation isn't the main culprit just now, because foreign workers are still cheaper than robots. But someday that will change and will bring the outsourcing bandwagon to a screeching halt. The lights-out unmanned factories of the future seem certain to be built close to the markets they serve.

As automation nears fulfillment of its manifest destiny, the struggle for employment will increasingly resemble a professional golf tournament in which only the lowest fifty scorers of the first two days are entitled to contend over the last two for shares of the prize money. Players who don't make the cut earn nothing, and ship out on Saturday morning—even as their more successful rivals are warming up for the third round of play—to gain an extra day or two of practice at the site of next week's tournament. The steady advance of automation will only make it harder for the average citizen to make the cut separating members of the workforce from the hard-core unemployed, of whom there will be increasingly many. At present, even some of those who do make it into the full-time workforce—the so-called working poor—find it impossible to subsist on the wages they earn.

Factory workers are by no means the only victims of outsourcing. Engineers, accountants, middle managers, and document translators—to name but a few—have likewise seen their jobs exported to the other side of the world. Some have even been obliged to train their replacements. Jobs go where labor is cheapest and move on when cheaper replacements are found. Many of the jobs that left the United States for Mexico under NAFTA have now moved on to the Orient, where the so-called China price prevails. Big-box discount stores like Wal-Mart scour the world for the lowest possible price and—after finding it in China—oblige first-world suppliers to meet it or lose their big-box accounts. Whoever wishes to sell to Wal-Mart, or to firms that compete with Wal-Mart, is more or less obliged to operate the Wal-Mart way. A few resist temporarily but are eventually driven by "market forces" to outsource their production—and increasingly their "backroom operations"—to distant lands.

China, incidentally, appears determined to remain an outsourcing mecca. Sherrod Brown, Democratic senator from Ohio, quotes several highly placed Chinese officials to the effect that increasing prosperity must not and will not be allowed to alter the servile nature of the Chinese workforce, most members of which toil for pennies an hour in conditions no organized labor force would tolerate.[8] He observes, for example, that "an estimated 10,000 workers perish in China's coal mines each year; in contrast, 42 U.S. coal miners died in 2001." China, which produces about 40 percent more coal per annum than the United States, with a far larger workforce, is the world's leading coal producer. Brown also summarizes what is known concerning the Chinese *laogai* system of forced labor camps, into which an estimated fifty million people have vanished since 1949.[9] Once part of the PRC's department of corrections, the camps now stand beside the Chinese Communist Party and the People's Liberation Army as integral parts of the Chicom industrial complex.

Yet even Chinese John Henrys are no match for automation. As soon as that first robot starts playing catch with Little Leaguers, the days of the manual laborer will be numbered. They will disappear gradually as successive generations of machines take over more and more tasks from their human rivals, leaving the machine owners (a.k.a. corporate shareholders) to claim—with perfect justification—an ever-increasing share of the rewards. Fewer and fewer of those rewards will find their way into the pockets of machine operators, who will in consequence find it harder and harder to earn a living. The process is already well under way, with no end in sight. The oft-lamented gap between rich and poor people, along with the one between rich and poor nations, can only continue to widen.

The shape of things to come has long been evident. Norbert Wiener, father of cybernetics, wrote as early as 1947 that "the automatic factory and the assembly line without human agents . . . gives the human race a new and most effective collection of mechanical slaves to perform its labor. Such mechanical labor has most of the economic properties of slave labor . . . any labor that accepts the conditions of competition accepts the conditions of slave labor, is essentially slave labor."[10] A few years later he added that "if these changes . . . come upon us in a haphazard and ill-organized way, we may well be in for the greatest period of unemployment we have yet seen."[11] On other occasions, he omitted the maybe: "It is perfectly clear that this will produce an unemployment situation, in comparison with which the . . . depression of the thirties will seem a pleasant joke."[12] The fact that it hasn't happened yet is no sign that it isn't going to. Like world-class chess, hand-eye coordination is taking longer to achieve by artificial means than the early automators anticipated.

The driving logic is starkly Malthusian. Truck drivers, herdsmen, territorial salesmen, heavy-equipment operators, and hoteliers—not unlike money managers—assume responsibility for specific income-bearing assets. Their mission, in each case, is to extract an appropriate return on investment. Managers who disappoint are quickly replaced. They easily can be, because automation and information technology are rapidly increasing the quantity of assets a single manager can effectively manage. Indeed, the rate at which that quantity increases consistently exceeds the rate at which the total accumulation of income-bearing assets seems able to grow. Hence, the number of asset managers required can only continue to decrease.

The Washington policy process seems to recognize only two legitimate responses to job shortages at home or abroad: enhanced skills and reduced compensation. While waiting for these measures to resolve a particular shortfall, Congress routinely resorts to deficit spending. That eases the (often devastating) symptoms of prolonged unemployment and underemployment while disguising the fact that enhanced skills and reduced compensation accomplish next to nothing.

The former fails because the vast majority of modern jobs require few if any real skills. A March 1999 report prepared by the Job Opportunities Task Force of Baltimore leaves little doubt about this. The task force interviewed almost every employer in the Baltimore metropolitan area about foreseeable job opportunities.[13] It concluded that in even the best of circumstances, roughly 62 percent of all job offerings would be of the low-wage, low-skill variety.

For reasons unknown, this finding came as a surprise. The market research firm engaged to conduct the necessary interviews undertook the project in the expectation that upon completion, similar services could be sold to other municipalities. Having demonstrated an ability to identify the most marketable skills in one area, the firm would surely be asked to render similar services to other local governments anxious to equip their students with more marketable skills. Teachers' unions awaited the survey results with bated breath. No one expected to discover that post-grammar-school education and job training are of little commercial value to much of the population.

Virtually anyone with an eighth-grade education and a willingness to follow instructions can fill most of the positions Baltimore employers then wanted—or expected to want—filled. The successful entry-level job candidate need demonstrate no knowledge of science, no skill at algebra, no smattering of history and geography, no writing or foreign-language skills, and—perhaps most surprising—no vestige of computer literacy.

On a typical day, about 26,500 full-time low-skill jobs were vacant in the Baltimore metropolitan area, while almost three times that number of individuals—fewer than half of whom were eligible to be counted among the officially unemployed—expressed a desire to find such jobs. As well they might. Plenty of low-skill jobs still pay rather well and offer decent working conditions. Almost a third then paid at least $11.25 per hour, or $22,500 per year—well above the prevailing national minimum wage of $5.15 per hour as well as the federal poverty guidelines figures of $5.22, $6.56, and $7.91 for a single parent with one, two, or three children, respectively.

Minority scholarships, computers in the classroom, and even state-funded training programs are powerless to cure the existing shortage of low-wage, low-skill jobs. Indeed, all the politically correct schemes, up to and including "the end of welfare as we know it," merely aggravate the situation by increasing the supply of qualified candidates—of whom there has never been any shortage—without increasing the demand for their services. The problem confronting metropolitan Baltimore is not a shortage of skills but a shortage of jobs in which to exercise the skills residents already possess.

Perhaps the most unfortunate aspect of the Baltimore study is the lack of others like it. Few if any other jurisdictions seem to know what kinds of jobs their public-school graduates are now, or soon will be, competing for. Nor has such information been gathered in the past. No one seems to know how the study's key findings—that 62 percent of the jobs in the region are low-skill jobs, and that the ratio of low-skill job seekers to low-skill job openings is roughly three to one—compare with corresponding figures in other jurisdictions, or in Baltimore's past.

A strikingly similar report did appear in England in 1997.[14] It found that 63 percent of total employment in Great Britain was in the six lowest-paid occupational groups: clerical/secretarial, personal/protective services, sales, handicrafts, plant and machine operatives, and "other" manual labor. Moreover, the median job in those categories required no formal education beyond level 4 in the British National Curriculum—a milestone intended to be reached by the typical eleven-year-old at the end of primary schooling. Only 4 percent of the employers interviewed nationwide felt that their business objectives were in any way jeopardized by the lack of literacy and numeracy among their current employees. "Once the vast majority of the adult population are functionally literate," the study concluded, "any link between literacy, numeracy, and economic performance is very hard to demonstrate." In short, inadequate schooling is not the cause of Britain's perennial job shortage, and better schooling will not cure it. Neither, as it turns out, will reduced compensation.

Conventional economic wisdom has always had trouble understanding how anyone willing and able to work could have trouble earning a living. Will labor markets not clear as others do, when wages are allowed to descend to their "natural" levels? The question ignores the fact that in a given time and place, the "full employment" wage may not be enough to live on. When reliable trucks and farm tractors became available, during the 1920s and 1930s, plow and draft horses could no longer earn their keep. They avoided starvation only by being sold to glue factories and dog-food suppliers. A similar disaster can befall human workers, and will soon after that first robot starts playing catch with Little Leaguers. Untold millions will discover that they can no longer survive on what they are able to earn. The so-called iron law of wages, which asserts that workers must be paid enough to keep on working and reproducing, does not apply to plow horses, or to others for whom there is no work.

Something of the sort actually happened, according to Michael Dunkerley, to farmers and day laborers during the early years of the Roman Empire. As endless streams of slaves began pouring into the homeland from newly conquered territories, he explains in *The Jobless Economy?*, ordinary citizens found it increasingly difficult to make ends meet.[15] Displaced workers eventually coalesced into the feared "Roman mob," which generations of emperors were obliged to placate with bread and circuses. Both Dunkerley and Jeremy Rifkin argued persuasively during the 1990s that automation and globalization are—like ancient Roman slaves—rapidly eliminating jobs in important sectors of the global economy, thereby necessitating the modern equivalent of bread and circuses.[16] Both authors are largely forgotten today, in part because the policy process favors inaction.

Although they cannot be sold or rented, jobs can be rationed in the way that food staples and gasoline were rationed during World War II. Juliet Schor points out in *The Overworked American* that a number of firms experimented rather successfully with a thirty-hour workweek during the 1930s, after Senator Hugo Black of Alabama introduced a bill in the Senate mandating such reform.[17] Although the Senate passed Black's bill, the House struck it down in favor of more comprehensive New Deal legislation. A thirty-hour workweek, if enacted today, might create as many as forty million new jobs in the United States while obliging employees to accept a reduced standard of living. Although the resulting standard could still be high, by global and historical norms, the transition to it would be more difficult than during the 1930s, due to the debt burdens so many Americans now bear.

The licensed employment agencies proposed earlier could ration civilian jobs in a very different way, by patterning them after jobs in the military. Enlisted men and women qualify for (admittedly meager) pensions and (remarkably comprehensive) VA health care after just twenty years of service. As in the military, agency employees compiling distinguished records—or possessing exceptional skills—could be rewarded with extra years on the job, coupled with pay increases and enhanced retirement benefits. If the number of people eligible for employment in the corporate sector of the economy is limited, their level of compensation can be increased. And if they are encouraged to embrace realistic expectations, their overall satisfaction can be increased as well.

Any early-retirement scheme would oblige employees to accept a large fraction of their compensation in deferred form. How else can they hope—after a mere twenty years in the workforce—to earn the greater part of their eventual retirement benefits? Although many will denounce any such reform as "creeping socialism," an appropriately crafted version might attract broad-based support on the grounds that it is market based and no viable alternative exists.

Like the licensed farm cooperatives proposed earlier, such agencies would grow to be vastly profitable member-owned and member-operated corporations. They would largely eliminate the need for corporate personnel departments, since they would assume most if not all of the record keeping, pension management, benefit administration, and disciplinary responsibilities such departments currently perform. They would also, in all probability, find it necessary to operate schools akin to the ones at which enlisted personnel—few of whom are extensively educated—are trained to deploy the highly sophisticated weapons of modern warfare. Such schools have long ranked among the nation's most successful, mainly because graduates progress more or less directly into higher-paying jobs.

Educators tend to forget that morale is as important to students as it is to soldiers and athletes. It is also very contagious. A soldier, student, or athlete surrounded by discouragement is likely to become infected. Because students communicate extensively among themselves, student body morale requires the shared conviction that investments made in the classroom will bear fruit later on. Military schools create that conviction by imparting immediately useful skills to students motivated to acquire them. For many years, a substantial fraction of the nation's civilian welders, crane operators, and aircraft mechanics—to name but a few—learned their trades in the military. The schools from which they graduated had the luxury of admitting only a few more candidates than were needed to fill the anticipated positions, and the liberty to "wash out" all who failed to "qualify" for such positions. How many civilian faculties, especially in inner-city high schools, yearn for such luxury and liberty?

By restricting the quantity of labor available to corporate enterprise, employment agencies of the proposed sort could offer better wages, along with relatively continuous progress from entry-level work up to more demanding tasks and eventually out into secure retirement. Promotion would be as incremental, as performance based, and as apparently fair as it currently is in the military. Moreover, by admitting teenagers (at least on a part-time basis) as soon as they can demonstrate functional literacy and numeracy, such agencies could greatly increase morale and performance in the nation's public schools. The result would be a younger, healthier, and more motivated workforce. At a guess, corporate hiring policies would evolve to exempt only the most promising college graduates from agency employment by hiring them directly onto corporate payrolls. Run-of-the-mill college students would be encouraged to enter the workforce through agency-organized summer jobs, internships, and/or work-study programs.

Entry-level employees would of course inherit the most menial tasks, before earning promotion to more responsible work. As skill positions become available, the likelier candidates within an agency would be offered places in schools designed to qualify them for those specific positions. Less likely candidates would have fewer opportunities to obtain such schooling. Yet even the most menial positions become endurable when a mere twenty years in them leads to secure pensions with guaranteed benefits.

Twenty-first-century leaders must reconcile themselves to the likelihood that good jobs will never again be as plentiful as of old. Many college graduates already recognize that their degrees aren't really needed for the work they do, and they suspect that time will continue to erode the commercial value of such degrees. A college education will never cease to enrich the life of anyone fortunate enough to possess one, but it will no longer do so through enhanced earning power. In the unending struggle to better one's own condition, education will be reduced to a niche strategy—like advanced artistic or athletic training—of economic value only to those born with the rare and innate talents needed to exploit such training.

Charles Murray, co-author of *The Bell Curve*, has long maintained that the United States spends too much of its education budget on the least gifted and least motivated students, few of whom care to exploit the opportunity public education represents, and too little on those with the talent and inclination to do so.[18] By taking the least teachable students out of the classroom and putting them to work in jobs that require little schooling, the proposed employment agencies might allow teachers to devote more time to students not now receiving the attention they need. Imagine the possibilities! Strawberries picked by mall rats instead of illegal aliens!

Retirement from an employment agency need never and should never necessitate retirement from the workforce. Agency retirees can and should be at liberty to open their own small businesses—lawn services, beauty shops, bars, luncheonettes, perhaps even tax preparation and bookkeeping firms—and be given every incentive to defer some or all retirement benefits (with interest) until the need for them arises. Retirees need only be prevented from competing—by supplying for-profit corporations with cheap labor—with the agencies obligated to support them in old age.

The organizational details of the proposed farm cooperatives and employment agencies need not concern us here. The intended purpose is—in each case—to allow members to earn more money by doing less work, in the manner of oligopolists the world over. Success will depend—again in each case—on strict enforcement of the enabling legislation, which must provide effective penalties for the unlicensed invasion of markets in which the licensees alone are authorized to trade.

Neither agricultural cooperatives nor employment agencies can hope to achieve their intended purposes under the aegis of free trade. With or without such institutions, U.S. jobs and incomes will remain hostage to competition from abroad as long as global corporations remain free to import foreign-made products into the country free of charge. Only under cover of job-protecting import duties can the proposed employment agencies hope to perform as intended. And because farmers now represent only about 1 percent of the U.S. population, agricultural cooperatives cannot expect to prosper unless the bulk of the population has money to spend. The time has come to reverse the nation's well-intentioned but ill-advised commitment to free trade. Only when that is done can U.S. farm and factory workers—the vast majority of whom work both hard and smart—expect to be paid a living wage.

Simply by imposing a 25 percent import duty on foreign-made automobiles and automobile parts, like the one we already impose on foreign-made trucks and truck parts, the United States could breathe new life into the auto industry. Similar duties on foreign garments, textiles, and steel could do the same for those industries. Duties on other imports would furnish U.S. manufacturers—many of whom would prefer not to outsource their production—with the countervailing power needed to oppose Wal-Mart's strong-arm tactics.

Such duties would, to be sure, elevate the prices U.S. consumers pay. But they would elevate the median income even more. U.S. leaders need to overcome their irrational fear of tariff protection—

somehow instilled in them by economists they only pretend to understand—to reverse the ongoing (and very real) decline of the U.S. economy. Until that happens, the public will continue to be placated with meaningless worker-protection agreements appended to treaties like NAFTA and CAFTA in lieu of effective measures. Will the Mexican government enforce an agreement committing Mexican firms to pay a minimum wage or to provide humane working conditions? Could it do so if it wanted? Will the Chinese government enforce such agreements? The record—including statements by Chinese leaders—suggests otherwise.

The first lesson to be learned from the systematic study of competition concerns the linkage between rules and outcomes. The rules of basketball favor tall, gifted athletes; those of football favor bulky, gifted athletes; those of chess favor mental gymnasts; and so on. Every set of rules favors one group of competitors over all others. It's unavoidable, particularly as it applies to market competition in the laboratory. Douglas Davis and Charles Holt observe, on page 5 of their book *Experimental Economics,* "One of the most prominent lessons of laboratory research [in economics] is the importance of trading rules and institutions to market outcomes." We have already commented on this fact with respect to Vernon Smith's "double auction" format—which he adopted precisely because it seemed most likely to confirm the predictions of neoclassical theory—and it seems to remain valid in other market experiments as well.

Today, as during most of American history, the rules of commerce are made by moneyed elites. And, as of old, the rules they make mostly favor themselves. Corporate profits have seldom been higher. Shareholders and CEOs have never been better compensated. The wealthy have seldom been wealthier, or paid fewer taxes. On the other hand, through no obvious fault of their own, people who have worked all their lives for pensions, health care, and Social Security are in the process of losing all three. They haven't shared the wealth of the last quarter century, and they won't begin to until they amass countervailing power of the sort that meaningful tariff protection, licensed farm cooperatives, and licensed employment agencies could so quickly provide.

Seldom does the opportunity arise to depose a ruling elite. The last one in the United States occurred during the Great Depression, when the elites of the day—unchecked by the Coolidge and Hoover administrations—seriously overplayed their hand. This they did by inflating a Wall Street bubble, insisting that nothing was fundamentally wrong with the struggling U.S. economy, advocating unlimited patience, repeating ad nauseam that "prosperity is just around the corner," and otherwise insulting the dignity and intelligence of the so-called "man in the street." Enraged by the proliferation of breadlines and shantytowns, the public demanded prompt remedial action. With communism, socialism, and anarchy steadily gaining adherents, New Dealers were able to push through a worried Congress the banking reforms of 1933, workers' compensation, unemployment insurance, the Wagner Act (giving workers the right to form unions), federal jobs programs, the Social Security system, and more.

In time, the ruling elites will again overplay their hand. The mounting federal deficits, massive trade imbalances, global warming, environmental degradation, impending energy shortage, loss of arable land, and uncontrolled population growth are all unsustainable. Sooner or later, a confluence of unsustainable trends will precipitate a crisis the ruling elite can't spin its way out of. Public outrage will again place them on the defensive—where they haven't been since the election of Ronald Reagan—and mandate a new wave of populist reform.

The New Deal reforms, like the populist reforms of the early twentieth century, were cobbled together with little rhyme or reason in response to suddenly pressing needs. President-elect Roosevelt had no clue, in November 1932, how his administration would go about trying to save American capitalism. His campaign had stressed budget balancing, which—heaven be praised—he was already starting to think better of. Orthodox economic thought endorsed the inaction of the Hoover administration, while the leading heterodox schools of the day—mainly Marxism, socialism, and American institutionalism—offered little if any practical guidance. The publication of Keynes's *General Theory of Employment, Interest, and Money* was still four years in the future. Of necessity, New Deal reform of the U.S. economy was going to be an ad hoc undertaking.

The coming wave of populist reform has a chance to be more coherent and lasting. There is still time to assemble, from parts already in existence, a comprehensive blueprint for a postindustrial economy that does not abandon the working poor to the fate of the plow horse. Protective tariffs, licensed farm cooperatives, and licensed employment agencies—or their equivalent in countervailing power—could contribute much toward that end. So, too, of course, could tax reform, education reform, budget reform, immigration reform, transportation reform, environmental reform, and health-care reform, to name but a few.

Unfortunately, any blueprint assembled in Washington today—by either party—would be predicated on conventional economic wisdom. As a result, it would be doomed to repeat the mistakes of the recent past. Tariff protection, licensed farm cooperatives, licensed employment agencies, and most if not all other sources of countervailing power are incompatible with conventional economic wisdom. There is no room for them in the perfectly competitive model, or elsewhere in the canons of economic orthodoxy. Their justification, if any, must come from the emerging science of competition, which alone can validate the highly unorthodox principle of economics that declares oligopoly and monopoly prices to be all but identical. Once validated, that principle alone will justify confidence that the reforms proposed above—or their equivalent in countervailing power—could pull the entire global economy back from the abyss toward which U.S. policy is presently propelling it.

Epilogue

To the ordinary citizen, science is something done by people in white lab coats, in remote locations, at the behest of the military-industrial complex. Graybeards remember when it won World War II, cured polio, and enabled a few possessors of "the right stuff" to stroll briefly on the moon. Their grandchildren know that science is taught in school, involves math, and has produced a pill for erectile dysfunction. They hope it will one day cure cancer, halt global warming, and increase the number of songs they can download onto their iPods. But few can explain the difference between real science and creation science, Scientology, or what the second Bush administration is pleased to call "sound science." Fewer still can distinguish between medical science and alternative medicine, or explain why a used-car dealer from Mississippi is likely to be mistaken when claiming to have discovered an unlimited supply of low-cost, environmentally friendly energy.

Seldom has anyone explained what science is—and is not—as simply and well as Richard Feynman in his 1974 commencement address to the students at Caltech.[1] Science, he said on that occasion, is nothing more than a method developed over the years for separating ideas that work from ideas that don't. Anyone who observes the same natural phenomena day after day, such as the ebb and flow of the tides or the barking of dogs in a village street, will begin to develop ideas about them. Try it and see. There's nothing scientific about having ideas. Everyone does that. Science, said Feynman, begins when somebody figures out a way to test an idea to see if it works or not.

The idea that the tides recede at noon and midnight is easily tested. It might work approximately, once or twice a year on a particular stretch of coastline, but usually it won't. A better idea is needed. The notion that dogs bark only when approached by strange men is equally easy to test. It, too, fails, because dogs often bark at stray cats and squirrels. And so on. It takes a special kind of imagination to think up ideas (hypotheses) that can survive even the simplest testing.

Scientific imagination, as Feynman elsewhere noted, is confined to a straitjacket.[2] Scientists cannot allow themselves to "seriously imagine things which are obviously in contradiction to the known laws of nature." No geologist, for instance, can allow himself to imagine a chemical change to take place—under any combination of subterranean temperatures and pressures—in ways that contradict the known laws of chemistry. Nor can an astrophysicist allow herself to imagine a nuclear reaction to take place in another galaxy that contradicts the known laws of atomic fission and fusion, as established here on earth in accelerator-collider experiments. And so on. It was Feynman's considered opinion that "the problem of creating something [an idea] which is new, but which is consistent with everything which has been seen before, is one of extreme difficulty."

Some sciences are as old as botany and astronomy, others as new as chaos and complexity. Some seem unusually pure or applied, others particularly theoretical or experimental in nature. Many observers distinguish between "hard" and "soft" sciences, while most deem the "physical" sciences more reliable than the "biological" and "social" ones. And so on. Dozens of ways have been found for comparing one science with another. Physicist Murray Gell-Mann, in *The Quark and the Jaguar*, follows the lead of nineteenth-century French philosopher Auguste Comte in ranking the sciences from the most to the least fundamental.[3]

Physics is more fundamental than chemistry, because its laws include those of quantum mechanics, which govern the reactions in which various elements combine to form compounds like water (H_2O), table salt (NaCl), and methane gas (CH_4). Chemistry is more fundamental than biology, because plants and animals are composed of atoms and molecules, which are subject to the laws of chemistry. Biology is more fundamental than psychology, or any of the other social sciences.

Gell-Mann considers elementary particle physics and cosmology to be the most fundamental sciences, and those devoted to the study of (complex) living things to be among the least. That doesn't mean one is more important than the other. It merely means that those who study complex living things must respect the laws of the more fundamental sciences when formulating their own. In contrast, those who study cosmology and particle physics are free—by the very nature of what they do—to disregard the possibility that some biochemist in the lab, or primatologist in the field, might observe something that contradicts an established law of physics.

After expressing doubt that mathematics is really a science at all—on the ground that the numbers and operations studied do not exist anywhere in nature—Gell-Mann goes on to say that math is still more fundamental than anything else, because mathematical models can be constructed of anything nature has seen fit to assemble, along with much that it never has and probably never will. The question to be considered here, of course, concerns the emerging science of competition. Where does it rank in Gell-Mann's hierarchy? Is it a fundamental science like astronomy, or a derivative one like geology? Is it still too soon to tell?

On the whole, the new science seems to be rather fundamental, since any tree game—or close relative thereof—can be played by inanimate computers, with little regard for the laws of any other science. Tournaments can be held to determine which strategies defeat which others, and who may claim to be (or to own) the current world champion player of any particular game.

The precise relationship between physics and the new science must remain obscure until it can be decided whether or not information—the lifeblood of strategic competition—is a physical quantity. Far-fetched though it may seem, the possibility that bits of information can be condensed into energy, and thence into matter, seems to be gaining adherents among those best qualified to comment on such notions.[4] Whatever the result may be—and debate on the subject has barely begun—it seems crystal clear that the science of competition is more fundamental than any of the social sciences, including the one the Bank of Sweden is pleased to describe as "economic science."

Economists have always bemoaned, with some justification, the fact that the experimental method is unavailable to them. After all, one cannot reenact the 1930s to see if the U.S. economy would have responded better to another four to eight years of Hooverian laissez-faire than to New Deal interference with the market mechanism. But the experimental method is obviously not unavailable to the emerging science of competition. Every contest between a man and a computer, or between two computers, constitutes a new and potentially revealing experiment. It is a pity perhaps that so many such experiments go unrecorded.

It is not too late to begin organizing tournaments—along the lines of the Axelrod tournaments—to identify effective ways of playing market games of the sort considered in Chapter 16. Although it may never be possible to identify unbeatable strategies for playing such games, it is surely possible to separate the ones that prosper in a given tournament from those that falter. And because each new generation of strategies will be obliged to defeat the best of the old, the winners of each annual tournament will be the best of all time at the game in which they excel. One may safely predict that from the very first, the winning strategies will handily defeat anything to be found in the orthodox economics literature.

Such a prediction can safely be made because orthodox strategies invariably require superhuman predictive powers—powers that would have permitted anyone possessing them to romp undefeated through both Axelrod tournaments. Until someone demonstrates an ability to incorporate such powers into a computer program, it must be assumed that no such powers exist.

Alone among the sciences, economics has adopted a unique (and uniquely unsound) modus operandi. In its twentieth-century approaches to auction bidding and financial markets, as in its nineteenth-century approach to markets in general, it began by bestowing the name "efficiency" on the select market outcomes, and then composed narratives explaining why, in each case, free-market competition would foreclose all other outcomes.

Twentieth-century scholarship has reduced those narratives to mathematical theories, complete in every detail, in which every conclusion is deduced by impeccable logic from purportedly self-evident axioms. Led by Alfred Marshall, the Cambridge school of economic thought accomplished the feat for the so-called neoclassical model of commercial interaction, leaving the task essentially finished by the time of Marshall's death in 1924.[5] Then, beginning soon after World War II, their successors more or less duplicated the feat for auctions and finance. It was an impressive performance. The ancient Greeks—inventors of the axiomatic method—would surely have approved.

Recall, however, that the courts have ruled creationism not to be a science on the ground that its conclusions are predetermined. It involves no search for the truth, because the truth is already known. Whatever search there is seeks only corroborative evidence, wherewith to persuade the uninitiated. The same criticism applies with equal force to orthodox economic theory. It is no more a science than creation science, because it, too, has already identified the truth. Small wonder that economists so often stand accused of self-deception.

Feynman devoted a substantial portion of his 1974 commencement address to the subject of scientific integrity. Scientists, he said, have a responsibility to other scientists—and perhaps to the public as well—not to fool themselves. "After you've not fooled yourself," he assured his listeners, "it's easy not to fool other scientists." But not fooling yourself is far from easy because, liking your own ideas, "you are the easiest one to fool." Scientists have been learning for generations—indeed are still learning—ways of avoiding self-deception. One such way, he hastened to add, is to divulge every reason you can think of why your conclusions are only tentative and may yet be proven wrong.

The history of science is replete with instances of self-deception. The cold-fusion fiasco of the late 1980s was the most recent to achieve notoriety, and the canals on Mars surely the most famous. But there have been plenty of other occasions on which reputable scientists observed, measured, and wrote peer-reviewed articles about phenomena that were later proven not to exist. Physicist Robert Park has written an entire book on erroneous or "voodoo" science, of which he recognizes four separate and distinct categories.[6] Among them is pathological science, first described by Nobel Prize–winning chemist Irving Langmuir during the 1930s.

The perpetrators of pathological science are guilty not of fraud but of self-deception. Enamored of their own ideas, and fully expecting their experiments to confirm their theories, they find confirmation where none exists and—entirely too often—rush into print with results that are easily disproved. Such behavior is irresponsible, because it creates unnecessary work for others. Yet those who engage in it are seldom accused of dishonesty. Lust for fame and fortune perhaps, along with honest beguilement with intriguing ideas, but seldom dishonesty. Few ideas have proven more beguiling, historically, than those associated with orthodox, neoclassical, laissez-faire economics.

There is a scene in Bertolt Brecht's play *Galileo* in which the master and his assistants are preparing to test the Copernican notion that the earth revolves about the sun.[7] Galileo explains to the others that as a matter of discipline, their purpose must be to prove the earth stationary. Only if the ascertainable facts render that position untenable may they allow themselves to find in Copernicus's favor. In fact, says Galileo, "if we find anything which would suit us, that thing will we eye with particular distrust." For only by such discipline can scientists discharge their obligation *as scientists* to resist fooling themselves. Seldom if ever has economic science acknowledged any such responsibility. Rarely do its practitioners "eye with particular distrust" anything they are disposed to believe.

On the contrary, it would be more nearly correct to say that no proposition has been too preposterous to serve as an axiom of economic theory if it led to the desired conclusions, and no contrary fact has been too obvious or too firmly established to ignore if it seemed to discredit such conclusions. The assumption that transportation costs are nil—without which Ricardo's principle of comparative advantage becomes invalid—is but one example of an obvious falsehood that serves as an axiom (albeit an unacknowledged one) of mainstream economic theory. The assumption that ordinary economic agents possess what Frank Knight described as "perfect foresight" and "foreknowledge free from uncertainty" is another such. The fact that the accounting conventions employed in estimating Harberger triangles misrepresent "profits reinvested" in a business as "costs of doing business" is habitually ignored by those who claim to have verified the allocative efficiency of free-market economies. And so on.

Like moviegoers, economists have opted for the story with the happiest ending—the one in which near-perfect competition begets near-perfect allocative efficiency in this most nearly perfect of all possible worlds. Their efforts to uncover reasons for doubting that story have been desultory in the extreme. Perhaps the emergence of a genuine science of competition, directly contradicting core teachings of orthodox economic theory, will stimulate more realistic expectations of free-market competition. Although such expectations are by no means the only benefit that further progress in the emerging science of competition seems likely to bring, it would be hard to imagine one of which the twenty-first century is likely to have greater need.

Notes

Preface

1. Harold W. Kuhn, "Extensive Games," *Proceedings of the National Academy of Sciences,* vol. 36 (1950), pp. 570–76. See also H. W. Kuhn, "Extensive Games and the Problem of Information," in *Contributions to the Theory of Games II* (Princeton, NJ: Princeton University Press, 1953), pp. 193–216.
2. In a letter to J. E. Switzer, reprinted in Derek J. de Solla Price, *Science Since Babylon* (New Haven, CT: Yale University Press, 1962).
3. This inability is extensively documented in Robert Park's splendid book *Voodoo Science: The Road from Foolishness to Fraud* (Oxford, UK: Oxford University Press, 2000) and—to a lesser extent—in Chris Mooney's *The Republican War on Science* (New York: Basic Books, 2005).

Chapter 1: Man Versus Machine

1. Brief descriptions of these little-known games, along with assessments of ongoing efforts to solve them, can be found in the final chapter of *Chips Challenging Champions: Games, Computers and Artificial Intelligence*, eds. Jonathan Schaeffer and Jaap van den Herick (Amsterdam: Elsevier Science B. V., 2002), pp. 321–55.
2. Feng-hsiung Hsu, *Behind Deep Blue* (Princeton, NJ: Princeton University Press, 2002); and Jonathan Schaeffer, *One Jump Ahead: Challenging Human Supremacy in Checkers* (Berlin: Springer, 1997).
3. Arthur Samuel, "Programming Computers to Play Games," in *Advances in Computers*, ed. F. Alt (New York: Academic Press, 1960), vol. 1, pp. 165–92.

4. Edward O. Thorp, *Beat the Dealer: A Winning Strategy for the Game of Twenty-one* (New York: Blaisdell, 1962).
5. The terms "strategy" and "game plan" are essentially synonymous, the main difference being that a concise definition will eventually be given for strategies, while game plans will remain undefined.
6. See William Poundstone, *Fortune's Formula: The Untold Story of the Scientific Betting System That Beat the Casinos and Wall Street* (New York: Hill and Wang, 2005).
7. Gerald Tesauro, "Programming Backgammon Using Self-Teaching Neural Nets," in *Chips Challenging Champions: Games, Computers and Artificial Intelligence*, pp. 223–41.
8. Arthur Samuel, "Some Studies in Machine Learning Using the Game of Checkers," *IBM Journal of Research and Development* 3 (1959), pp. 210–29.
9. Steven Quartz and Terrence J. Sejnowski, *Liars, Lovers, and Heroes: What the New Brain Science Reveals About How We Become Who We Are* (New York: William Morrow, 2002), chap. 5.
10. The standard reference on differential games continues to be Rufus Isaacs, *Differential Games* (New York: Wiley, 1965). For an introduction to combinatorial games, consult Elwyn R. Berlekamp, John H. Conway, and Richard K. Guy, *Winning Ways for Your Mathematical Plays* (London: Academic Press, 1982), vols. 1 and 2. Also Richard J. Nowakowski, ed., *Games of No Chance*, MSRI Publications, vol. 29 (New York: Cambridge University Press, 1996).
11. As used among programming professionals, the term "hacker" is not derogatory. On the contrary, it is a term of approbation, applied to colleagues who do what they do as much for satisfaction as for money. Hackers describe the pranksters who break into university computer systems to change everyone's grades, and the like, as "crackers." See Pekka Himanen, *The Hacker Ethic and the Spirit of the Information Age* (New York: Random House, 2001).

Chapter 2: The Art and Science of Competition

1. Jane Goodall, *The Chimpanzees of Gombe* (Cambridge, MA: Belknap Press, 1986), pp. 530–34.
2. See en.wikipedia.org/Neanderthal (accessed April 13, 2007).
3. Sun Tzu, *The Art of War* (New York: Writers' Club Press, 2002), p. 53.
4. Ibid., p. 16.
5. Carl von Clausewitz, *On War* (Princeton, NJ: Princeton University Press, 1976), p. 193.
6. A sizeable collection of Clausewitz's better-known quotes, including this one, will be found at www.clausewitz.com/CWZHOME/CWZbiz.htm.
7. Al Ries and Jack Trout, *Marketing Warfare* (New York: McGraw-Hill, 1986).
8. W. W. Rouse Ball, *Mathematical Recreations and Essays* (London: Macmillan, 1939), p. 104.

9. Charles L. Bouton, "Nim, a Game with a Complete Mathematical Theory," *Annals of Mathematics*, 2nd ser., 3 (1902), pp. 35–39.
10. Godfrey H. Hardy and E. M. Wright, *An Introduction to the Theory of Numbers* (Oxford, UK: Oxford University Press, 1938), pp. 117–19.
11. Meaning that, after each move by White save the first, the two most recently placed nickels lie on a common diameter of the circular tabletop, at equal distances from the center.
12. Kenneth S. Howard, *Classic Chess Problems by Pioneer Composers* (New York: Dover, 1970), p. 53.
13. Godfrey H. Hardy, *A Mathematician's Apology* (New York: Cambridge University Press, 1992), p. 108.
14. John von Neumann and Oskar Morgenstern, *Theory of Games and Economic Behavior* (Princeton, NJ: Princeton University Press, 1944).
15. Meaning relatively, if not absolutely, "drag-free."

Chapter 3: Tree Games and Backward Induction

1. John von Neumann, "Zur Theorie der Gesselshaftsspeile," *Math. Annalen* 100 (1928), pp. 295–320.
2. The complete tree for tic-tac-toe contains $9! = 362,880 = 9 \times 8 \times 7 \times 6 \times 5 \times 4 \times 3 \times 2 \times 1$ paths leading to an equal number of leaves, because the first mark may be placed in any of nine cells, the second in any of the remaining eight cells, and so on. The reduced tree reflects the fact that the first mark must occupy one of only three distinct positions (a side, corner, or central cell)—meaning that the root node *O* has only three downstream neighbors—the second one of at most five, and so on.
3. He called them "games in extensive form."
4. John McDonald, *The Game of Business* (Garden City, NY: Doubleday, 1975).
5. Alfred P. Sloan, *My Years with General Motors* (New York: Doubleday, 1964).
6. Hsu, *Behind Deep Blue*, p. 225.
7. See William Poundstone, *Prisoner's Dilemma* (New York: Doubleday, 1992), chap. 3.

Chapter 4: Models and Paradigms

1. It is often said that the best econometric models have successfully predicted eleven of the last seven recessions.
2. The 1961 edition of *Webster's New World Dictionary* defines a truism to be "a statement the truth of which is obvious and well known," and suggests as synonyms "platitude" and "commonplace."
3. Hardy, *A Mathematician's Apology*, p. 107.

4. Many if not most useful models consist of equations. But numbers and sets of numbers are the building blocks from which equations are constructed.
5. *Econometrica* 11, no. 1 (April 1943), pp. 1–12.
6. George Stanic and Jeremy Kilpatrick, eds., *A History of School Mathematics* (Reston, VA: National Council of Teachers of Mathematics, 2003), vol. 1, pp. 647–71.
7. Dorothy M. Livingston, *The Master of Light: A Biography of A. A. Michelson* (Chicago: University of Chicago Press; reprint edition, 1979), chap. 5.
8. An eminently readable account of these "classical" non-Euclidean geometries will be found in David Hilbert and S. Cohn-Vossen, *Geometry and the Imagination* (New York: Chelsea, 1952).
9. Lobachevsky's measurements are cited in N. V. Efimov, *Higher Geometry* (Moscow: Mir Publishers, 1980), p. 35.
10. David Hilbert, *Foundations of Geometry* (Chicago: Open Court Press, 1971).
11. An excellent account of ancient Greek science will be found in the opening chapters of Isaac Asimov's *New Guide to Science* (New York: Basic Books, 1984).
12. Stephen P. Ellner and John Guckenheimer, *Dynamic Models in Biology* (Princeton, NJ: Princeton University Press, 2006), p. 81.
13. By this it is meant that the entire curve lies above the given horizontal, but that it intersects any slightly higher horizontal at exactly two points.
14. For those who may have forgotten, the equation of a Gaussian bell curve is $y = \exp[-\frac{1}{2}((x - \mu)/\sigma)^2]$. The parameters α and σ may be determined one from the other by exploiting the fact that $\alpha\sigma = 1/\sqrt{2\pi} \approx 0.3989\ldots$
15. Two kinds of infinitude arise in applied mathematics. Although there are infinitely many proper fractions—numbers like ⅔, ⅘, and ⅞, in which the number on the bottom (the denominator) exceeds the one on the top (the numerator)—the proper fractions represent a tiny knowable elite within the population of numbers that lie between 0 and 1.

Chapter 5: Two-Sided Competition

1. This is obvious if the two lie in a common row and/or column. If they lie in different rows *and* columns, let x be the number at the intersection of the row containing the circled number and the column containing the one enclosed in a square. The circled number cannot then exceed x, which cannot exceed the one enclosed in a square.
2. Ted Williams, with David Pietrusza, *My Life in Pictures* (Kingston, NY: Total Sports Publishing, 2001), p. 149.
3. Melvin Dresher, *Games of Strategy* (Englewood Cliffs, NJ: Prentice-Hall, 1961), chap. 9.
4. Ibid., chap. 10.
5. Ernest Cockayne, "Plane Pursuit with Curvature Constraints," *SIAM Journal of Applied Math* 15 (1967), pp. 1511–16.

6. Rufus Isaacs, *Differential Games* (New York: John Wiley, 1965).
7. William Poundstone, *Prisioner's Dilemma* (New York: Doubleday, 1992), p. 103.
8. Ibid., p. 173.

Chapter 6: Many-Sided Competition

1. Garrett Hardin, "The Tragedy of the Commons," *Science* 162 (1968), pp. 1243–48.
2. Philip Straffin, *Game Theory and Strategy* (Washington, DC: The Mathematical Association of America, 1993), p. 146.
3. Steven Brams and Philip Straffin, "Prisoners' Dilemma and Professional Sports Drafts," *The American Mathematical Monthly* 86 (1979), pp. 80–88.
4. Adam Smith, *An Inquiry into the Nature and Causes of the Wealth of Nations* (Buffalo, NY: Prometheus Books, 1991), p. 349, italics added.
5. Ibid.
6. Steven Brams and Peter Fishburn, *Approval Voting* (Boston: Birkhauser, 1982).
7. Donald G. Saari, *Chaotic Elections!* (Providence, RI: American Mathematical Society, 2001), p. 17.
8. Ibid., p. 20.
9. Ibid., p. 27.
10. Ibid., p. 26
11. John von Neumann and Oskar Morgenstern, *Theory of Games and Economic Behavior* (Princeton, NJ: Princeton University Press, 1944), chap. 6.
12. Ibid., chap. 10.
13. Lloyd Shapley and Martin Shubik, "A Method for Evaluating the Distribution of Power in a Committee System," *American Political Science Review* 48 (1954), pp. 787–92.
14. Sylvia Nasar, *A Beautiful Mind* (New York: Simon and Schuster, 1998), p. 364.
15. After Vilfredo Pareto (1848–1923), an Italian industrialist and social scientist.
16. Drew Fudenberg and David K. Levine, *The Theory of Learning in Games* (Cambridge, MA: MIT Press, 1998).
17. David H. Stern, "Some Notes on Oligopoly Theory and Experiments," in *Essays in Mathematical Economics in Honor of Oskar Morgenstern*, ed. Martin Shubik (Princeton, NJ: Princeton University Press, 1967).
18. James Case, *Economics and the Competitive Process* (New York: New York University Press, 1979), chaps. 4–6.
19. U.S. Department of Transportation, *Theory of Operations, National Intelligent Transportation Systems Architecture* (Washington, DC, 1998).
20. *New York Times*, April 22, 2005, opinion.

Chapter 7: Competition in the Wild

1. Konrad Lorenz, *On Aggression* (New York: Harcourt, Brace & World, 1966).
2. Shirley C. Strum, *Almost Human: A Journey into the World of Baboons* (New York: W. W. Norton, 1987).
3. Goodall, *The Chimpanzees of Gombe*, p. 274.
4. John Maynard Smith and George R. Price, "The Logic of Animal Conflict," *Nature* 246 (1973), pp. 15–18.
5. John Maynard Smith, *Evolution and the Theory of Games* (Cambridge, UK: Cambridge University Press, 1982).
6. Konrad Lorenz, *King Solomon's Ring* (New York: HarperCollins, 1982).
7. Dian Fossey, *Gorillas in the Mist* (Boston: Houghton Mifflin, 1983), p. 69.
8. Joan Roughgarden, *Theory of Population Genetics and Evolutionary Ecology* (New York: Macmillan, 1979).
9. Karl Sigmund, *Games of Life* (New York: Penguin Books, 1995), pp. 45–46.
10. Adrian Desmond and James Moore, *Darwin: The Life of a Tormented Evolutionist* (New York: Warner Books, 1991), chap. 30.

Chapter 8: Auctions

1. Kenneth S. Deffeyes, *Beyond Oil* (New York: Hill and Wang, 2005).
2. See www.gomr.mms.gov/homepg/regulate/environ/history_louisiana.html.
3. Lawrence Friedman, "A Competitive Bidding Strategy," *Operations Research* 4 (Feb. 1956), pp. 105–106.
4. E. C. Capen, R. V. Clapp, and W. M. Campbell, "Competitive Bidding in High Risk Situations," *Journal of Petroleum Technology* (June 1971).
5. E. C. Capen and R. V. Clapp, "Conflicting Bidding Models in a High Stakes Game," paper presented to ORSA national meeting, Las Vegas, Nov. 16–19, 1975.
6. Michael H. Rothkopf, "A Model of Rational Competitive Bidding," *Management Science* 15 (1969), pp. 362–73. See also Robert Wilson, "Competitive Bidding with Asymmetric Information," *Management Science* 13 (1967): 816–20. Both Rothkopf and Wilson wrote technical reports on the subject some years before the foregoing papers appeared in the open literature.
7. There is no end to the variety of ways mathematicians have devised, over the centuries, for computing the mean or average of a collection of numbers. There are arithmetic means, geometric means, logarithmic means, harmonic means, root-mean-squares, and more. Each may be weighted or unweighted, trimmed or untrimmed, and combined with others to obtain yet other means. The mean of several means is again a mean. In the language of mathematics, the familiar average is the unweighted, untrimmed arithmetic mean.
8. Shmuel Oren and A. C. Williams, "Optimal Bidding in Sequential Auctions," *Operations Research* 23 (1975).

9. *Encyclopaedia Britannica* (1991), p. 395.
10. On auctions for book publication rights, see John P. Dessauer, *Book Publishing* (New York: Bowker, 1981); on baseball's free-agency market, see James Cassing and Richard W. Douglas, "Implications of the Auction Mechanism in Baseball's Free Agent Draft," *Southern Economic Journal* 47 (1980), pp. 110–21, and Barry Blecherman and Colin Cammerer, "Is There a Winner's Curse in the Market for Baseball Players?" mimeograph, Brooklyn Polytechnic Institute, Brooklyn, NY; on corporate takeovers, see Richard Roll, "The Hubris Hypothesis in Corporate Takeovers," *Journal of Business* 59 (1986), pp. 197–216; and on real-estate auctions, see Orley C. Ashenfelter and David Genesove, "Testing for Price Anomalies in Real Estate Auctions," *American Economic Review: Papers and Proceedings* 82 (May 1992), pp. 501–505.
11. Max H. Bazerman and William F. Samuelson, "I Won the Auction but Don't Want the Prize," *Journal of Conflict Resolution* 27 (1983), pp. 618–34.
12. John H. Kagel and Dan Levin, *Common Value Auctions and the Winner's Curse* (Princeton, NJ: Princeton University Press, 2002), chap. 2.
13. Ibid., chap. 6.
14. This rule change was used in the New Zealand radio spectrum auctions of 1990, for instance.
15. Paul Milgrom, *Putting Auction Theory to Work* (Cambridge, UK: Cambridge University Press, 2004), p. 30.
16. Alvin E. Roth, "New Physicians: A Natural Experiment in Market Organization," *Science* 250 (1990), pp. 1524–28; and "A Natural Experiment in the Organization of Entry Level Labor Markets: Regional Markets for New Physicians and Surgeons in the UK," *American Economic Review* 81 (1991), pp. 415–40.
17. Excerpt from the vice president's speech at the beginning of FCC auction 4, August 1969.
18. Paul Klemperer, *Auctions: Theory and Practice* (Princeton, NJ: Princeton University Press, 2004), p. 166.

Chapter 9: Competition in Financial Markets

1. Peter L. Bernstein, *Capital Ideas: The Improbable Origins of Modern Wall Street* (New York: Free Press, 1992).
2. As quoted in ibid., p. 161.
3. Benjamin Graham and David Dodd, *Security Analysis: Principles and Techniques* (New York: McGraw-Hill, various editions).
4. John Eatwell, Murray Milgate, and Peter Newman, eds. *The New Palgrave: A Dictionary of Economics* (London: Macmillan, 1987).
5. Benoit Mandelbrot and Richard L. Hudson, *The (Mis)behavior of Markets: A Fractal View of Risk, Ruin, and Reward* (New York: Basic Books, 2004).

6. See Bernstein, *Capital Ideas*, p. 225.
7. Fischer Black, "On Robert C. Merton," *MIT Sloan Management Review* (1988), p. 28.
8. "On Market Timing and Investment Performance," *Journal of Business* 54, no. 3 (1981), p. 378.
9. Robert C. Merton, private communication.
10. Burton Malkiel, *A Random Walk Down Wall Street* (New York: Norton, various editions).
11. Gunduz Caginalp, David Porter, and Vernon Smith, "Financial Bubbles: Excess Cash, Momentum, and Incomplete Information," *Journal of the Psychology of Financial Markets* 2 (2001), pp. 80–99.
12. Robert J. Shiller, *Irrational Exuberance* (Princeton, NJ: Princeton University Press, 2000).
13. John L. Kelly, "A New Interpretation of Information Rate," *Bell System Technical Journal* 91 (1956), pp. 917–26.
14. William Poundstone, *Fortune's Formula: The Untold Story of the Scientific Betting System That Beat the Casinos and Wall Street* (New York: Hill and Wang, 2005).
15. Edward O. Thorp, *Beat the Dealer: A Winning Strategy for the Game of Twenty-one*, rev. ed. (New York: Random House, 1966).
16. Paul A. Samuelson, "Lifetime Portfolio Selection by Dynamic Stochastic Programming," *Review of Economics and Statistics*, Aug. 1969, pp. 239–46.
17. Ibid., "The Fallacy of Maximizing the Geometric Mean in Long Sequences of Investing or Gambling," *Proceedings of the National Academy of Sciences* 68: pp. 2493–96.
18. William Poundstone, *Fortune's Formula*, op. cit. p. 329.
19. Ibid., p. 327.
20. Ibid., p. 329.

Chapter 10: Orthodox Economic Thought

1. *The Penguin Dictionary of Economics*, 4th ed. (London: Penguin Books, 1987).
2. Simon Kuznets, *National Income and Its Composition, 1919–1938* (New York: National Bureau of Economic Research, 1941).
3. Simon Kuznets, *National Product Since 1869* (New York: National Bureau of Economic Research, 1946).
4. William Stanley Jevons, *The Theory of Political Economy* (London: Macmillan, 1871), p. 22.
5. *Economist*, July 15, 2006, pp. 67–69.
6. John R. MacArthur, *The Selling of Free Trade* (New York: Hill and Wang, 2000), p. 284.
7. To be found at www.economy.com.

8. It is widely believed in the macroeconomic community that low levels of unemployment tend to accelerate the rate of inflation, while high levels tend to decelerate it. So there should be a "happy medium" at which inflation proceeds at a constant rate. This non-accelerating inflation rate of unemployment is frequently and inaccurately described as the "natural rate of unemployment" by careless economists and unscrupulous advocates of particular policies.
9. *Economist*, July 15, 2006, p. 67.
10. Ibid., p. 69.
11. Some will object that scientific issues are not, or ought not to be, emotional issues. This is of course nonsense—scientists become as emotionally attached to their pet theories as barroom drunks. Being aware of the danger does not always protect them from falling prey to it.
12. In the building trades, the vertical planks in a staircase are called risers, while the horizontal ones are called treads.

Chapter 11: Economic Competition

1. Italics added. See Paul A. Samuelson and William Nordhaus, *Economics*, 16th ed. (Boston: Irwin/McGraw-Hill, 1998), p. 29.
2. Alan S. Blinder, Elie R. D. Canetti, David E. Lebow, and Jeremy B. Rudd, *Asking About Prices: A New Approach to Understanding Price Stickiness* (New York: Russell Sage Foundation, 1998).
3. There is a long-standing tradition in economics that questionnaire data are unreliable. Only the prominence of the lead investigator—Alan Blinder holds a named chair in economics at Princeton and is a former vice chairman of the Board of Governors of the Federal Reserve and a former member of the president's Council of Economic Advisers—got the present study into print.
4. James Crotty, "Why There Is Chronic Excess Capacity—the Market Failures Issue," *Challenge*, Nov.–Dec. 2002, pp. 21–44.
5. After Vilfredo Pareto (1848–1923), Walras's successor at Lausanne.
6. As quoted by Richard Parker, in his biography of John Kenneth Galbraith (New York: Farrar, Straus and Giroux, 2005), p. 197.
7. Arrow shared the 1972 prize with John R. Hicks; Debreu received his in 1983.
8. Mark W. Hendrickson, ed., *The Morality of Capitalism*, 2nd rev. ed. (Irvington-on-Hudson, NY: Foundation for Economic Education, 1996).
9. Ibid., italics added.
10. Laurent Maudit, "Economics Students Denounce the Lack of Pluralism in the Teaching Offered," *Le Monde*, June 21, 2000.
11. The adjective "flagrant," according to *Webster's New Collegiate Dictionary* (1961), applies to that which "can neither escape notice nor be condoned."
12. Alan Greenspan, "The Assault on Integrity," *The Objectivist Newsletter*, vol. 2, no. 3, 1963.

13. Ken Auletta, *World War 3.0: Microsoft and Its Enemies* (New York: Random House, 2001), p. 297.
14. "Why Does Cereal Cost So Much?" *Consumer Reports*, Nov. 1992, pp. 689–92.
15. Paul Ormerod, *The Death of Economics* (New York: St. Martin's Press, 1994), p. 77.

Chapter 12: Evidence Pro and Con

1. James T. Hong and Charles R. Plott, "Rate Filing Policies for Inland Water Transportation," *Bell Journal of Economics* 13, no. 1 (1982), pp. 1–19.
2. As quoted by John Eatwell, Murray Milgate, and Peter Newman, eds., *The Invisible Hand* (New York: W. W. Norton, 1989), p. 102.
3. It should be pointed out in this connection that—whether by accident or by design—Chamberlin chose supply and demand curves (depicted in figure 12.1) which correspond to a market in which collusion is unrewarding, and that most of the ones employed by Smith do likewise. What might they have discovered had they chosen differently?
4. Arnold C. Harberger, "Monopoly and Resource Allocation," *American Economic Review* 44 (1954), pp. 77–87.
5. William G. Shepherd, *The Economics of Industrial Organization*, 2nd ed. (Englewood Cliffs, NJ: Prentice-Hall, 1985).
6. Keith Cowling and Dennis C. Mueller, "The Social Costs of Monopoly Power," *Economic Journal* 88 (Aug. 1978), pp. 727–48.
7. Victor J. and Carol Horton Tremblay, *The U.S. Brewing Industry: Data and Economic Analysis* (Cambridge, MA: MIT Press, 2005), p. 162.
8. Vernon L. Smith, "Experimental Economics: Induced Value Theory," *American Economic Review: Papers and Proceedings* 66 (1976), pp. 274–79.
9. Hong and Plott, "Rate Filing Policies."
10. Dennis Mueller, 1998, private communication.
11. See en.wikipedia.org/wiki/Capacity_utilization (accessed April 13, 2007).
12. Central Intelligence Agency, *The World Factbook* (Washington, DC, 2002), p. 80.
13. As quoted by Robert L. Heilbroner in *The Worldly Philosophers*, op cit., p. 213.

Chapter 13: Free Trade

1. Paul A. Samuelson, *Economics: An Introductory Analysis*, 4th ed. (New York: McGraw-Hill, 1958), p. 672.
2. Stanislaw Ulam, *Adventures of a Mathematician* (Berkeley: University of California Press, 1991), p. 86.
3. Todd G. Buchholz, *New Ideas from Dead Economists* (New York: Penguin, 1999), p. 71.

4. Milton Friedman and Rose Friedman, *Free to Choose* (New York: Harcourt Brace Jovanovich, 1980), p. 39.
5. Italics added.
6. James Buchanan, *Essays on the Political Economy* (Honolulu: University of Hawaii Press, 1989), p. 52.
7. Paul Ormerod, *The Death of Economics* (New York: St. Martin's Press, 1994), p. 8.
8. Pew Research Center poll, Nov. 2005.
9. Peter T. Marsh, *Bargaining on Europe: Britain and the First Common Market, 1860–1892* (New Haven, CT: Yale University Press, 1999), p. 6.
10. As quoted in ibid., p. 13.
11. Ibid., p. 16.
12. William J. Gill, *Trade Wars Against America* (New York: Praeger, 1990), p. 39.
13. Ravi Batra, *The Pooring of America: Competition and the Myth of Free Trade* (New York: Collier Books, 1993), p. 38.
14. Ibid., p. 42.
15. Lester Thurow, *Head to Head: The Coming Economic Battle Among Japan, Europe, and America* (New York: William Morrow, 1992), chap. 5.
16. Paul Bairoch, *Economics and World History* (Chicago: University of Chicago Press, 1993).
17. Elhanan Helpman, *The Mystery of Economic Growth* (Cambridge, MA: Belknap Press of Harvard University Press, 2004).
18. Kevin O'Rourke, "Tariffs and Growth in the Late Nineteenth Century," *Economic Journal* 110 (2000), pp. 456–83.
19. Michael Clemens and Jeffrey G. Williamson, "Why Did the Tariff-Growth Correlation Reverse After 1950?" NBER Working Paper 9181 (2002).
20. Francisco Rodríguez and Dani Rodrik, "Trade Policy and Economic Growth: A Skeptic's Guide to the Cross-National Evidence," *NBER Macroeconomic Annual 2000* 15, pp. 261–325.
21. The remark is variously attributed to both Mark Twain and Artemus Ward.
22. Raymond W. Baker, *Capitalism's Achilles Heel: Dirty Money and How to Renew the Free-Market System* (New York: John Wiley & Sons, 2005).
23. Ibid., p. 165.
24. The new division might consist of a single employee and a single telephone.
25. P. J. O'Rourke, *Eat the Rich: A Treatise on Economics* (New York: Atlantic Monthly Press, 1998), pp. 116–19.
26. Scott Adams, *The Dilbert Future: Thriving on Stupidity in the 21st Century* (New York: HarperBusiness, 1997), p. 160.
27. Buchholz, *New Ideas from Dead Economists*, p. 75.
28. Ibid.

Chapter 14: Heterodox Economic Thought

1. Richard Brookhiser, *What Would the Founders Do?* (New York: Basic Books, 2006). On page 84, he asserts, "All the founders were familiar with Smith, but . . ."
2. James Fallows, *Looking at the Sun* (New York: Pantheon Books, 1994), p. 197.
3. William O. Henderson, *Friedrich List: Economist and Visionary* (London: Frank Cass, 1983).
4. Fallows, *Looking at the Sun*, chap. 4.
5. See Richard Parker, *John Kenneth Galbraith: His Life, His Politics, His Economics* (New York: Farrar, Straus and Giroux, 2005), p. 699, *n*18.
6. Thorstein Veblen, "Why Is Economics Not an Evolutionary Science?" *Quarterly Journal of Economics* 12, no. 4. (1898).
7. Louis Brandeis, "The Living Law," *Illinois Law Review* 10 (1916).
8. Todd G. Buchholz, *New Ideas from Dead Economists* (New York: Plume Books, 1989), p. 192.
9. Alfred Eichner, ed., *A Guide to Post Keynesian Economics* (White Plains, NY: M. E. Sharpe, 1978), p. 9.
10. Ibid., pp. 11–16.
11. Donella H. Meadows, Dennis L. Meadows, Jorgen Randers, and William W. Behrens III, *The Limits to Growth* (New York: Universe Books, 1972).
12. Donella H. Meadows, Dennis L. Meadows, and Jorgen Randers, *Beyond the Limits* (Post Mills, VT: Chelsea Green Pub. Co., 1992); Donella H. Meadows, Jorgen Randers, and Dennis L. Meadows, *Limits to Growth: The Thirty-Year Update* (White River Junction, VT: Chelsea Green Pub. Co., 2004).
13. Jared Diamond, *Collapse: How Societies Choose to Fail or Succeed* (New York: Penguin Books, 2005), p. 81.
14. Herman E. Daly, *Beyond Growth: The Economics of Sustainable Development* (Boston: Beacon Press, 1996), p. 7.
15. "Materially closed" means that, unlike energy, which is constantly entering the biosphere from the sun and being radiated away into outer space by the warmth of the earth's surface, matter is unable either to enter or to exit the system.
16. The interview is posted at www.thenation.com under the title "A Conversation with Robert Rubin," by William Greider, dated July 14, 2006. The resulting article, titled "Born-Again Rubinomics," is posted nearby.
17. Fallows, *Looking at the Sun*, p. 180.

Chapter 15: Spontaneous Cooperation

1. Robert Axelrod, *The Evolution of Cooperation* (New York: Basic Books, 1984).
2. This and the following section closely follow the account given in ibid., chap. 3.

3. Tony Ashworth, *Trench Warfare, 1914–1918: The Live and Let Live System* (New York: Holmes & Meier, 1980).
4. Steven D. Levitt and Stephen J. Dubner, *Freakonomics: A Rogue Economist Explores the Hidden Side of Everything* (New York: William Morrow, 2005), p. 135.
5. Axelrod, *Evolution of Cooperation.*
6. John H. Holland, *Hidden Order: How Adaptation Builds Complexity* (Reading, MA: Helix Books, 1995), chap. 2.
7. Robert Axelrod, "The Evolution of Strategies in the Iterated Prisoners' Dilemma," in *Genetic Algorithms and Simulated Annealing*, ed. Lawrence Davis (London: Pitman, 1987), pp. 32–43.
8. Anatol Rapoport and Albert M. Chammah, *Prisoner's Dilemma* (Ann Arbor: University of Michigan Press, 1965), p. 201.
9. Paul A. Samuelson and William Nordhaus, *Economics*, pp. 196–204.

Chapter 16: Imperfect Competition

1. For the last time ever, as fate would have it. The franchise moved to Minnesota in 1961.
2. Edward H. Chamberlin, *The Theory of Monopolistic Competition* (Cambridge, MA: Harvard University Press, 1933).
3. Paul A. Samuelson, *Economics: An Introductory Analysis*, 4th ed. (New York: McGraw-Hill, 1958), pp. 456–57.
4. William Breit and Roger L. Ransom, *Academic Scribblers*, 2nd ed. (New York: CBS College Publishers, 1982), p. 254.
5. Auletta, *World War 3.0*, p. 297.
6. Robert L. Heilbroner and Lester C. Thurow, *Economics Explained*, updated ed. (New York: Simon and Schuster, 1987), p. 181.
7. Ibid.
8. At that time, the market for kerosene (used mainly in lamps and lanterns) was still much larger than the emerging ones for gasoline and lubricants.
9. Donald Dewey, *The Antitrust Experiment in America* (New York: Columbia University Press, 1990), pp. 23 and 132.
10. Orthodox theory explains these gaps in terms of "opportunity costs," which are the premiums society pays manufacturers to remain in their present lines of business. By this argument, the extra $40 Microsoft long charged for a copy of Windows was a bribe for not turning its Redmond, Washington, campus into a brewery or big-box store.
11. The exception being the natural price leader who succeeds in driving a competitor or two out of the market.
12. Al Ries and Jack Trout, *Marketing Warfare* (New York: McGraw-Hill, 1986), p. 40.
13. *Atlantic*, May 2006, p. 53.

14. That the decline need not be rapid is witnessed by the many brands of beer that were bought up in the 1950s, 1960s, and 1970s which have never been advertised since, yet which continue to sell briskly in certain parts of the country.
15. James Case, *Economics and the Competitive Process* (New York: New York University Press, 1979).
16. The fact that WE can never profitably exceed WE*, together with the symmetry of the situation, suggests that THEY is unlikely to exceed WE* multiplied by the number of opponents.
17. As quoted by Douglas Davis and Charles Holt in *Experimental Economics* (Princeton, NJ: Princeton University Press, 1993), p. 4.
18. Cournot used as an example well water, which, then as now, brought a healthy price in much of Europe.
19. Benjamin Franklin, *Writings* (New York: Library of America, 1987).

Chapter 17: Policy Implications

1. John Kenneth Galbraith, *American Capitalism: The Concept of Countervailing Power* (New Brunswick, NJ: Transaction Publishers, 1993), pp. 42–43.
2. The facts in this and the following paragraph are as reported by Richard Parker in *John Kenneth Galbraith: His Life, His Politics, His Economics* (New York: Farrar, Straus and Giroux, 2005), pp. 235–38.
3. The capacities would have to apply to individual crops: so much corn, so much wheat, so much soy, so much milk, so many hogs, so many chickens, so many tons of beef, and so on. After a few years, an authorization exchange program could be phased in so that farmers wishing to plant more corn and less soy, or raise more chickens and fewer hogs, could trade with others having complementary wishes. Provision would also have to be made for over- and under-production due to good and bad weather. Because the United States has only a million farms left, the information-processing requirements need not exceed the capacity of a single (possibly desktop) computer.
4. Kenneth Deffeyes, *Hubbert's Peak* (Princeton, NJ: Princeton University Press, 2001), p. 4.
5. Gailbraith, *American Capitalism*, p. x.
6. *American Scientist*, vol. 81, no. 5, p. 458.
7. Some have said that such factories can never be completely unmanned, because they will always have to contain a man and at least one dog. The man will be there to feed the dogs, and the dogs will be there to make sure that no man—their keeper included—lays a hand on the machinery.
8. Sherrod Brown, *Myths of Free Trade: Why American Trade Policy Has Failed* (New York: The New Press), p. 125.
9. Ibid., p. 121.
10. Norbert Wiener, "A Scientist Rebels," *Atlantic Monthly*, Jan. 1947, p. 27. The term "cybernetics" comes from the Greek word *kybernētēs*, meaning steers-

man or governor, and refers to the science of goal-seeking, or self-regulating, devices.

11. Norbert Wiener, *I Am a Mathematician: The Later Life of a Prodigy* (New York: Doubleday, 1956), pp. 295–96.
12. Norbert Wiener, *The Human Use of Human Beings: Cybernetics and Society* (Boston: Houghton Mifflin, 1950), p. 220.
13. In 1999, the metropolitan area included both Baltimore City and Baltimore County, along with Anne Arundel, Carroll, Harford, and Howard counties. Kent County, on the other side of the Chesapeake Bay, has since been added.
14. Peter Robinson, *Literacy, Numeracy, and Economic Performance* (London: Centre for Economic Performance, 1997).
15. Michael Dunkerley, *The Jobless Economy?: Computer Technology in the World of Work* (Cambridge, UK: Polity Press, 1996).
16. Jeremy Rifkin, *The End of Work: The Decline of the Global Labor Force and the Dawn of the Post-market Era* (New York: Putnam, 1995).
17. Juliet Schor, *The Overworked American* (New York: Basic Books, 1991), pp. 154–55.
18. Charles Murray and Richard Herrnstein, *The Bell Curve* (New York: The Free Press, 1994), pp. 434–35.

Epilogue

1. As quoted in Ralph Leighton, *Surely You're Joking, Mr. Feynman!* (New York: W. W. Norton, 1985), p. 338.
2. Richard P. Feynman, *The Feynman Lectures on Physics* (Reading, MA: Addison-Wesley, 1963), vol. 2, pp. 20–21.
3. Murray Gell-Mann, *The Quark and the Jaguar* (New York: W. H. Freeman, 1994), chap. 9.
4. Feynman explains his reasons in *Feynman Lectures on Computation* (New York: Westview Press, 1996), chap. 5.
5. The neoclassical model of commercial interaction differs from contemporary versions of the PCM in presuming that capital and labor are—as they then appeared to be—incapable of crossing national boundaries.
6. Robert Park, *Voodoo Science: The Road from Foolishness to Fraud* (New York: Oxford University Press, 2000), pp. 9–10.
7. Bertolt Brecht, *Galileo* (New York: Grove Press, 1966), scene 8.

Index